Technological Slavery

The Collected Writings of Theodore J. Kaczynski, a.k.a. "The Unabomber"

Introduction by Dr. David Skrbina

Feral House

Technological Slavery is a revised and enlarged version of the book, *Road to Revolution*, published in an English edition of 400 copies, and also in a French edition in 2008 by Éditions Xenia of Vevey Switzerland.

North American publishing rights for *Technological Slavery* were granted to Feral House by its owner, Technological Slavery LLC.

10 9 8 7 6 5 4 3 2 1

Feral House
1240 W. Sims Way, Suite 124
Port Townsend WA 98368
www.FeralHouse.com

Design by Bill Smith

To the memory of Joy Richards,
with love.

From the Publisher

Feral House has not published this book to justify the crimes committed by Mr. Kaczynski. But we do feel that there is a great deal of legitimate thought in this book, and the First Amendment allows readers to judge whether or not this is the case.

Technophiles like Ray Kurzweil and Bill Joy also expressed their regard for Theodore Kaczynski's writing:

"Like many of my colleagues, I felt that I could easily have been the Unabomber's next target. He is clearly a Luddite, but simply saying this does not dismiss his argument.... As difficult as it is for me to acknowledge, I saw some merit in the reasoning in [Kaczynski's writing]."

—Bill Joy, founder of Sun Microsystems,
in "Why the Future Doesn't Need Us," Wired *magazine*

Theodore J. Kaczynski does not receive remuneration for his essays in this book, and a portion of all profits will be donated to the American Red Cross.

TABLE OF CONTENTS

AUTHOR'S NOTE TO THE SECOND EDITION

This book was first published by Editions Xenia under the title *The Road to Revolution*. Unfortunately, the Xenia edition was riddled with errors, most of which were not my fault. In this second edition, if the publisher has done his work properly, the errors have been corrected and the book has been improved in other ways.

I want to make clear that I have no control over the cover design of this book and no control over the way it is advertised and promoted. I expect it to be advertised and promoted in ways that I will find offensive. Moreover, I do not like the new title of the book. Nevertheless, I have cooperated in the creation of this new edition and consented to the change of title because I think it is important to make the book available in its present corrected and improved form.

Ted Kaczynski
December 8, 2009

FOREWORD

I have to begin by saying that I am deeply dissatisfied with this book. It should have been an organized and systematic exposition of a series of related ideas. Instead, it is an unorganized collection of writings that expound the ideas unsystematically. And some ideas that I consider important are not even mentioned. I simply have not had the time to organize, rewrite, and complete the contents of this book.

The principal reason why I have not had time is that agencies of the United States government have created unnecessary legal difficulties for me. To mention only the most important of these difficulties, the United States Attorney for the Eastern District of California has formally proposed to round up and confiscate the original and every copy of everything I have ever written and turn over all such papers to my alleged "victims" through a fictitious sale that will allow the "victims" to acquire all of the papers without having to pay anything for them. Under this plan, the government would even confiscate papers that I have given to libraries, including papers that have been on library shelves for several years. The documents in which the United States Attorney has put forward this proposal are available to the public: They are Document 704 and Document 713, Case Number CR-S-96-259 GEB, United States District Court for the Eastern District of California.

At this writing, I have the assistance of lawyers in resisting the government's actions in regard to my papers. But I have learned from hard experience that it is unwise to leave everything in the hands of lawyers; one is well advised to research the legal issues oneself, keep track of what the lawyers are doing, and intervene when necessary. Such work is time-consuming, especially when one is confined in a maximum-security prison and therefore has only very limited access to law books.

I would have preferred to delay publication of the present book until I'd had time to prepare its contents properly, but it seemed advisable to publish before the government took action to confiscate all my papers. I have, moreover, another reason to avoid delay: The Federal Bureau of Prisons has proposed new regulations that would allow prison wardens to cut off almost all communications between allegedly "terrorist" prisoners and the outside world. The proposed regulations are published in the Federal Register, Volume 71, Number 63, pages 16520-25.

I have no idea when the new regulations may be approved, but if and when that happens it is all too possible that my communications will be cut off. Obviously it is important for me to publish while I can still communicate relatively freely, and that is why this book has to appear now in an unfinished state.

The version of "Industrial Society and its Future" that appears in this book differs from the original manuscript only in trivial ways; spelling, punctuation, capitalization, and the like have been corrected or improved here and there. As far as I know, all earlier versions of "Industrial Society and its Future" published in English or French contain numerous errors, such as the omission of parts of sentences and even of whole sentences, and some of these errors are serious enough so that they change or obscure the meaning of an entire paragraph.

What is much more serious is that at least one completely spurious article has been published under my name. I recently received word from a correspondent in Spain that an article titled "*La Rehabilitación del Estado por los Izquierdistas*" ("The Rehabilitation of the State by the Leftists") had been published and attributed to me. But I most certainly did not write such an article. So the reader should not assume that everything published under my name has actually been written by me. Needless to say, all writings attributed to me in the present book are authentic.

I would like to thank Dr. David Skrbina for having asked questions and raised arguments that spurred me to formulate and write down certain ideas that I had been incubating for years.

I owe thanks to a number of other people also. At the end of "The Truth About Primitive Life" I have thanked by name (and with their permission) several people who provided me with materials for that essay, and some of those people have helped me enormously in other ways as well. In particular, I owe a heavy debt of gratitude to Facundo Bermudez, Marjorie Kennedy, and Patrick Scardo. I owe special thanks to my Spanish correspondent who writes under the pseudonym "Último Reducto," and to a female friend of his, both of whom provided stimulating argument; and Último Reducto moreover has ably translated many of my writings into Spanish. I hesitate to name others to whom I owe thanks, because I'm not sure that they would want to be named publicly.

For the sake of clarity, I want to state here in summary form the four main points that I've tried to make in my writings.

Technological progress is carrying us to inevitable disaster. There may be physical disaster (for example, some form of environmental catastrophe), or there may be disaster in terms of human dignity (reduction of the human race to a degraded and servile condition). But disaster of one kind or another will certainly result from continued technological progress.

This is not an eccentric opinion. Among those frightened by the probable consequences of technological progress are Bill Joy, whose article "Why the Future Doesn't Need Us"[1] is now famous, Martin Rees, author of the book *Our Final Century*,[2] and Richard A. Posner, author of *Catastrophe: Risk and Response*.[3] None of these three is by any stretch of the imagination radical or predisposed to find fault with the existing structure of society. Richard Posner is a conservative judge of the United States Court of Appeals for the Seventh Circuit. Bill Joy is a well-known computer wizard, and Martin Rees is the Astronomer Royal of Britain. These last two men, having devoted their lives to technology, would hardly be likely to fear it without having good reason to do so.

Joy, Rees, and Posner are concerned mainly with physical disaster and with the possibility or indeed the likelihood that human beings will be supplanted by machines. The disaster that technological progress implies for human dignity has been discussed by men like Jacques Ellul and Lewis Mumford, whose books are widely read and respected. Neither man is considered to be out on the fringe or even close to it.

Only the collapse of modern technological civilization can avert disaster. Of course, the collapse of technological civilization will itself bring disaster. But the longer the technoindustrial system continues to expand, the worse will be the eventual disaster. A lesser disaster now will avert a greater one later.

The development of the technoindustrial system cannot be controlled, restrained, or guided, nor can its effects be moderated to any substantial degree.

This, again, is not an eccentric opinion. Many writers, beginning with Karl Marx, have noted the fundamental importance of technology in determining the course of society's development. In effect, they have recognized that it is technology that rules society, not the other way around. Ellul especially has emphasized the autonomy of technology, i.e., the fact that modern technology has taken on a life of its own and is not subject to human control. Ellul, moreover, was not the first to formulate this conclusion. Already in 1934 the Mexican thinker Samuel Ramos[4] clearly stated the principle of technological autonomy, and this insight was adumbrated as early as the 1860s by Samuel Butler. Of course, no one questions the obvious fact that human individuals or groups can control technology in the sense that at a given point in time they can decide what to do with a particular item of technology. What the principle of technological autonomy asserts is that the overall development of technology, and its long-term consequences for society, are not subject to human control. Hence, as long as modern technology continues to exist, there is little we can do to moderate its effects.

A corollary is that nothing short of the collapse of technological society can avert a greater disaster. Thus, if we want to defend ourselves against technology, the only action we can take that might prove effective is an effort to precipitate the collapse of technological society. Though this conclusion is an obvious consequence of the principle of technological autonomy, and though it possibly is implied by certain statements of Ellul, I know of no conventionally published writer who has explicitly recognized that our only way out is through the collapse of technological society. This seeming blindness to the obvious can only be explained as the result of timidity.

If we want to precipitate the collapse of technological society, then our goal is a revolutionary one under any reasonable definition of that term. What we are faced with, therefore, is a need for out-and-out revolution.

3. The political left is technological society's first line of defense against revolution. In fact, the left today serves as a kind of fire extinguisher that douses and quenches any nascent revolutionary movement. What do I mean by "the left"? If you think that racism, sexism, gay rights, animal rights, indigenous people's rights, and "social justice" in general are among the most important issues that the world currently faces, then you are a leftist as I use that term. If you don't like this application of the word "leftist," then you are free to designate the people I'm referring to by some other term. But, whatever you call them, the people who extinguish revolutionary movements are the people who are drawn indiscriminately to causes: racism, sexism, gay rights, animal rights, the environment, poverty, sweatshops, neocolonialism…it's all the same to them. These people constitute a subculture that has been labeled "the adversary culture."[5] Whenever a movement of resistance begins to emerge, these leftists (or whatever you choose to call them) come swarming to it like flies to honey until they outnumber the original members of the movement, take it over, and turn it into just another leftist faction, thereby emasculating it. The history of "Earth First!" provides an elegant example of this process.[6]

4. What is needed is a new revolutionary movement, dedicated to the elimination of technological society, that will take measures to exclude all leftists, as well as the assorted neurotics, lazies, incompetents, charlatans, and persons deficient in self-control who are drawn to resistance movements in America today. Just what form a revolutionary movement should take remains open to discussion. What is clear is that, for a start, people who are serious about addressing the problem of technology must establish systematic contact with one another and a sense of common purpose; they must strictly separate themselves from the "adversary culture"; they must be oriented toward practical action, without renouncing a priori the most extreme forms of action; and they must take as their goal nothing less than the dissolution of technological civilization.

---

[1] *Wired* magazine, April 2000.

[2] Published by William Heinemann, 2003.

[3] Oxford University Press, 2004.

[4] *El perfil del hombre y la cultura en México*, Décima Edición, Espasa-Calpe Mexicana, Mexico City, 1982 (originally published in 1934), pages 104-105.

[5] See Paul Hollander, *The Survival of the Adversary Culture.*

[6] The process is ably documented by Martha F. Lee, *Earth First!: Environmental Apocalypse*, Syracuse University Press, 1995.

A Revolutionary for Our Times

introduction by Dr. David Skrbina

We are steeped in a technological milieu. Technology surrounds us on all sides, envelops us, and, perhaps, suffocates us. It determines or shapes every course of action that we take in our daily lives—how we live, eat, sleep, get to work, where and how we work, how we entertain ourselves, how we run our government, how we conduct our wars. Technological considerations dictate what we can and cannot do, how we do it, and frequently even why we do it. Technology and its direct effects are in our air, our water, across our landscape, and in our bodies. In the developed nations of the 21st century, for all practical purposes, there is no escape from its pervasive effects.

Needless to say, this was not always the case. For the vast majority of our existence, humanity has lived without advanced technology. Ever since the genus Homo emerged from the African savannahs some 2 million years ago, humans have survived and thrived with only the crudest of tools. We lived as wanderers, typically in groups of 50 people or less, and only occasionally stopping to establish temporary encampments. Of the 2 million years of our existence we had controlled use of fire for perhaps only half that time. Durable, stone-tipped spears appeared only 100,000 years ago, and arrowheads, needles, and harpoons some 25,000 years ago—scarcely 1% of humanity's lifetime. We faced all the challenges and threats of nature with only the spear and the hand axe, wearing only crude furs and simple woven clothing, and, for some, with a campfire to keep warm and cook food. I will not idealize the primitive life; it was hard, brutal, sometimes violent, sometimes cruel. But it was the life humanity came to live.

Like it or not, our bodies and our minds are adapted by 2 million years of evolution to a primitive, low-tech existence. Yet today we are surrounded by ubiquitous, advanced, inscrutable technology. And therein lies our predicament.

How can we, creatures of nature, who have spent 99% of our existence using only the simplest of tools, thrive and live well in a high-tech world? Rationally, it seems impossible—and it *is* impossible. There is no good reason to expect that human beings, whose physiology is virtually unchanged since the Stone Age, could adapt well to such a radically altered lifestyle.

By way of illumination, compare the two-million-year lifetime of humanity with a 50-year-old man. Humans have been non-hunter-gatherers—that is, farm-, village- or city-dwellers—for only the past 10,000 years; this so-called civilized portion of history represents a mere 0.5% of our species' lifetime. On a scale of 50 years, then, this "modern" existence corresponds to just *three months*.

Let's say, hypothetically, we find a man born and raised as a nomadic hunter-gatherer in the wilds of sub-Saharan Africa, utterly unaffected by civilization and high technology. We wish to 'help' him by introducing him, progressively over three months, to all the benefits of modern life. So we take him, first, to a small farm, and show him how we grow domesticated crops and raise domesticated animals—organisms he has never seen in the wild. We introduce him to sowing, weeding, harvesting, animal husbandry. We allow him one month to adapt.

Then we take him to a small rural village. We show him writing, and teach him the basics of metals and ceramics. He interacts with a relatively large number of people every day, in relatively close quarters. He is subject to the rules of the village. We allow him a second month to adapt to this.

For the third month we take him on a tour of human cities: smaller first, then mid-sized, finally to a large modern metropolis. Over the course of his final 30 days he sees, in turn: complex wood and metal tools, guns, mechanical clocks, large buildings, ocean-going ships, railroads, cameras, refrigerators, bicycles, gasoline engines, telephones, light bulbs, cars, radios. On the *final day*, we show him, for the first time ever: jet airplanes, television, computers, nuclear reactors and nuclear weapons, integrated circuits, the space shuttle.

Then we turn him loose. We give him a few dollars, a small home in the suburbs, dress him up in a suit and a tie, and say, "Have a good life." "Be a good citizen," we say; "and don't do anything wrong. But don't worry, you'll adapt—we did!"

What shall we expect for our African friend? What are his prospects for the future? We humans, as a whole, are no better off than this 50-year-old hunter-gatherer. As individuals we are, of course, born and raised in a technological world, and so we think we can adapt. But our physical and mental selves are really locked in the past. We try to hide this past with fancy clothes and sophisticated language, and we arm ourselves with all varieties of clever technological aids. But our ancient hunter-gatherer selves are still there, deep inside, struggling to make sense of the world.

# # #

Empirically, the evidence points to one likely outcome: namely, that we humans are in fact *unable* to handle advanced technologies without causing massive disruption to our bodies, our psyche, and our environment.

Consider first our physical health. We suffer from a range of modern ills that have traditionally been very rare: obesity, cancers, accidental death and injury, deliberate death through high-tech weapons (including handguns) and warfare, global plagues like AIDS. Automobile accidents kill over 40,000 Americans every year, and about 1.3 million people globally—that's roughly 3,300 people killed *every day*. Nearly 44% of the American population is medicated.[1] A recent study suggests that 28% of all teenagers suffer chronic headaches, with 40% of these occurring daily.[2] Even the mundane daily computer use that many of us experience imposes its own risks: carpal-tunnel syndrome, eyestrain, back and joint pain, headache, toxic chemicals on keyboards and monitors, and the general ill health that results from sedentary behavior.

Modern foods are killing us: pesticides, chemical fertilizers, growth hormones, radically new genetically-modified crops, too much sugar, too much fat, too much *meat*. Primitive humans rarely ate meat, but when they did it was typically freshly-killed, always wild game, and usually after putting in several exhausting hours of chase, on foot, with sticks or handmade spears.[3] We moderns eat something like 3.5 pounds per week—a half-pound per day, every day—of domesticated, fat-laden, hormone-injected, antibiotic-laced, high-tech factory-farmed animal flesh. Little surprise that cancer and other ailments result.[4]

There is also the potential for direct, violent physical harm. Terrorists achieve their ends through the use of high technology—especially those residing in the halls of government. Virtually all major terrorist threats, including biochemical agents, bio-toxins, nuclear weapons, and other WMDs, are the direct result of advanced industrial technology. The claim that the 9/11 attacks were "low-tech" is a lie; the hijackers made good use of one of the most advanced products of modern technology, the jet airliner.

Psychologically, we suffer widely from illnesses that, to the best of our knowledge, were rarely seen in ancient times: clinical depression, insomnia, suicide, bipolar disorders, dementia, anxiety, and numerous byproducts of extreme mental stress. Nearly 15% of the US population has a personality disorder.[5] Some 26% can be classified as mentally ill.[6] The use of anti-psychotic drugs among children is soaring, both in the US and the UK; British rates increased from 3.9 to 7.7 per 10,000 children over 13 years, whereas American rates ran significantly higher yet: from 23 to 45 per 10,000, over just five years.[7]

Attention deficit disorder and autism have been linked to television and video games, and studies have argued that they are quite literally addictive.[8] So too the Internet. A 2006 Stanford University study found that "more than one out of eight Americans exhibited at least one possible sign of problematic Internet use," including finding it "hard to stay away," concealing nonessential use, using it as an escape mechanism, and harming relationships—all classic signs of addiction.[9] More broadly, researchers now find that a whole range of psychological ailments correlates closely with daily computer usage.[10] And social psychologists have long suspected that many of our modern era's senseless and brutal crimes stem from an assortment of social stresses, exacerbated by industrial technology.[11]

Even the putative *benefits* of technology often turn out to be nonexistent, or to have some nasty strings attached. The Internet, which brings a flood of information into every household and allows for instantaneous, mass communication, comes with severe side effects. Evidence is building that it is literally rewiring our brains' cognitive circuits, resulting in a diminished ability to focus and concentrate on longer and more demanding tasks, such as reading substantive articles or books. Journalist Nicholas Carr recently observed[12] that "over the past few years I've had an uncomfortable sense that someone, or something, has been tinkering with my brain…I'm not thinking the way I used to think. …Now my concentration often starts to drift after two or three pages. I get fidgety, lose the thread…" He lays the blame on Internet "power browsing," which places highest priority on *efficiency* and *immediacy*, causing everything else to take a back seat—in particular, deep reflection and sustained concentration.

Cell phones, which offer continuous and immediate contact with nearly everyone, continue to raise red flags. They are suspected of damaging our cellular DNA,[13] correlate with an increase in anxiety among teens,[14] pose risks to pregnant women and unborn fetuses,[15] and increase the risk of brain cancer and malignant tumors.[16] Other studies attempt to dispute these findings, but it is clear that cell phone radiation is producing at least some detrimental effects on our bodies.

Technology in schools provides yet another classic example. Computers and other high-tech learning aids were, for many years, hyped as the Holy Grail of improved academic performance. They have even been promoted for use by young children and infants. Now we find, instead, that computers and iPods are increasingly used for cheating and plagiarism.[17] High-speed, ultra-short messaging, as with Twitter, threatens emotional and moral development.[18] Text messaging in general now appears to damage language skills.[19] Educational technology for infants, such as 'Baby Einstein' and related video tools, is now found to not only *not* help children, but is actually detrimental.[20] The death blow to the pro-tech lobby came in 2007, with the publication of a major study by the US government. A review of 16 leading ed-tech products, covering more than 9,400 students in 132 schools, showed *no increase in achievement scores*.[21] As a consequence, schools are now bailing out. A *New York Times* article[22] quotes a local school board president: "After seven years, there was literally no evidence it had any impact on student achievement—none." Given the costs and health risks, it's no wonder schools are now seriously reconsidering their technology plans.

Finally, when we look outside the human sphere, to nature, we find disastrous problems: unprecedented species extinction, destruction of forests, resource depletion, global climate change. The toxic byproducts of industrial society are found in the bodies of arctic seals. Costa Rican tree frogs suffer from acid rain produced in New York. Global warming alters age-old weather patterns and threatens to disrupt every ecosystem on the planet. Nuclear reactor wastes will remain deadly for millennia. And the exploding global population is a direct result of highly advanced agricultural and health-care technologies.

Of these concerns, climate change is perhaps the most troubling. A 2009 report by a UN-affiliated think tank projects that, without drastic mitigation actions, climate change will cause "much of civilization to collapse," for large portions of the world.[23] Here we have the ultimate irony: a technological civilization created and powered by fossil fuels, which ends up being so disruptive to the global climate that it destroys itself. Along the way we will have eliminated thousands of other species, and put our own existence at risk. Perhaps a kind of cosmic justice is at work after all.

# # #

From an objective standpoint, then, the situation seems clear: In advanced technology we are dealing with something—a set of tools, a structure, a mindset, a force, a power—which is damaging all aspects of our lives, and seriously undermining the health of the planet. And, for all practical purposes, it is beyond our rational control.

Modern technology, then, even though it is the product of natural beings and developed from the materials of nature, is a profoundly *unnatural* phenomenon. Nothing in humanity's evolutionary past, or in the Earth's evolutionary past, has equipped us to deal with the consequences of this phenomenon. And yet we, and all the world, are confronted with its effects every minute of the day.

There is no doubt that modern technology poses a profound dilemma for humanity. A recent textbook stated the following: "That technology represents a problem of major importance, requiring analysis and interpretation, needs no argument. ... It is the controlling power of our age, affecting and shaping virtually all aspects of human existence in this century." And I think many people—most people—have an intuitive sense that this is true: that the 'problem of technology' is very real, and very serious.

A recent poll of 69,000 people in North America revealed that a *majority*, 51%, can be classified as "technological pessimists," meaning that they are at best indifferent to modern technology, and at worst outright hostile toward it.[24] This is a huge number—something in excess of 100 million adults in North America alone. We know from experience that Europeans tend to be even more skeptical about such things, and thus they are likely to have an even higher number of pessimists. So there seems to be a widespread and deep-seated feeling that something is wrong with our technological age.

So what shall we do? We are faced with a whole range of threats to our wellbeing, and all of them—literally, all major problems confronting humanity—are created or enabled by advanced technology. Shall we just sit here and take it, stoically? Shall we wring our hands, bemoaning the fact that the system is too large, too impenetrable, too unmovable to change? Shall we ask our leaders for help? Shall we pray to God? Shall we wait for the scientists and technologists to save us? What irony—to look to technology to save us *from itself!*

These are a few of the issues that we will raise in this book. They are complex, far-reaching, and vitally important for our collective future. As difficult as it may be, it is a discussion that we cannot avoid.

# # #

The occasion for the discussion at hand is, of course, the work of Theodore Kaczynski. Convicted of the Unabomber crimes in 1998, Kaczynski is now spending the remainder of his life in a high-security supermax prison in Colorado. The Unabomber case received worldwide attention, due in part to the inability of the FBI to track him down after 17 years of trying, and in part to the unique motivation of the person or group known as "FC." FC's primary demand, to which the FBI eventually agreed, was to allow publication in a major newspaper or journal of a lengthy anti-technology manifesto entitled "Industrial Society and its Future" (ISAIF). The *Washington Post* published a nearly complete version of ISAIF on September 19, 1995; roughly 1.2 million copies were sold that day. Soon thereafter, Theodore's brother, David Kaczynski, recognized the style and content of the manifesto and contacted the FBI. Theodore, then age 53, was arrested at his small wooden home in rural Montana on April 3, 1996. On April 15 he was on the cover of *Time* magazine, and the whole world saw the man that had eluded capture for so long.

This book was never intended to be a biography, but it is worth recalling a few basic facts of Kaczynski's life story. He was born in Chicago on May 22, 1942. From his early childhood it was clear that he was an academic standout, and he excelled at school. Skipping two grades, he left high school for Harvard at age 16. By 1962, at age 20, Kaczynski had completed his Bachelor's degree in mathematics. He headed to graduate school at the University of Michigan at Ann Arbor, where, over the next five years, he earned Master's and PhD degrees in math. In 1967 he acquired a teaching job at the prestigious University of California at Berkeley; it was a position he held for just two years. By 1971 he had decided to buy some land near

Lincoln, Montana and make a homestead there. He worked odd jobs and was periodically seen in nearby towns, but by and large kept to himself.

Under different circumstances, we might never have heard from Kaczynski again. But this was not to be. In one of his letters to me, he recounts how both recreationists and the Forest Service continually pressed in on him—to the point where a peaceful life was no longer possible. This invasion constituted a kind of war, and Kaczynski began to defend himself.

It was not until a few years later, in mid-1978, that the first so-called Unabomber attack occurred. Between 1978 and 1985 there were eight mail- or package-bombings, including one on an airplane, which resulted in a total of 20 injuries. All were connected with universities or airlines, hence the name given by the FBI: 'un-a-bomber.'

The first fatality occurred in December 1985, when computer storeowner Hugh Scrutton was killed by a package bomb left in his parking lot. Between 1987 and 1995 there were five more attacks, killing two (advertising executive Thomas Mosser and California Forestry Association president Gilbert Murray) and injuring three. The ISAIF manifesto was published five months after the final attack, and Kaczynski was arrested seven months after that.

In the 14 years since his imprisonment, the public has heard and read many things about Kaczynski, but nothing from Kaczynski himself until now. This book is the first comprehensive and unedited collection of his writings.

This book will *not* address the many sensational issues surrounding Kaczynski: the details of the Unabomber case, Kaczynski's personal history, his so-called "troubled past," the "psychology of a murderer," or the ineptitudes of the American criminal justice system.[25] This book does not advocate violence, bomb-making, murder, or any other heinous acts that one might fear finding here. It does not even discuss violence except very indirectly, as one potential but undefined aspect of the "revolution against technology."

The entire focus of this book is the *problem of technology*: where we stand today, what kind of imminent future we are facing, and what we ought to do about it.

The challenge to the reader is to make a firm separation between the Unabomber crimes and a rational, in-depth, no-holds-barred discussion of the threat posed by modern technology. Kaczynski has much to offer to this discussion even if we accept that he was guilty of certain reprehensible crimes. We do ourselves no favors by ignoring him. His ideas have no less force, his arguments are none the weaker, simply because they issue from a maximum-security cell.

Kaczynski's writings revolve around a core argument against modern technology. To briefly recap that argument:

- Human beings evolved under primitive, low-tech conditions. This is our natural state of existence.
- Present technological society is radically different than our natural state, and imposes unprecedented stresses upon us, and on nature.
- Technologically-induced stress is bad now and will get much worse, leading to a condition where humans will be completely manipulated and molded to serve the needs of the system. Such a state of affairs is undignified, abhorrent, disastrous for nature, and profoundly dehumanizing.
- The technological system cannot be fixed or reformed so as to avoid this dehumanized future.
- Therefore, the system must be brought to an end.

The logic is sound. However, we are free to challenge any of the premises. Perhaps we did not evolve under low-tech conditions—maybe God created humans 6,000 years ago. Perhaps modern technology is, in some sense, not an aberrant condition but is really our "natural state." Perhaps the stresses of modern life will not get worse. Perhaps reform is possible. Perhaps revolution, though justified, is futile. These are just some of the responses we might make to Kaczynski's argument, and in defense of the status quo. All these points will be touched on in this book; I hope that some progress will be made.

# # #

As will become apparent, Kaczynski is a careful, insightful thinker who makes forceful arguments against technology—arguments that are not easily refuted. In spite of this, even at the peak of the Unabomber trial, one rarely heard anything of these arguments. Instead we were treated to an interesting spectacle: a near-universal assault on his character and actions, without a shred of meaningful discussion of his ideas. This shameful, deliberate act of mindlessness was typically "justified" in three ways—none of which are rational. These tactics need to be firmly buried, so that a real inquiry can proceed.

First: "He's a murderer, and we must not dignify a murderer by discussing his ideas." Based on his plea bargain, we indeed must accept that Kaczynski did deliver the fatal mail bombs. For that he is rightly punished with a life sentence in a federal penitentiary. His tactics were deplorable, and I for one do not endorse such actions.

And yet, in any civilized society even the most nefarious of prisoners has some rights. Freedom of speech is one of these. Every prisoner in any modern nation should have the right to communicate to outsiders, to express his or her ideas, and even to publish books or artwork, provided they hold to the same broad restrictions of any citizen. American prisoners cannot profit from their work—this is the famous "Son of Sam" law—but that is not at issue here. Kaczynski gets not one dime of profit from this book. But he cannot be denied the legal or moral right to express his views.

Furthermore, every document that Kaczynski receives or sends out is reviewed in detail by personnel from the US Federal Bureau of Prisons. We need have no apprehensions about him communicating secret plans to destroy the world, or to kill again.

But do we dignify Kaczynski unduly? I recall a similar concern in late 2005, when a documentary ran on American public television about Mark David Chapman, the killer of John Lennon. Similar complaints were raised: "We dignify this criminal too much by even mentioning his name"; "We should never hear his voice"; "We should never read a word of what he says," and so on. Many opposed the documentary, and yet it was produced, and aired. And nothing was to be gained except sheer voyeurism. There was no deep message, no residual value in hearing Chapman speak. It was pure pop culture. And yet it aired, because he has a right to speak, and we have a right to know. How much more important to hear from Kaczynski—not just the mail-bomber who eluded the FBI for 17 years, but a man with ideas that challenge the core of our modern worldview, and even offer a kind of salvation.

That said, we could clearly opt to close our eyes and ears to the man. But this solves nothing. We are still left facing the same issues, and having to answer the same difficult questions. In dealing with his writings perhaps we do dignify him. But more importantly, we dignify our children, the natural world, and ourselves—because it is these that will bear the consequences of our actions.

Second: "Sure, technology causes problems, but we've got no choice. What are we supposed to do, go live in a cave?" The point here, presumably, is that technological society is an irrevocable reality, and any discussion to the contrary is a complete waste of time. To this I can only say: (a) If you really think that you have no choice, then the debate is over. Kaczynski has won. If you have no choice, you have no freedom. You are little better than a slave to the system. You may be a *comfortable* slave—an Uncle Tom, if you will—but this is an utterly undignified existence. And (b), if by *cave* we mean a life *without* technology, then this is ludicrous, and impossible. For the 2 million years of our existence we have used tools—technology—to survive. It cannot be otherwise. The whole question is, what *level* of technology shall we use? We can choose simple, natural, manageable, biodegradable tools, or we can choose complex, enslaving, toxic tools.

If the cave imagery is intended as a shorthand notion for a simple, low-tech lifestyle, then I respond, yes, this is precisely what we need. We modern people think life unlivable without electricity, the Internet, air conditioning, and indoor plumbing. Obviously it was not always like this. The greatest accomplishments of humanity occurred without computers, without electricity, without plumbing. Think of it—life without computers! What barbarians those Renaissance men must have been! Those ancient Greeks—brute animals! And yet the Greeks, for example, though living with only the most basic of tools, were able to create one of the greatest societies in history. The whole point of technology, of society, is, after all, to *have a good life*; and a good life requires almost nothing at all.

The third common tactic was to raise a series of red herrings—to discuss everything about the man *except* his "crazy" ideas. His arguments no doubt pose a threat to the system, and thus many people, especially those in positions of power, are very anxious to repudiate Kaczynski and his ideas—preferably, in such a way as to avoid actually addressing them. The arguments are not easily defeated, especially by simple-minded politicians, jealous or jaded intellectuals, or apologists for big business, so they tend to mount superficial or trivial attacks. They will talk about his mental state, his upbringing, the legal circus—anything to distract the public from substantive inquiry. In this way, Kaczynski's dangerous ideas are safely hidden out of sight. Virtually every mass media discussion of either Kaczynski or ISAIF is guilty of this ploy; even at the height of the media frenzy, the most one could hope for would be to hear or read a few snippets from the manifesto. [26] The cover story in *Time* the week after Kaczynski's arrest is a perfect case in point: not a word on the substance of his thinking.[27]

One instance that was especially egregious, if only because one would have expected better, was the largely inane critique of the manifesto by Kirkpatrick Sale in *Nation*.[28] Given a rare opportunity to provide an in-depth assessment of the piece in a high-visibility venue, Sale fumbled badly. He spends an inordinate amount of time on trivial, incidental, or pointless issues, belaboring the Unabomber's "wooden," "plodding," and "leaden" writing style, and his lack of pure originality ("thinks he's the first person who ever worked out such ideas")—as if such things have any bearing at all on the arguments at hand.

In fact Kaczynski's writing style is perfectly suited to the task. He is clear, precise, and articulate. He writes in a commonsense manner, largely free of technical terms. When he does introduce precise terms, he is generally careful to define them. He is respectful of the reader. He writes to a broad audience. He is methodical and meticulous. Clarity and precision are of utmost importance, befitting the severity of the situation.

Kaczynski's originality is not really in dispute. It is true that many of the themes he addresses have been discussed by others, but this fact takes nothing away from the force of his arguments. Quite the contrary—it only *strengthens* his position. He follows in a long line of important thinkers who had grave concerns about technology, and its potential to disrupt society. The earliest of these was Lao Tzu, the venerable Chinese philosopher of 2,500 years ago, who observed: "The more sharpened tools the people have / the more benighted the state." Sharp tools cut through the social fabric, separating people from themselves and from the world. Such tools cast us all into a dark time, from which we are unable to see our way ahead. We build them at our own risk.

Shortly afterward, Plato was making the first connection between *techne* and *logos*, and warning us about even so benign a technology as *writing*:

> *This invention will produce forgetfulness in the minds of those who learn to use it, because they will not practice their memory. ... [Writing is] an elixir not of memory, but of* reminding... *[It offers us] the* appearance *of wisdom, not true wisdom...* (Phaedrus, *275a)*

Such early reflections led, in time, to Rousseau's full-blown critique of technology in his *Discourse on the Arts and Sciences* (1750), and to Henry David Thoreau's anti-technological musings in *Walden* (1850). Not long thereafter, British essayist Samuel Butler felt compelled to issue the first unequivocal attack against the technological system:

> *Day by day, the machines are gaining ground upon us; day by day we are becoming more subservient to them... The time will come when the machines will hold the real supremacy over the world and its inhabitants... Our opinion is that* war to the death *should be instantly proclaimed against them. Every machine of every sort should be destroyed by the well-wisher of his species. (*Darwin Among the Machines, *1863)*[29]

Noted philosophers like Scheler, Whitehead, and Heidegger published stinging critiques. Orwell's *Road to Wigan Pier* (1937) concludes with a penetrating and insightful attack on mechanization and the "machine society." Of special significance to Kaczynski, and the whole technology debate, is Jacques Ellul's 1954 masterpiece *The Technological Society*; his portrayal of technology as a monistic, self-driving force in the world that is able to invade all aspects of human existence, deeply undermining our freedom in the process, was as ground-breaking as it was troubling. In the 1960s and 70s, radical thinkers like Marcuse and Illich called for virtual revolution against the system.[30] Through the present day, some elements of the so-called green anarchist movement attempt to do the same—see R. Scarce (2006).

Thus, even though Kaczynski addresses many issues which others before him have raised, he carries the analysis to a new level of intensity. His uniqueness is expressed in a number of ways. First is his relentless focus on technology itself as the root cause of our predicament; he is adamant that, directly or indirectly, modern technology is the sole basis for our most pressing contemporary problems. Second, he assigns highest value to the dignity and autonomy, or freedom, of the human being; it is these things that are chiefly threatened by technology. Third, he explicitly calls for revolution against the system, in a way that no prior critic has done. And revolution is not merely some whimsical afterthought—it is a core element of his overall critique. Fourth, he is very authoritative in his research, citing in a careful and scholarly manner the relevant ideas that support his claims. He does not make idle statements, or offer appeals to emotion, or engage in hyperbole. Finally, Kaczynski is very pragmatic. This is not just theory for him. The situation demands action, and he offers specific plans to assist the transition to a post-technological world.

# # #

With these pseudo-criticisms and diversionary tactics out of the way, a true inquiry can proceed. In order to move ahead and seriously tackle the problem of technology, there are three main issues that we should bear in mind:

(1) What is the present state of affairs? (in terms of human stress and indignity, environmental damage, etc). How bad are things at the moment?

(2) What is our likely future in the near term; say, in the next few decades? Will things get better? Stay the same? Get worse? Get *much* worse?

(3) What can, or should, we do about it?

Most people, being more or less adapted to modern society, would likely rate present conditions as a mixed bag: some good, some bad, some problems we need to work on but nothing imminently pressing. The near-term future they would see as more of the same—a few improvements, a few new problems, overall slightly better, perhaps. This automatically implies a conservative course of action: Carry on with the status quo, don't rock the boat, be a 'cooperator,' work hard, follow the rules, vote, hoist the flag of nationalism when called to. No major catastrophes coming, and in any case we have the government, the scientists, and corporate self-interest to take care of any problems that may arise. This view, according to Kaczynski, is naively optimistic—*dangerously* optimistic. It fails to respond to the exponentially growing power of technology, and its rapidly increasing ability to assert control over life on this planet.

Faced with persistent technological crises, there is also the common attitude of 'no pain, no gain': "Yes, there are inevitable problems with technology, but they are a necessary part of the learning process. Without the pain of the mistakes we could not enjoy the gains that technology offers." This line of thinking would be fine, if (a) the pains were predictable, limited, and manageable; (b) they were fairly and justly distributed; and (c) the 'gains' were in fact true improvements on the human condition. Kaczynski argues, rightly I think, that all three of these assumptions are false. And not just 'a little false,' but radically false—false in a deeply deceiving fashion.

Kaczynski's answers to the central questions are quite clear. In my exchange of letters with him, I pressed him on these points in order to better understand his reasoning, and to examine any weaknesses. These questions are, in fact, core issues that we all should ask ourselves. Furthermore, they do not end. This is an inquiry that must be ongoing, and responsive to the changing nature of technology itself. An answer one day may well be exposed as inadequate or fallacious the next.

One hundred years ago, Henry Ford could not begin to anticipate the highway deaths, urban sprawl, wars over oil, and global warming that his automobiles would bring. The inventors of television could not anticipate that it would lead to obesity, ill health, lower academic performance, and attention deficit disorder. The inventors of aerosol propellants (chlorofluorocarbons) could not know that they would destroy the planetary ozone layer. Early coal miners could not know that their product would disrupt the climate of the entire planet. These were not simple mistakes, mere oversights; they are an unavoidable aspect of advanced technology. We can *never* know what the consequences will be, and the more powerful and more ubiquitous the technology, the greater the risk. If global warming destroys the Earth's ability to sustain life as we know it, then all the wonderful gains of the industrial age will be utterly worthless.

Paraphrasing Lao Tzu: *the sharper the tools, the darker the times*. We live in an age of *very sharp tools*. Consequently, it is also a very dark time. But tools cut both ways. Can they even, perhaps, be turned against themselves? Does the technological system contain the seeds of its own destruction? This may be our only hope.

We are clearly in dire need of a substantive inquiry into the problem of technology. In recent years we have seen just the beginning of what may lie ahead—a potentially catastrophic future. If most people are not yet convinced that drastic action is warranted, it is only because the worst outcomes have yet to be realized. On the other hand, if we wait until the crisis is obvious to all, it will be far too late. What can we do, *now*, to regain human dignity, defend the planet, and give ourselves the best chance for long-term survival? This is the question that presses upon us with the greatest urgency. We ignore it at our peril. •

NOTES:

[1] From the report "Health: US 2004," by the US Health and Human Services Department. See AP news story, 2 December 2004.

[2] In the bibliography, see Powers et al. (2003), and Split and Neuman (1999).

[3] Though evidence suggests that humans also scavenged dead animals killed by other predators. But doing this, of course, still meant fighting off the competition, including perhaps the predator who made the kill. One can imagine that this still involved considerable risk, effort, and skill, especially when armed with only sticks and stones.

[4] For the connection between modern meat consumption and cancer, see: Chao et al. (2005); Nothlings et al. (2005); Norat et al. (2005); Xu et al. (2007); Egeberg et al. (2008); Allen et al. (2008). Also, the industrial production of domesticated meat has an astonishingly negative impact on the global environment. It produces 22% of human-induced global warming gases, more than the total transport sector combined (see *Lancet* study, 13 September 2007). According to the UN's FAO agency, livestock directly or indirectly utilize an amazing 30% of the earth's entire land surface area. And they represent fully 20% of the total land animal biomass ("Livestock a major threat to environment," 29 November 2006). This cannot but have a catastrophic long-term impact on the planet.

[5] Grant et al. (2004).

[6] Kessler et al. (2004).

[7] Reported by the AP (5 May 2008)—"Anti-psychotic drug use soars among US and UK kids." See also Rani et al. (2008).

[8] Christakis et al. (2004), and Kubey and Csikszentmihalyi (2002). Regarding the possible connection between television and autism, see Waldman et al. (2006). Autism in fact seems to be more prevalent than commonly thought; recent estimates suggest that about one of every 150 children (0.7%) has some form of this disorder, significantly higher than previous estimates (see Rice, 2007).

[9] Aboujaoude et al. (2006).

[10] Nakazawa et al. (2002).

[11] See, for example, the AP story of 5 April, 2007 ("Technology may fuel recorded assaults"), citing evidence that rape and other sexual assaults are on the increase due to the ability to record and transmit images of such acts.

[12] Carr (2008): "Is Google making us stupid?"

[13] Reuters news story, 21 December 2004: "Mobile phone radiation harms DNA."

[14] *Los Angeles Times*, 25 May 2006: "Cell phone use may signal teen anxiety."

[15] Story by G. Lean, in the British newspaper *Independent* (18 May 2008). "Women who use mobile phones when pregnant are more likely to give birth to children with behavioral problems." He adds,

"using the handsets just two or three times a day was enough to raise the risk" of hyperactivity and emotional problems. For the full report, see Divan et al. (2008).

[16] As reported in the *Independent*, "using a mobile phone for more than 10 years increases the risk of getting brain cancer" (7 October 2007). Long-term users "are twice as likely to get a malignant tumor on the side of the brain where they hold the handset." See Hardell et al. (2007). See also AP story, "Cancer expert warns employees on cell phones" (24 July 2008).

[17] AP news story, 27 April 2007: "Schools say iPods becoming tool for cheaters."

[18] CNN news story, 14 April 2009: "Scientists warn of Twitter dangers."

[19] See Reuters news story (27 April 2007) on a report of the Irish government: "Text messaging harms written language." Teens were found to be "unduly reliant on short sentences, simple tenses, and a limited vocabulary."

[20] Zimmerman et al. (2007). For each additional hour of video watched per day, infants understood six to eight *fewer* words, on average.

[21] Dynarski et al. (2007). Among their main findings: "Test scores were not significantly higher in classrooms using selected reading and mathematics software products."

[22] *New York Times*, 4 May 2007 (page A1). Headline: "Seeing no progress, some schools drop laptops."

[23] *State of the Future Report* (2009), by The Millennium Project. As an added bonus, it now appears that the very same emissions that cause global warming also lower the IQ of unborn children. See the article in *Time* magazine (23 July 2009: "Study links exposure to pollution with lower IQ"), or Perera et al. (2009).

[24] Forrester Research study, "The State of Consumers and Technology: Benchmark 2005" (3 August 2005).

[25] That said, I would like to make one brief observation on his mental condition. At his trial Kaczynski was diagnosed by a government-appointed psychologist as 'paranoid-schizophrenic.' The inference for the public was that this man was half-insane, delusional, and incapable of rational thinking—"of course, he sent mail bombs, he must be mad." I am not a psychologist, but in a couple hundred pages of correspondence with Kaczynski, spread out over more than six years, I have found him to be stable, rational, lucid, and extremely articulate. I have seen no indication of any mental illness. This picture, however, does not serve the interests of the government, who would like to portray anyone attempting to undermine industrial technology as insane—insanity by definition.

As further evidence, Kaczynski has passed on comments from two staff psychologists—Drs. Watterson and Morrison—at his prison. According to Kaczynski they repeatedly denied finding evidence of serious mental illness, and called the schizophrenia diagnosis "ridiculous," "wildly improbable," and merely a "political diagnosis."

[26] There were of course a few exceptions, including: Wright (1995), Fulano (1996), Akai (1997), Finnegan (1998), and Coatimundi (1998).

[27] Gibbs (1996).

[28] Sale (1995).

[29] See also his essay "Mechanical creation" (1865).

[30] Marcuse (1964) and Illich (1973, 1974).

Industrial Society

1

and Its Future (ISAIF)

Introduction

1. The Industrial Revolution and its consequences have been a disaster for the human race. They have greatly increased the life expectancy of those of us who live in "advanced" countries, but they have destabilized society, have made life unfulfilling, have subjected human beings to indignities, have led to widespread psychological suffering (in the Third World to physical suffering as well) and have inflicted severe damage on the natural world. The continued development of technology will worsen the situation. It will certainly subject human beings to greater indignities and inflict greater damage on the natural world, it will probably lead to greater social disruption and psychological suffering, and it may lead to increased physical suffering even in "advanced" countries.

2. The industrial-technological system may survive or it may break down. If it survives, it MAY eventually achieve a low level of physical and psychological suffering, but only after passing through a long and very painful period of adjustment and only at the cost of permanently reducing human beings and many other living organisms to engineered products and mere cogs in the social machine. Furthermore, if the system survives, the consequences will be inevitable: There is no way of reforming or modifying the system so as to prevent it from depriving people of dignity and autonomy.

3. If the system breaks down the consequences will still be very painful. But the bigger the system grows the more disastrous the results of its breakdown will be, so if it is to break down it had best break down sooner rather than later.

4. We therefore advocate a revolution against the industrial system. This revolution may or may not make use of violence; it may be sudden or it may be a relatively gradual process spanning a few decades. We can't predict any of that. But we do outline in a very general way the measures that those who hate the industrial system should take in order to prepare the way for a revolution against that form of society. This is not to be a POLITICAL

revolution. Its object will be to overthrow not governments but the economic and technological basis of the present society.

5. In this article we give attention to only some of the negative developments that have grown out of the industrial-technological system. Other such developments we mention only briefly or ignore altogether. This does not mean that we regard these other developments as unimportant. For practical reasons we have to confine our discussion to areas that have received insufficient public attention or in which we have something new to say. For example, since there are well-developed environmental and wilderness movements, we have written very little about environmental degradation or the destruction of wild nature, even though we consider these to be highly important.

The Psychology of Modern Leftism

6. Almost everyone will agree that we live in a deeply troubled society. One of the most widespread manifestations of the craziness of our world is leftism, so a discussion of the psychology of leftism can serve as an introduction to the discussion of the problems of modern society in general.

7. But what is leftism? During the first half of the 20th century leftism could have been practically identified with socialism. Today the movement is fragmented and it is not clear who can properly be called a leftist. When we speak of leftists in this article we have in mind mainly socialists, collectivists, "politically correct" types, feminists, gay and disability activists, animal rights activists and the like. But not everyone who is associated with one of these movements is a leftist. What we are trying to get at in discussing leftism is not so much a movement or an ideology as a psychological type, or rather a collection of related types. Thus, what we mean by "leftism" will emerge more clearly in the course of our discussion of leftist psychology. (Also, see paragraphs 227-230.)

8. Even so, our conception of leftism will remain a good deal less clear than we would wish, but there doesn't seem to be any remedy for this. All we are trying to do here is indicate in a rough and approximate way the two psychological tendencies that we believe are the main driving force of modern leftism. We by no means claim to be telling the WHOLE truth about leftist psychology. Also, our discussion is meant to apply to modern leftism only. We leave open the question of the extent to which our discussion could be applied to the leftists of the 19th and early 20th centuries.

9. The two psychological tendencies that underlie modern leftism we call *feelings of inferiority* and *oversocialization*. Feelings of inferiority are characteristic of modern leftism as a whole, while oversocialization is characteristic only of a certain segment of modern leftism; but this segment is highly influential.

Feelings of Inferiority

10. By "feelings of inferiority" we mean not only inferiority feelings in the strict sense but a whole spectrum of related traits: low self-esteem, feelings of powerlessness, depressive tendencies, defeatism, guilt, self-hatred, etc. We argue that modern leftists tend to have some such feelings (possibly more or less repressed), and that these feelings are decisive in determining the direction of modern leftism.

11. When someone interprets as derogatory almost anything that is said about him (or about groups with whom he identifies), we conclude that he has inferiority feelings or low self-esteem. This tendency is pronounced among minority-rights activists, whether or not they belong to the minority groups whose rights they defend. They are hypersensitive about the words used to designate minorities and about anything that is said concerning minorities. The terms "Negro," "oriental," "handicapped," or "chick" for an African, an Asian, a disabled person or a woman originally had no derogatory connotation. "Broad" and "chick" were merely the feminine equivalents of "guy," "dude" or "fellow." The negative connotations have been attached to these terms by the activists themselves. Some animal rights activists have gone so far as to reject the word "pet" and insist on its replacement by "animal companion." Leftish anthropologists go to

great lengths to avoid saying anything about primitive peoples that could conceivably be interpreted as negative. They want to replace the word "primitive" by "nonliterate." They seem almost paranoid about anything that might suggest that any primitive culture is inferior to our own. (We do not mean to imply that primitive cultures ARE inferior to ours. We merely point out the hypersensitivity of leftish anthropologists.)

12. Those who are most sensitive about "politically incorrect" terminology are not the average black ghetto-dweller, Asian immigrant, abused woman or disabled person, but a minority of activists, many of whom do not even belong to any "oppressed" group but come from privileged strata of society. Political correctness has its stronghold among university professors, who have secure employment with comfortable salaries, and the majority of whom are heterosexual white males from middle to upper-class families.

13. Many leftists have an intense identification with the problems of groups that have an image of being weak (women), defeated (American Indians), repellent (homosexuals), or otherwise inferior. The leftists themselves feel that these groups are inferior. They would never admit to themselves that they have such feelings, but it is precisely because they do see these groups as inferior that they identify with their problems. (We do not mean to suggest that women, Indians, etc., ARE inferior; we are only making a point about leftist psychology.)

14. Feminists are desperately anxious to prove that women are as strong and as capable as men. Clearly they are nagged by a fear that women may NOT be as strong and as capable as men.

15. Leftists tend to hate anything that has an image of being strong, good and successful. They hate America, they hate Western civilization, they hate white males, they hate rationality. The reasons that leftists give for hating the West, etc., clearly do not correspond with their real motives. They SAY they hate the West because it is warlike, imperialistic, sexist, ethnocentric and so forth, but where these same faults appear in socialist countries or in primitive cultures, the leftist finds excuses for them, or at best he GRUDGINGLY admits that they exist; whereas he ENTHUSIASTICALLY points out (and often greatly exaggerates) these faults where they appear in Western civilization. Thus it is clear that these faults are not the leftist's real motive for hating America and the West. He hates America and the West because they are strong and successful.

16. Words like "self-confidence," "self-reliance," "initiative," "enterprise," "optimism," etc., play little role in the liberal and leftist vocabulary. The leftist is anti-individualistic, pro-collectivist. He wants society to solve everyone's problems for them, satisfy everyone's needs for them, take care of them. He is not the sort of person who has an inner sense of confidence in his ability to solve his own problems and satisfy his own needs. The leftist is antagonistic to the concept of competition because, deep inside, he feels like a loser.

17. Art forms that appeal to modern leftish intellectuals tend to focus on sordidness, defeat and despair, or else they take an orgiastic tone, throwing off rational control as if there were no hope of accomplishing anything through rational calculation and all that was left was to immerse oneself in the sensations of the moment.

18. Modern leftish philosophers tend to dismiss reason, science, objective reality and to insist that everything is culturally relative. It is true that one can ask serious questions about the foundations of scientific knowledge and about how, if at all, the concept of objective reality can be defined. But it is obvious that modern leftish philosophers are not simply cool-headed logicians systematically analyzing the foundations of knowledge. They are deeply involved emotionally in their attack on truth and reality. They attack these concepts because of their own psychological needs. For one thing, their attack is an outlet for hostility, and, to the extent that it is successful, it satisfies the drive for power. More importantly, the leftist hates science and rationality because they classify certain beliefs as true (i.e., successful, superior) and other beliefs as false (i.e., failed, inferior). The leftist's feelings of inferiority run so deep that he cannot tolerate any classification of some things as successful or superior and other things as failed or inferior. This also underlies the rejection by many leftists of the concept of mental illness and of the utility of IQ tests. Leftists are antagonistic to genetic explanations of human abilities or behavior because such explanations tend to make some persons appear superior or inferior to others. Leftists prefer to give society the credit or blame for an individual's ability or lack of it. Thus if a person is "inferior" it is not his fault, but society's, because he has not been brought up properly.

19. The leftist is not typically the kind of person whose feelings of inferiority make him a braggart, an egotist, a bully, a self-promoter, a ruthless competitor. This kind of person has not wholly lost faith in himself. He has a deficit in his sense of power and self-worth, but he can still conceive of

himself as having the capacity to be strong, and his efforts to make himself strong produce his unpleasant behavior.[1] But the leftist is too far gone for that. His feelings of inferiority are so ingrained that he cannot conceive of himself as individually strong and valuable. Hence the collectivism of the leftist. He can feel strong only as a member of a large organization or a mass movement with which he identifies himself.

20. Notice the masochistic tendency of leftist tactics. Leftists protest by lying down in front of vehicles, they intentionally provoke police or racists to abuse them, etc. These tactics may often be effective, but many leftists use them not as a means to an end but because they PREFER masochistic tactics. Self-hatred is a leftist trait.

21. Leftists may claim that their activism is motivated by compassion or by moral principles, and moral principle does play a role for the leftist of the oversocialized type. But compassion and moral principle cannot be the main motives for leftist activism. Hostility is too prominent a component of leftist behavior; so is the drive for power. Moreover, much leftist behavior is not rationally calculated to be of benefit to the people whom the leftists claim to be trying to help. For example, if one believes that affirmative action is good for black people, does it make sense to demand affirmative action in hostile or dogmatic terms? Obviously it would be more productive to take a diplomatic and conciliatory approach that would make at least verbal and symbolic concessions to white people who think that affirmative action discriminates against them. But leftist activists do not take such an approach because it would not satisfy their emotional needs. Helping black people is not their real goal. Instead, race problems serve as an excuse for them to express their own hostility and frustrated need for power. In doing so they actually harm black people, because the activists' hostile attitude toward the white majority tends to intensify race hatred.

22. If our society had no social problems at all, the leftists would have to INVENT problems in order to provide themselves with an excuse for making a fuss.

23. We emphasize that the foregoing does not pretend to be an accurate description of everyone who might be considered a leftist. It is only a rough indication of a general tendency of leftism.

Oversocialization

24. Psychologists use the term "socialization" to designate the process by which children are trained to think and act as society demands. A person is said to be well socialized if he believes in and obeys the moral code of his society and fits in well as a functioning part of that society. It may seem senseless to say that many leftists are oversocialized, since the leftist is perceived as a rebel. Nevertheless, the position can be defended. Many leftists are not such rebels as they seem.

25. The moral code of our society is so demanding that no one can think, feel and act in a completely moral way. For example, we are not supposed to hate anyone, yet almost everyone hates somebody at some time or other, whether he admits it to himself or not. Some people are so highly socialized that the attempt to think, feel and act morally imposes a severe burden on them. In order to avoid feelings of guilt, they continually have to deceive themselves about their own motives and find moral explanations for feelings and actions that in reality have a non-moral origin. We use the term "oversocialized" to describe such people.[2]

26. Oversocialization can lead to low self-esteem, a sense of powerlessness, defeatism, guilt, etc. One of the most important means by which our society socializes children is by making them feel ashamed of behavior or speech that is contrary to society's expectations. If this is overdone, or if a particular child is especially susceptible to such feelings, he ends by feeling ashamed of HIMSELF. Moreover the thought and the behavior of the over-socialized person are more restricted by society's expectations than are those of the lightly socialized person. The majority of people engage in a significant amount of naughty behavior. They lie, they commit petty thefts, they break traffic laws, they goof off at work, they hate someone, they say spiteful things or they use some underhanded trick to get ahead of the other guy. The oversocialized person cannot do these things, or if he does do them he generates in himself a sense of shame and self-hatred. The oversocialized person cannot even experience, without guilt, thoughts or feelings that are contrary to the accepted morality; he cannot think "unclean" thoughts. And socialization is not just a matter of morality; we are socialized to conform to many norms of behavior that do not fall under the heading of morality. Thus the oversocialized person is kept on a psychological leash and

spends his life running on rails that society has laid down for him. In many oversocialized people this results in a sense of constraint and powerlessness that can be a severe hardship. We suggest that oversocialization is among the more serious cruelties that human beings inflict on one another.

27. We argue that a very important and influential segment of the modern left is oversocialized and that their oversocialization is of great importance in determining the direction of modern leftism. Leftists of the oversocialized type tend to be intellectuals or members of the upper middle class. Notice that university intellectuals[3] constitute the most highly socialized segment of our society and also the most left-wing segment.

28. The leftist of the oversocialized type tries to get off his psychological leash and assert his autonomy by rebelling. But usually he is not strong enough to rebel against the most basic values of society. Generally speaking, the goals of today's leftists are NOT in conflict with the accepted morality. On the contrary, the left takes an accepted moral principle, adopts it as its own, and then accuses mainstream society of violating that principle. Examples: racial equality, equality of the sexes, helping poor people, peace as opposed to war, nonviolence generally, freedom of expression, kindness to animals. More fundamentally, the duty of the individual to serve society and the duty of society to take care of the individual. All these have been deeply rooted values of our society (or at least of its middle and upper classes[4]) for a long time. These values are explicitly or implicitly expressed or presupposed in most of the material presented to us by the mainstream communications media and the educational system. Leftists, especially those of the oversocialized type, usually do not rebel against these principles but justify their hostility to society by claiming (with some degree of truth) that society is not living up to these principles.

29. Here is an illustration of the way in which the oversocialized leftist shows his real attachment to the conventional attitudes of our society while pretending to be in rebellion against it. Many leftists push for affirmative action, for moving black people into high-prestige jobs, for improved education in black schools and more money for such schools; the way of life of the black "underclass" they regard as a social disgrace. They want to integrate the black man into the system, make him a business executive, a lawyer, a scientist just like upper middle-class white people. The leftists will reply that the last thing they want is to make the black man into a copy of the white man; instead, they want to preserve African-American culture.

But in what does this preservation of African-American culture consist? It can hardly consist in anything more than eating black-style food, listening to black-style music, wearing black-style clothing and going to a black-style church or mosque. In other words, it can express itself only in superficial matters. In all ESSENTIAL respects most leftists of the oversocialized type want to make the black man conform to white middle-class ideals. They want to make him study technical subjects, become an executive or a scientist, spend his life climbing the status ladder to prove that black people are as good as white. They want to make black fathers "responsible," they want black gangs to become nonviolent, etc. But these are exactly the values of the industrial-technological system. The system couldn't care less what kind of music a man listens to, what kind of clothes he wears or what religion he believes in as long as he studies in school, holds a respectable job, climbs the status ladder, is a "responsible" parent, is nonviolent and so forth. In effect, however much he may deny it, the oversocialized leftist wants to integrate the black man into the system and make him adopt its values.

30. We certainly do not claim that leftists, even of the over-socialized type, NEVER rebel against the fundamental values of our society. Clearly they sometimes do. Some oversocialized leftists have gone so far as to rebel against one of modern society's most important principles by engaging in physical violence. By their own account, violence is for them a form of "liberation." In other words, by committing violence they break through the psychological restraints that have been trained into them. Because they are oversocialized these restraints have been more confining for them than for others; hence their need to break free of them. But they usually justify their rebellion in terms of mainstream values. If they engage in violence they claim to be fighting against racism or the like.

31. We realize that many objections could be raised to the foregoing thumbnail sketch of leftist psychology. The real situation is complex, and anything like a complete description of it would take several volumes even if the necessary data were available. We claim only to have indicated very roughly the two most important tendencies in the psychology of modern leftism.

32. The problems of the leftist are indicative of the problems of our society as a whole. Low self-esteem, depressive tendencies and defeatism are not restricted to the left. Though they are especially noticeable in the left, they are widespread in our society. And today's society tries to socialize us to

a greater extent than any previous society. We are even told by experts how to eat, how to exercise, how to make love, how to raise our kids and so forth.

The Power Process

33. Human beings have a need (probably based in biology) for something that we will call the *power process*. This is closely related to the need for power (which is widely recognized) but is not quite the same thing. The power process has four elements. The three most clear-cut of these we call goal, effort and attainment of goal. (Everyone needs to have goals whose attainment requires effort, and needs to succeed in attaining at least some of his goals.) The fourth element is more difficult to define and may not be necessary for everyone. We call it autonomy and will discuss it later (paragraphs 42-44).

34. Consider the hypothetical case of a man who can have anything he wants just by wishing for it. Such a man has power, but he will develop serious psychological problems. At first he will have a lot of fun, but by and by he will become acutely bored and demoralized. Eventually he may become clinically depressed. History shows that leisured aristocracies tend to become decadent. This is not true of fighting aristocracies that have to struggle to maintain their power. But leisured, secure aristocracies that have no need to exert themselves usually become bored, hedonistic and demoralized, even though they have power. This shows that power is not enough. One must have goals toward which to exercise one's power.

35. Everyone has goals; if nothing else, to obtain the physical necessities of life: food, water and whatever clothing and shelter are made necessary by the climate. But the leisured aristocrat obtains these things without effort. Hence his boredom and demoralization.

36. Non-attainment of important goals results in death if the goals are physical necessities, and in frustration if non-attainment of the goals is compatible with survival. Consistent failure to attain goals throughout life results in defeatism, low self-esteem or depression.

37. Thus, in order to avoid serious psychological problems, a human being needs goals whose attainment requires effort, and he must have a reasonable rate of success in attaining his goals.

Surrogate Activities

38. But not every leisured aristocrat becomes bored and demoralized. For example, the emperor Hirohito, instead of sinking into decadent hedonism, devoted himself to marine biology, a field in which he became distinguished. When people do not have to exert themselves to satisfy their physical needs they often set up artificial goals for themselves. In many cases they then pursue these goals with the same energy and emotional involvement that they otherwise would have put into the search for physical necessities. Thus the aristocrats of the Roman Empire had their literary pretensions; many European aristocrats a few centuries ago invested tremendous time and energy in hunting, though they certainly didn't need the meat; other aristocracies have competed for status through elaborate displays of wealth; and a few aristocrats, like Hirohito, have turned to science.

39. We use the term "surrogate activity" to designate an activity that is directed toward an artificial goal that people set up for themselves merely in order to have some goal to work toward, or, let us say, merely for the sake of the "fulfillment" that they get from pursuing the goal. Here is a rule of thumb for the identification of surrogate activities. Given a person who devotes much time and energy to the pursuit of goal X, ask yourself this: If he had to devote most of his time and energy to satisfying his biological needs, and if that effort required him to use his physical and mental faculties in a varied and interesting way, would he feel seriously deprived because he did not attain goal X? If the answer is no, then the person's pursuit of a goal X is a surrogate activity. Hirohito's studies in marine biology clearly constituted a surrogate activity, since it is pretty certain that if Hirohito had had to spend his time working at interesting non-scientific tasks in order to obtain the necessities of life, he would not have felt deprived because he didn't know all about the anatomy and life-cycles of marine animals. On the other hand the pursuit of sex and love (for example) is not a surrogate activity, because most people, even if their existence were otherwise satisfactory, would feel deprived if they passed their lives without ever having a relationship with a member of the opposite sex. (But pursuit of an excessive amount of sex, more than one really needs, can be a surrogate activity.)

40. In modern industrial society only minimal effort is necessary to satisfy one's physical needs. It is enough to go through a training program

to acquire some petty technical skill, then come to work on time and exert the very modest effort needed to hold a job. The only requirements are a moderate amount of intelligence and, most of all, simple OBEDIENCE. If one has those, society takes care of one from cradle to grave. (Yes, there is an underclass that cannot take the physical necessities for granted, but we are speaking here of mainstream society.) Thus it is not surprising that modern society is full of surrogate activities. These include scientific work, athletic achievement, humanitarian work, artistic and literary creation, climbing the corporate ladder, acquisition of money and material goods far beyond the point at which they cease to give any additional physical satisfaction, and social activism when it addresses issues that are not important for the activist personally, as in the case of white activists who work for the rights of nonwhite minorities. These are not always PURE surrogate activities, since for many people they may be motivated in part by needs other than the need to have some goal to pursue. Scientific work may be motivated in part by a drive for prestige, artistic creation by a need to express feelings, militant social activism by hostility. But for most people who pursue them, these activities are in large part surrogate activities. For example, the majority of scientists will probably agree that the "fulfillment" they get from their work is more important than the money and prestige they earn.

41. For many if not most people, surrogate activities are less satisfying than the pursuit of real goals (that is, goals that people would want to attain even if their need for the power process were already fulfilled). One indication of this is the fact that, in many or most cases, people who are deeply involved in surrogate activities are never satisfied, never at rest. Thus the money-maker constantly strives for more and more wealth. The scientist no sooner solves one problem than he moves on to the next. The long-distance runner drives himself to run always farther and faster. Many people who pursue surrogate activities will say that they get far more fulfillment from these activities than they do from the "mundane" business of satisfying their biological needs, but that is because in our society the effort required to satisfy the biological needs has been reduced to triviality. More importantly, in our society people do not satisfy their biological needs AUTONOMOUSLY but by functioning as parts of an immense social machine. In contrast, people generally have a great deal of autonomy in pursuing their surrogate activities.

Autonomy

42. Autonomy as a part of the power process may not be necessary for every individual. But most people need a greater or lesser degree of autonomy in working toward their goals. Their efforts must be undertaken on their own initiative and must be under their own direction and control. Yet most people do not have to exert this initiative, direction and control as single individuals. It is usually enough to act as a member of a SMALL group. Thus if half a dozen people discuss a goal among themselves and make a successful joint effort to attain that goal, their need for the power process will be served. But if they work under rigid orders handed down from above that leave them no room for autonomous decision and initiative, then their need for the power process will not be served. The same is true when decisions are made on a collective basis if the group making the collective decision is so large that the role of each individual is insignificant.[5]

43. It is true that some individuals seem to have little need for autonomy. Either their drive for power is weak or they satisfy it by identifying themselves with some powerful organization to which they belong. And then there are unthinking, animal types who seem to be satisfied with a purely physical sense of power (the good combat soldier, who gets his sense of power by developing fighting skills that he is quite content to use in blind obedience to his superiors).

44. But for most people it is through the power process—having a goal, making an AUTONOMOUS effort and attaining the goal—that self-esteem, self-confidence and a sense of power are acquired. When one does not have adequate opportunity to go through the power process the consequences are (depending on the individual and on the way the power process is disrupted) boredom, demoralization, low self-esteem, inferiority feelings, defeatism, depression, anxiety, guilt, frustration, hostility, spouse or child abuse, insatiable hedonism, abnormal sexual behavior, sleep disorders, eating disorders, etc.[6]

Sources of Social Problems

45. Any of the foregoing symptoms can occur in any society, but in modern industrial society they are present on a massive scale. We aren't the first to mention that the world today seems to be going crazy. This sort of thing is not normal for human societies. There is good reason to believe that primitive man suffered from less stress and frustration and was better satisfied with his way of life than modern man is. It is true that not all was sweetness and light in primitive societies. Abuse of women was common among the Australian aborigines, transsexuality was fairly common among some of the American Indian tribes. But is does appear that GENERALLY SPEAKING the kinds of problems that we have listed in the preceding paragraph were far less common among primitive peoples than they are in modern society.

46. We attribute the social and psychological problems of modern society to the fact that that society requires people to live under conditions radically different from those under which the human race evolved and to behave in ways that conflict with the patterns of behavior that the human race developed while living under the earlier conditions. It is clear from what we have already written that we consider lack of opportunity to properly experience the power process as the most important of the abnormal conditions to which modern society subjects people. But it is not the only one. Before dealing with disruption of the power process as a source of social problems we will discuss some of the other sources.

47. Among the abnormal conditions present in modern industrial society are excessive density of population, isolation of man from nature, excessive rapidity of social change and the breakdown of natural small-scale communities such as the extended family, the village or the tribe.

48. It is well known that crowding increases stress and aggression. The degree of crowding that exists today and the isolation of man from nature are consequences of technological progress. All preindustrial societies were predominantly rural. The Industrial Revolution vastly increased the size of cities and the proportion of the population that lives in them, and modern agricultural technology has made it possible for the Earth to support a far denser population than it ever did before. (Also, technology exacerbates the effects of crowding because it puts increased disruptive powers in people's hands. For example, a variety of noise-making devices: power mowers, radios,

motorcycles, etc. If the use of these devices is unrestricted, people who want peace and quiet are frustrated by the noise. If their use is restricted, people who use the devices are frustrated by the regulations. But if these machines had never been invented there would have been no conflict and no frustration generated by them.)

49. For primitive societies the natural world (which usually changes only slowly) provided a stable framework and therefore a sense of security. In the modern world it is human society that dominates nature rather than the other way around, and modern society changes very rapidly owing to technological change. Thus there is no stable framework.

50. The conservatives are fools: They whine about the decay of traditional values, yet they enthusiastically support technological progress and economic growth. Apparently it never occurs to them that you can't make rapid, drastic changes in the technology and the economy of a society without causing rapid changes in all other aspects of the society as well, and that such rapid changes inevitably break down traditional values.

51. The breakdown of traditional values to some extent implies the breakdown of the bonds that hold together traditional small-scale social groups. The disintegration of small-scale social groups is also promoted by the fact that modern conditions often require or tempt individuals to move to new locations, separating themselves from their communities. Beyond that, a technological society HAS TO weaken family ties and local communities if it is to function efficiently. In modern society an individual's loyalty must be first to the system and only secondarily to a small-scale community, because if the internal loyalties of small-scale communities were stronger than loyalty to the system, such communities would pursue their own advantage at the expense of the system.

52. Suppose that a public official or a corporation executive appoints his cousin, his friend or his coreligionist to a position rather than appointing the person best qualified for the job. He has permitted personal loyalty to supersede his loyalty to the system, and that is "nepotism" or "discrimination," both of which are terrible sins in modern society. Would-be industrial societies that have done a poor job of subordinating personal or local loyalties to loyalty to the system are usually very inefficient. (Look at Latin America.) Thus an advanced industrial society can tolerate only those small-scale communities that are emasculated, tamed and made into tools of the system.[7]

53. Crowding, rapid change and the breakdown of communities have been widely recognized as sources of social problems. But we do not believe they are enough to account for the extent of the problems that are seen today.

54. A few preindustrial cities were very large and crowded, yet their inhabitants do not seem to have suffered from psychological problems to the same extent as modern man. In America today there still are uncrowded rural areas, and we find there the same problems as in urban areas, though the problems tend to be less acute in the rural areas. Thus crowding does not seem to be the decisive factor.

55. On the growing edge of the American frontier during the 19th century, the mobility of the population probably broke down extended families and small-scale social groups to at least the same extent as these are broken down today. In fact, many nuclear families lived by choice in such isolation, having no neighbors within several miles, that they belonged to no community at all, yet they do not seem to have developed problems as a result.

56. Furthermore, change in American frontier society was very rapid and deep. A man might be born and raised in a log cabin, outside the reach of law and order and fed largely on wild meat; and by the time he arrived at old age he might be working at a regular job and living in an ordered community with effective law enforcement. This was a deeper change than that which typically occurs in the life of a modern individual, yet it does not seem to have led to psychological problems. In fact, 19th century American society had an optimistic and self confident tone, quite unlike that of today's society.[8]

57. The difference, we argue, is that modern man has the sense (largely justified) that change is IMPOSED on him, whereas the 19th century frontiersman had the sense (also largely justified) that he created change himself, by his own choice. Thus a pioneer settled on a piece of land of his own choosing and made it into a farm through his own effort. In those days an entire county might have only a couple of hundred inhabitants and was a far more isolated and autonomous entity than a modern county is. Hence the pioneer farmer participated as a member of a relatively small group in the creation of a new, ordered community. One may well question whether the creation of this community was an improvement, but at any rate it satisfied the pioneer's need for the power process.

58. It would be possible to give other examples of societies in which there has been rapid change and/or lack of close community ties without the kind of massive behavioral aberration that is seen in today's industrial society. We contend that the most important cause of social and psychological problems in modern society is the fact that people have insufficient opportunity to go through the power process in a normal way. We don't mean to say that modern society is the only one in which the power process has been disrupted. Probably most if not all civilized societies have interfered with the power process to a greater or lesser extent. But in modern industrial society the problem has become particularly acute. Leftism, at least in its recent (mid- to late-20th century) form, is in part a symptom of deprivation with respect to the power process.

Disruption of the Power Process in Modern Society

59. We divide human drives into three groups: (1) those drives that can be satisfied with minimal effort; (2) those that can be satisfied but only at the cost of serious effort; (3) those that cannot be adequately satisfied no matter how much effort one makes. The power process is the process of satisfying the drives of the second group. The more drives there are in the third group, the more there is frustration, anger, eventually defeatism, depression, etc.

60. In modern industrial society natural human drives tend to be pushed into the first and third groups, and the second group tends to consist increasingly of artificially created drives.

61. In primitive societies, physical necessities generally fall into group 2: They can be obtained, but only at the cost of serious effort. But modern society tends to guarantee the physical necessities to everyone[9] in exchange for only minimal effort, hence physical needs are pushed into group 1. (There may be disagreement about whether the effort needed to hold a job is "minimal"; but usually, in lower- to middle-level jobs, whatever effort is required is merely that of OBEDIENCE. You sit or stand where you are told to sit or stand and do what you are told to do in the way you are told to do it. Seldom do you have to exert yourself seriously, and in any case you have hardly any autonomy in work, so that the need for the power process is not well served.)

62. Social needs, such as sex, love and status, often remain in group 2 in modern society, depending on the situation of the individual.[10] But, except for people who have a particularly strong drive for status, the effort required to fulfill the social drives is insufficient to satisfy adequately the need for the power process.

63. So certain artificial needs have been created that fall into group 2, hence serve the need for the power process. Advertising and marketing techniques have been developed that make many people feel they need things that their grandparents never desired or even dreamed of. It requires serious effort to earn enough money to satisfy these artificial needs, hence they fall into group 2. (But see paragraphs 80-82.) Modern man must satisfy his need for the power process largely through pursuit of the artificial needs created by the advertising and marketing industry,[11] and through surrogate activities.

64. It seems that for many people, maybe the majority, these artificial forms of the power process are insufficient. A theme that appears repeatedly in the writings of the social critics of the second half of the 20th century is the sense of purposelessness that afflicts many people in modern society. (This purposelessness is often called by other names such as "anomie" or "middle-class vacuity.") We suggest that the so-called "identity crisis" is actually a search for a sense of purpose, often for commitment to a suitable surrogate activity. It may be that existentialism is in large part a response to the purposelessness of modern life.[12] Very widespread in modern society is the search for "fulfillment." But we think that for the majority of people an activity whose main goal is fulfillment (that is, a surrogate activity) does not bring completely satisfactory fulfillment. In other words, it does not fully satisfy the need for the power process. (See paragraph 41.) That need can be fully satisfied only through activities that have some external goal, such as physical necessities, sex, love, status, revenge, etc.

65. Moreover, where goals are pursued through earning money, climbing the status ladder or functioning as part of the system in some other way, most people are not in a position to pursue their goals AUTONOMOUSLY. Most workers are someone else's employee and, as we pointed out in paragraph 61, must spend their days doing what they are told to do in the way they are told to do it. Even most people who are in business for themselves have only limited autonomy. It is a chronic complaint of small-business persons and entrepreneurs that their hands

are tied by excessive government regulation. Some of these regulations are doubtless unnecessary, but for the most part government regulations are essential and inevitable parts of our extremely complex society. A large portion of small business today operates on the franchise system. It was reported in the Wall Street Journal a few years ago that many of the franchise-granting companies require applicants for franchises to take a personality test that is designed to EXCLUDE those who have creativity and initiative, because such persons are not sufficiently docile to go along obediently with the franchise system. This excludes from small business many of the people who most need autonomy.

66. Today people live more by virtue of what the system does FOR them or TO them than by virtue of what they do for themselves. And what they do for themselves is done more and more along channels laid down by the system. Opportunities tend to be those that the system provides, the opportunities must be exploited in accord with the rules and regulations[13], and techniques prescribed by experts must be followed if there is to be a chance of success.

67. Thus the power process is disrupted in our society through a deficiency of real goals and a deficiency of autonomy in the pursuit of goals. But it is also disrupted because of those human drives that fall into group 3: the drives that one cannot adequately satisfy no matter how much effort one makes. One of these drives is the need for security. Our lives depend on decisions made by other people; we have no control over these decisions and usually we do not even know the people who make them. ("We live in a world in which relatively few people—maybe 500 or 1,000—make the important decisions," Philip B. Heymann of Harvard Law School, quoted by Anthony Lewis, New York Times, April 21, 1995.) Our lives depend on whether safety standards at a nuclear power plant are properly maintained; on how much pesticide is allowed to get into our food or how much pollution into our air; on how skillful (or incompetent) our doctor is; whether we lose or get a job may depend on decisions made by government economists or corporation executives; and so forth. Most individuals are not in a position to secure themselves against these threats to more than a very limited extent. The individual's search for security is therefore frustrated, which leads to a sense of powerlessness.

68. It may be objected that primitive man is physically less secure than modern man, as is shown by his shorter life expectancy; hence modern man

suffers from less, not more than the amount of insecurity that is normal for human beings. But psychological security does not closely correspond with physical security. What makes us FEEL secure is not so much objective security as a sense of confidence in our ability to take care of ourselves. Primitive man, threatened by a fierce animal or by hunger, can fight in self-defense or travel in search of food. He has no certainty of success in these efforts, but he is by no means helpless against the things that threaten him. The modern individual on the other hand is threatened by many things against which he is helpless; nuclear accidents, carcinogens in food, environmental pollution, war, increasing taxes, invasion of his privacy by large organizations, nationwide social or economic phenomena that may disrupt his way of life.

69. It is true that primitive man is powerless against some of the things that threaten him; disease for example. But he can accept the risk of disease stoically. It is part of the nature of things, it is no one's fault, unless it is the fault of some imaginary, impersonal demon. But threats to the modern individual tend to be MAN-MADE. They are not the results of chance but are IMPOSED on him by other persons whose decisions he, as an individual, is unable to influence. Consequently he feels frustrated, humiliated and angry.

70. Thus primitive man for the most part has his security in his own hands (either as an individual or as a member of a SMALL group), whereas the security of modern man is in the hands of persons or organizations that are too remote or too large for him to be able personally to influence them. So modern man's drive for security tends to fall into groups 1 and 3; in some areas (food, shelter, etc.) his security is assured at the cost of only trivial effort, whereas in other areas he CANNOT attain security. (The foregoing greatly simplifies the real situation, but it does indicate in a rough, general way how the condition of modern man differs from that of primitive man.)

71. People have many transitory drives or impulses that are necessarily frustrated in modern life, hence fall into group 3. One may become angry, but modern society cannot permit fighting. In many situations it does not even permit verbal aggression. When going somewhere one may be in a hurry, or one may be in a mood to travel slowly, but one generally has no choice but to move with the flow of traffic and obey the traffic signals. One may want to do one's work in a different way, but usually one can work only according to the rules laid down by one's employer. In many other ways as

well, modern man is strapped down by a network of rules and regulations (explicit or implicit) that frustrate many of his impulses and thus interfere with the power process. Most of these regulations cannot be dispensed with, because they are necessary for the functioning of industrial society.

72. Modern society is in certain respects extremely permissive. In matters that are irrelevant to the functioning of the system we can generally do what we please. We can believe in any religion we like (as long as it does not encourage behavior that is dangerous to the system). We can go to bed with anyone we like (as long as we practice "safe sex"). We can do anything we like as long as it is UNIMPORTANT. But in all IMPORTANT matters the system tends increasingly to regulate our behavior.

73. Behavior is regulated not only through explicit rules and not only by the government. Control is often exercised through indirect coercion or through psychological pressure or manipulation, and by organizations other than the government, or by the system as a whole. Most large organizations use some form of propaganda[14] to manipulate public attitudes or behavior. Propaganda is not limited to "commercials" and advertisements, and sometimes it is not even consciously intended as propaganda by the people who make it. For instance, the content of entertainment programming is a powerful form of propaganda. An example of indirect coercion: There is no law that says we have to go to work every day and follow our employer's orders. Legally there is nothing to prevent us from going to live in the wild like primitive people or from going into business for ourselves. But in practice there is very little wild country left, and there is room in the economy for only a limited number of small business owners. Hence most of us can survive only as someone else's employee.

74. We suggest that modern man's obsession with longevity, and with maintaining physical vigor and sexual attractiveness to an advanced age, is a symptom of unfulfillment resulting from deprivation with respect to the power process. The "mid-life crisis" also is such a symptom. So is the lack of interest in having children that is fairly common in modern society but almost unheard-of in primitive societies.

75. In primitive societies life is a succession of stages. The needs and purposes of one stage having been fulfilled, there is no particular reluctance about passing on to the next stage. A young man goes through the power process by becoming a hunter, hunting not for sport or for fulfillment but to get meat that is necessary for food. (In young women the process is more

complex, with greater emphasis on social power; we won't discuss that here.) This phase having been successfully passed through, the young man has no reluctance about settling down to the responsibilities of raising a family. (In contrast, some modern people indefinitely postpone having children because they are too busy seeking some kind of "fulfillment." We suggest that the fulfillment they need is adequate experience of the power process—with real goals instead of the artificial goals of surrogate activities.) Again, having successfully raised his children, going through the power process by providing them with the physical necessities, the primitive man feels that his work is done and he is prepared to accept old age (if he survives that long) and death. Many modern people, on the other hand, are disturbed by the prospect of physical deterioration and death, as is shown by the amount of effort they expend trying to maintain their physical condition, appearance and health. We argue that this is due to unfulfillment resulting from the fact that they have never put their physical powers to any practical use, have never gone through the power process using their bodies in a serious way. It is not the primitive man, who has used his body daily for practical purposes, who fears the deterioration of age, but the modern man, who has never had a practical use for his body beyond walking from his car to his house. It is the man whose need for the power process has been satisfied during his life who is best prepared to accept the end of that life.

76. In response to the arguments of this section someone will say, "Society must find a way to give people the opportunity to go through the power process." This won't work for those who need autonomy in the power process. For such people the value of the opportunity is destroyed by the very fact that society gives it to them. What they need is to find or make their own opportunities. As long as the system GIVES them their opportunities it still has them on a leash. To attain autonomy they must get off that leash.

How Some People Adjust

77. Not everyone in industrial-technological society suffers from psychological problems. Some people even profess to be quite satisfied with society as it is. We now discuss some of the reasons why people differ so greatly in their response to modern society.

78. First, there doubtless are innate differences in the strength of the drive for power. Individuals with a weak drive for power may have relatively little need to go through the power process, or at least relatively little need for autonomy in the power process. These are docile types who would have been happy as plantation darkies in the Old South. (We don't mean to sneer at the "plantation darkies" of the Old South. To their credit, most of the slaves were NOT content with their servitude. We do sneer at people who ARE content with servitude.)

79. Some people may have some exceptional drive, in pursuing which they satisfy their need for the power process. For example, those who have an unusually strong drive for social status may spend their whole lives climbing the status ladder without ever getting bored with that game.

80. People vary in their susceptibility to advertising and marketing techniques. Some people are so susceptible that, even if they make a great deal of money, they cannot satisfy their constant craving for the shiny new toys that the marketing industry dangles before their eyes. So they always feel hard-pressed financially even if their income is large, and their cravings are frustrated.

81. Some people have low susceptibility to advertising and marketing techniques. These are the people who aren't interested in money. Material acquisition does not serve their need for the power process.

82. People who have medium susceptibility to advertising and marketing techniques are able to earn enough money to satisfy their craving for goods and services, but only at the cost of serious effort (putting in overtime, taking a second job, earning promotions, etc.). Thus material acquisition serves their need for the power process. But it does not necessarily follow that their need is fully satisfied. They may have insufficient autonomy in the power process (their work may consist of following orders) and some of their drives may be frustrated (e.g., security, aggression). (We are guilty of oversimplification in paragraphs 80-82 because we have assumed that the desire for material acquisition is entirely a creation of the advertising and marketing industry. Of course it's not that simple.)[11]

83. Some people partly satisfy their need for power by identifying themselves with a powerful organization or mass movement. An individual lacking goals or power joins a movement or an organization, adopts its goals as his own, then works toward these goals. When some of the goals are attained, the individual, even though his personal efforts have

played only an insignificant part in the attainment of the goals, feels (through his identification with the movement or organization) as if he had gone through the power process. This phenomenon was exploited by the Fascists, Nazis and Communists. Our society uses it too, though less crudely. Example: Manuel Noriega was an irritant to the U.S. (goal: punish Noriega). The U.S. invaded Panama (effort) and punished Noriega (attainment of goal). The U.S. went through the power process and many Americans, because of their identification with the U.S., experienced the power process vicariously. Hence the widespread public approval of the Panama invasion; it gave people a sense of power.[15] We see the same phenomenon in armies, corporations, political parties, humanitarian organizations, religious or ideological movements. In particular, leftist movements tend to attract people who are seeking to satisfy their need for power. But for most people identification with a large organization or a mass movement does not fully satisfy the need for power.

84. Another way in which people satisfy their need for the power process is through surrogate activities. As we explained in paragraphs 38-40, a surrogate activity is an activity that is directed toward an artificial goal that the individual pursues for the sake of the "fulfillment" that he gets from pursuing the goal, not because he needs to attain the goal itself. For instance, there is no practical motive for building enormous muscles, hitting a little white ball into a hole or acquiring a complete series of postage stamps. Yet many people in our society devote themselves with passion to bodybuilding, golf or stamp-collecting. Some people are more "other-directed" than others, and therefore will more readily attach importance to a surrogate activity simply because the people around them treat it as important or because society tells them it is important. That is why some people get very serious about essentially trivial activities such as sports, or bridge, or chess, or arcane scholarly pursuits, whereas others who are more clear-sighted never see these things as anything but the surrogate activities that they are, and consequently never attach enough importance to them to satisfy their need for the power process in that way. It only remains to point out that in many cases a person's way of earning a living is also a surrogate activity. Not a PURE surrogate activity, since part of the motive for the activity is to gain the physical necessities and (for some people) social status and the luxuries that advertising makes them want. But many people put into their work far more effort than is necessary to earn whatever money and status they require,

and this extra effort constitutes a surrogate activity. This extra effort, together with the emotional investment that accompanies it, is one of the most potent forces acting toward the continual development and perfecting of the system, with negative consequences for individual freedom. (See paragraph 131.) Especially, for the most creative scientists and engineers, work tends to be largely a surrogate activity. This point is so important that it deserves a separate discussion, which we shall give in a moment (paragraphs 87-92).

85. In this section we have explained how many people in modern society do satisfy their need for the power process to a greater or lesser extent. But we think that for the majority of people the need for the power process is not fully satisfied. In the first place, those who have an insatiable drive for status, or who get firmly "hooked" on a surrogate activity, or who identify strongly enough with a movement or organization to satisfy their need for power in that way, are exceptional personalities. Others are not fully satisfied with surrogate activities or by identification with an organization. (See paragraphs 41, 64.) In the second place, too much control is imposed by the system through explicit regulation or through socialization, which results in a deficiency of autonomy, and in frustration due to the impossibility of attaining certain goals and the necessity of restraining too many impulses.

86. But even if most people in industrial-technological society were well satisfied, we (FC) would still be opposed to that form of society, because (among other reasons) we consider it demeaning to fulfill one's need for the power process through surrogate activities or through identification with an organization, rather than through pursuit of real goals.

The Motives of Scientists

87. Science and technology provide the most important examples of surrogate activities. Some scientists claim that they are motivated by "curiosity" or by a desire to "benefit humanity." But it is easy to see that neither of these can be the principal motive of most scientists. As for "curiosity," that notion is simply absurd. Most scientists work on highly specialized problems that are not the object of any normal curiosity. For example, is an astronomer, a mathematician or an entomologist curious about the properties of isopropyltrimethylmethane? Of course not. Only a chemist is curious about such a thing, and he is curious about it only

because chemistry is his surrogate activity. Is the chemist curious about the appropriate classification of a new species of beetle? No. That question is of interest only to the entomologist, and he is interested in it only because entomology is his surrogate activity. If the chemist and the entomologist had to exert themselves seriously to obtain the physical necessities, and if that effort exercised their abilities in an interesting way but in some nonscientific pursuit, then they wouldn't give a damn about isopropyltrimethylmethane or the classification of beetles. Suppose that lack of funds for postgraduate education had led the chemist to become an insurance broker instead of a chemist. In that case he would have been very interested in insurance matters but would have cared nothing about isopropyltrimethylmethane. In any case it is not normal to put into the satisfaction of mere curiosity the amount of time and effort that scientists put into their work. The "curiosity" explanation for the scientists' motive just doesn't stand up.

88. The "benefit of humanity" explanation doesn't work any better. Some scientific work has no conceivable relation to the welfare of the human race—most of archaeology or comparative linguistics for example. Some other areas of science present obviously dangerous possibilities. Yet scientists in these areas are just as enthusiastic about their work as those who develop vaccines or study air pollution. Consider the case of Dr. Edward Teller, who had an obvious emotional involvement in promoting nuclear power plants. Did this involvement stem from a desire to benefit humanity? If so, then why didn't Dr. Teller get emotional about other "humanitarian" causes? If he was such a humanitarian then why did he help to develop the H-bomb? As with many other scientific achievements, it is very much open to question whether nuclear power plants actually do benefit humanity. Does the cheap electricity outweigh the accumulating waste and the risk of accidents? Dr. Teller saw only one side of the question. Clearly his emotional involvement with nuclear power arose not from a desire to "benefit humanity" but from the personal fulfillment he got from his work and from seeing it put to practical use.

89. The same is true of scientists generally. With possible rare exceptions, their motive is neither curiosity nor a desire to benefit humanity but the need to go through the power process: to have a goal (a scientific problem to solve), to make an effort (research) and to attain the goal (solution of the problem). Science is a surrogate activity because scientists work mainly for the fulfillment they get out of the work itself.

90. Of course, it's not that simple. Other motives do play a role for many scientists. Money and status for example. Some scientists may be persons of the type who have an insatiable drive for status (see paragraph 79) and this may provide much of the motivation for their work. No doubt the majority of scientists, like the majority of the general population, are more or less susceptible to advertising and marketing techniques and need money to satisfy their craving for goods and services. Thus science is not a PURE surrogate activity. But it is in large part a surrogate activity.

91. Also, science and technology constitute a powerful mass movement, and many scientists gratify their need for power through identification with this mass movement. (See paragraph 83.)

92. Thus science marches on blindly, without regard to the real welfare of the human race or to any other standard, obedient only to the psychological needs of the scientists and of the government officials and corporation executives who provide the funds for research.

The Nature of Freedom

93. We are going to argue that industrial-technological society cannot be reformed in such a way as to prevent it from progressively narrowing the sphere of human freedom. But because "freedom" is a word that can be interpreted in many ways, we must first make clear what kind of freedom we are concerned with.

94. By "freedom" we mean the opportunity to go through the power process, with real goals not the artificial goals of surrogate activities, and without interference, manipulation or supervision from anyone, especially from any large organization. Freedom means being in control (either as an individual or as a member of a SMALL group) of the life-and-death issues of one's existence: food, clothing, shelter and defense against whatever threats there may be in one's environment. Freedom means having power; not the power to control other people but the power to control the circumstances of one's own life. One does not have freedom if anyone else (especially a large organization) has power over one, no matter how benevolently, tolerantly and permissively that power may be exercised. It is important not to confuse freedom with mere permissiveness (see paragraph 72).

95. It is said that we live in a free society because we have a certain number of constitutionally guaranteed rights. But these are not as important as they seem. The degree of personal freedom that exists in a society is determined more by the economic and technological structure of the society than by its laws or its form of government.[16] Most of the Indian nations of New England were monarchies, and many of the cities of the Italian Renaissance were controlled by dictators. But in reading about these societies one gets the impression that they allowed far more personal freedom than out society does. In part this was because they lacked efficient mechanisms for enforcing the ruler's will: There were no modern, well-organized police forces, no rapid long-distance communications, no surveillance cameras, no dossiers of information about the lives of average citizens. Hence it was relatively easy to evade control.

96. As for our constitutional rights, consider for example that of freedom of the press. We certainly don't mean to knock that right; it is a very important tool for limiting concentration of political power and for keeping those who do have political power in line by publicly exposing any misbehavior on their part. But freedom of the press is of very little use to the average citizen as an individual. The mass media are mostly under the control of large organizations that are integrated into the system. Anyone who has a little money can have something printed, or can distribute it on the Internet or in some such way, but what he has to say will be swamped by the vast volume of material put out by the media, hence it will have no practical effect. To make an impression on society with words is therefore almost impossible for most individuals and small groups. Take us (FC) for example. If we had never done anything violent and had submitted the present writings to a publisher, they probably would not have been accepted. If they had been accepted and published, they probably would not have attracted many readers, because it's more fun to watch the entertainment put out by the media than to read a sober essay. Even if these writings had had many readers, most of these readers would soon have forgotten what they had read as their minds were flooded by the mass of material to which the media expose them. In order to get our message before the public with some chance of making a lasting impression, we've had to kill people.

97. Constitutional rights are useful up to a point, but they do not serve to guarantee much more than what might be called the bourgeois conception of freedom. According to the bourgeois conception, a "free" man is essentially

an element of a social machine and has only a certain set of prescribed and delimited freedoms; freedoms that are designed to serve the needs of the social machine more than those of the individual. Thus the bourgeois's "free" man has economic freedom because that promotes growth and progress; he has freedom of the press because public criticism restrains misbehavior by political leaders; he has a right to a fair trial because imprisonment at the whim of the powerful would be bad for the system. This was clearly the attitude of Simón Bolívar. To him, people deserved liberty only if they used it to promote progress (progress as conceived by the bourgeois). Other bourgeois thinkers have taken a similar view of freedom as a mere means to collective ends. Chester C. Tan, *Chinese Political Thought in the Twentieth Century*, page 202, explains the philosophy of the Kuomintang leader Hu Han-Min: "An individual is granted rights because he is a member of society and his community life requires such rights. By community Hu meant the whole society or the nation." And on page 259 Tan states that according to Carsun Chang (Chang Chun-Mai, head of the State Socialist Party in China) freedom had to be used in the interest of the state and of the people as a whole. But what kind of freedom does one have if one can use it only as someone else prescribes? FC's conception of freedom is not that of Bolívar, Hu, Chang or other bourgeois theorists. The trouble with such theorists is that they have made the development and application of social theories their surrogate activity. Consequently the theories are designed to serve the needs of the theorists more than the needs of any people who may be unlucky enough to live in a society on which the theories are imposed.

98. One more point to be made in this section: It should not be assumed that a person has enough freedom just because he SAYS he has enough. Freedom is restricted in part by psychological controls of which people are unconscious, and moreover many people's ideas of what constitutes freedom are governed more by social convention than by their real needs. For example, it's likely that many leftists of the oversocialized type would say that most people, including themselves, are socialized too little rather than too much, yet the oversocialized leftist pays a heavy psychological price for his high level of socialization.

Some Principles of History

99. Think of history as being the sum of two components: an erratic component that consists of unpredictable events that follow no discernible pattern, and a regular component that consists of long-term historical trends. Here we are concerned with the long-term trends.

100. FIRST PRINCIPLE. If a SMALL change is made that affects a long-term historical trend, then the effect of that change will almost always be transitory—the trend will soon revert to its original state. (Example: A reform movement designed to clean up political corruption in a society rarely has more than a short-term effect; sooner or later the reformers relax and corruption creeps back in. The level of political corruption in a given society tends to remain constant, or to change only slowly with the evolution of the society. Normally, a political cleanup will be permanent only if accompanied by widespread social changes; a SMALL change in the society won't be enough.) If a small change in a long-term historical trend appears to be permanent, it is only because the change acts in the direction in which the trend is already moving, so that the trend is not altered but only pushed a step ahead.

101. The first principle is almost a tautology. If a trend were not stable with respect to small changes, it would wander at random rather than following a definite direction; in other words it would not be a long-term trend at all.

102. SECOND PRINCIPLE. If a change is made that is sufficiently large to alter permanently a long-term historical trend, then it will alter the society as a whole. In other words, a society is a system in which all parts are interrelated, and you can't permanently change any important part without changing all other parts as well.

103. THIRD PRINCIPLE. If a change is made that is large enough to alter permanently a long-term trend, then the consequences for the society as a whole cannot be predicted in advance. (Unless various other societies have passed through the same change and have all experienced the same consequences, in which case one can predict on empirical grounds that another society that passes through the same change will be likely to experience similar consequences.)

104. FOURTH PRINCIPLE. A new kind of society cannot be designed on paper. That is, you cannot plan out a new form of society in advance, then set it up and expect it to function as it was designed to do.

105. The third and fourth principles result from the complexity of human societies. A change in human behavior will affect the economy of a society and its physical environment; the economy will affect the environment and vice versa, and the changes in the economy and the environment will affect human behavior in complex, unpredictable ways; and so forth. The network of causes and effects is far too complex to be untangled and understood.

106. FIFTH PRINCIPLE. People do not consciously and rationally choose the form of their society. Societies develop through processes of social evolution that are not under rational human control.

107. The fifth principle is a consequence of the other four.

108. To illustrate: By the first principle, generally speaking an attempt at social reform either acts in the direction in which the society is developing anyway (so that it merely accelerates a change that would have occurred in any case) or else it has only a transitory effect, so that the society soon slips back into its old groove. To make a lasting change in the direction of development of any important aspect of a society, reform is insufficient and revolution is required. (A revolution does not necessarily involve an armed uprising or the overthrow of a government.) By the second principle, a revolution never changes only one aspect of a society, it changes the whole society; and by the third principle changes occur that were never expected or desired by the revolutionaries. By the fourth principle, when revolutionaries or utopians set up a new kind of society, it never works out as planned.

109. The American Revolution does not provide a counterexample. The American "Revolution" was not a revolution in our sense of the word, but a war of independence followed by a rather far-reaching political reform. The Founding Fathers did not change the direction of development of American society, nor did they aspire to do so. They only freed the development of American society from the retarding effect of British rule. Their political reform did not change any basic trend, but only pushed American political culture along its natural direction of development. British society, of which American society was an offshoot, had been moving for a long time in the direction of representative democracy. And prior to the War of Independence the Americans were already practicing a significant degree of representative

democracy in the colonial assemblies. The political system established by the Constitution was modeled on the British system and on the colonial assemblies. With major alterations, to be sure—there is no doubt that the Founding Fathers took a very important step. But it was a step along the road that the English-speaking world was already traveling. The proof is that Britain and all of its colonies that were populated predominantly by people of British descent ended up with systems of representative democracy essentially similar to that of the United States. If the Founding Fathers had lost their nerve and declined to sign the Declaration of Independence, our way of life today would not have been significantly different. Maybe we would have had somewhat closer ties to Britain, and would have had a Parliament and Prime Minister instead of a Congress and President. No big deal. Thus the American Revolution provides not a counterexample to our principles but a good illustration of them.

110. Still, one has to use common sense in applying the principles. They are expressed in imprecise language that allows latitude for interpretation, and exceptions to them can be found. So we present these principles not as inviolable laws but as rules of thumb, or guides to thinking, that may provide a partial antidote to naive ideas about the future of society. The principles should be borne constantly in mind, and whenever one reaches a conclusion that conflicts with them one should carefully reexamine one's thinking and retain the conclusion only if one has good, solid reasons for doing so.

Industrial-Technological Society Cannot Be Reformed

111. The foregoing principles help to show how hopelessly difficult it would be to reform the industrial system in such a way as to prevent it from progressively narrowing our sphere of freedom. There has been a consistent tendency, going back at least to the Industrial Revolution, for technology to strengthen the system at a high cost in individual freedom and local autonomy. Hence any change designed to protect freedom from technology would be contrary to a fundamental trend in the development of our society. Consequently, such a change either would be a transitory one—soon swamped by the tide of history—or, if large enough to be permanent, would alter the nature of our whole society. This by the first and second principles.

Moreover, since society would be altered in a way that could not be predicted in advance (third principle) there would be great risk. Changes large enough to make a lasting difference in favor of freedom would not be initiated because it would be realized that they would gravely disrupt the system. So any attempts at reform would be too timid to be effective. Even if changes large enough to make a lasting difference were initiated, they would be retracted when their disruptive effects became apparent. Thus, permanent changes in favor of freedom could be brought about only by persons prepared to accept radical, dangerous and unpredictable alteration of the entire system. In other words by revolutionaries, not reformers.

112. People anxious to rescue freedom without sacrificing the supposed benefits of technology will suggest naive schemes for some new form of society that would reconcile freedom with technology. Apart from the fact that people who make such suggestions seldom propose any practical means by which the new form of society could be set up in the first place, it follows from the fourth principle that even if the new form of society could be once established, it either would collapse or would give results very different from those expected.

113. So even on very general grounds it seems highly improbable that any way of changing society could be found that would reconcile freedom with modern technology. In the next few sections we will give more specific reasons for concluding that freedom and technological progress are incompatible.

Restriction of Freedom is Unavoidable in Industrial Society

114. As explained in paragraphs 65-67, 70-73, modern man is strapped down by a network of rules and regulations, and his fate depends on the actions of persons remote from him whose decisions he cannot influence. This is not accidental or a result of the arbitrariness of arrogant bureaucrats. It is necessary and inevitable in any technologically advanced society. The system HAS TO regulate human behavior closely in order to function. At work, people have to do what they are told to do, when they are told to do it and in the way they are told to do it, otherwise production would be thrown

into chaos. Bureaucracies HAVE TO be run according to rigid rules. To allow any substantial personal discretion to lower-level bureaucrats would disrupt the system and lead to charges of unfairness due to differences in the way individual bureaucrats exercised their discretion. It is true that some restrictions on our freedom could be eliminated, but GENERALLY SPEAKING the regulation of our lives by large organizations is necessary for the functioning of industrial-technological society. The result is a sense of powerlessness on the part of the average person. It may be, however, that formal regulations will tend increasingly to be replaced by psychological tools that make us want to do what the system requires of us. (Propaganda,[14] educational techniques, "mental health" programs, etc.)

115. The system HAS TO force people to behave in ways that are increasingly remote from the natural pattern of human behavior. For example, the system needs scientists, mathematicians and engineers. It can't function without them. So heavy pressure is put on children to excel in these fields. It isn't natural for an adolescent human being to spend the bulk of his time sitting at a desk absorbed in study. A normal adolescent wants to spend his time in active contact with the real world. Among primitive peoples the things that children are trained to do tend to be in reasonable harmony with natural human impulses. Among the American Indians, for example, boys were trained in active outdoor pursuits—just the sort of things that boys like. But in our society children are pushed into studying technical subjects, which most do grudgingly.

116. Because of the constant pressure that the system exerts to modify human behavior, there is a gradual increase in the number of people who cannot or will not adjust to society's requirements: welfare leeches, youth-gang members, cultists, anti-government rebels, radical environmentalist saboteurs, dropouts and resisters of various kinds.

117. In any technologically advanced society the individual's fate MUST depend on decisions that he personally cannot influence to any great extent. A technological society cannot be broken down into small, autonomous communities, because production depends on the cooperation of very large numbers of people and machines. Such a society MUST be highly organized and decisions HAVE TO be made that affect very large numbers of people. When a decision affects, say, a million people, then each of the affected individuals has, on the average, only a one-millionth share in making the decision. What usually happens in practice is that decisions are made by

public officials or corporation executives, or by technical specialists, but even when the public votes on a decision the number of voters ordinarily is too large for the vote of any one individual to be significant.[17] Thus most individuals are unable to influence measurably the major decisions that affect their lives. There is no conceivable way to remedy this in a technologically advanced society. The system tries to "solve" this problem by using propaganda to make people WANT the decisions that have been made for them, but even if this "solution" were completely successful in making people feel better, it would be demeaning.

118. Conservatives and some others advocate more "local autonomy." Local communities once did have autonomy, but such autonomy becomes less and less possible as local communities become more enmeshed with and dependent on large-scale systems like public utilities, computer networks, highway systems, the mass communications media and the modern health-care system. Also operating against autonomy is the fact that technology applied in one location often affects people at other locations far away. Thus pesticide or chemical use near a creek may contaminate the water supply hundreds of miles downstream, and the greenhouse effect affects the whole world.

119. The system does not and cannot exist to satisfy human needs. Instead, it is human behavior that has to be modified to fit the needs of the system. This has nothing to do with the political or social ideology that may pretend to guide the technological system. It is not the fault of capitalism and it is not the fault of socialism. It is the fault of technology, because the system is guided not by ideology but by technical necessity.[18] Of course the system does satisfy many human needs, but generally speaking it does this only to the extent that it is to the advantage of the system to do it. It is the needs of the system that are paramount, not those of the human being. For example, the system provides people with food because the system couldn't function if everyone starved; it attends to people's psychological needs whenever it can CONVENIENTLY do so, because it couldn't function if too many people became depressed or rebellious. But the system, for good, solid, practical reasons, must exert constant pressure on people to mold their behavior to the needs of the system. Too much waste accumulating? The government, the media, the educational system, environmentalists, everyone inundates us with a mass of propaganda about recycling. Need more technical personnel? A chorus of voices exhorts kids to study science. No one

stops to ask whether it is inhumane to force adolescents to spend the bulk of their time studying subjects that most of them hate. When skilled workers are put out of a job by technical advances and have to undergo "retraining," no one asks whether it is humiliating for them to be pushed around in this way. It is simply taken for granted that everyone must bow to technical necessity. And for good reason: If human needs were put before technical necessity there would be economic problems, unemployment, shortages or worse. The concept of "mental health" in our society is defined largely by the extent to which an individual behaves in accord with the needs of the system and does so without showing signs of stress.

120. Efforts to make room for a sense of purpose and for autonomy within the system are no better than a joke. For example, one company, instead of having each of its employees assemble only one section of a catalogue, had each assemble a whole catalogue, and this was supposed to give them a sense of purpose and achievement. Some companies have tried to give their employees more autonomy in their work, but for practical reasons this usually can be done only to a very limited extent, and in any case employees are never given autonomy as to ultimate goals—their "autonomous" efforts can never be directed toward goals that they select personally, but only toward their employer's goals, such as the survival and growth of the company. Any company would soon go out of business if it permitted its employees to act otherwise. Similarly, in any enterprise within a socialist system, workers must direct their efforts toward the goals of the enterprise, otherwise the enterprise will not serve its purpose as part of the system. Once again, for purely technical reasons it is not possible for most individuals or small groups to have much autonomy in industrial society. Even the small-business owner commonly has only limited autonomy. Apart from the necessity of government regulation, he is restricted by the fact that he must fit into the economic system and conform to its requirements. For instance, when someone develops a new technology, the small-business person often has to use that technology whether he wants to or not, in order to remain competitive.

The "Bad" Parts of Technology Cannot Be Separated from the "Good" Parts

121. A further reason why industrial society cannot be reformed in favor of freedom is that modern technology is a unified system in which all parts are dependent on one another. You can't get rid of the "bad" parts of technology and retain only the "good" parts. Take modern medicine, for example. Progress in medical science depends on progress in chemistry, physics, biology, computer science and other fields. Advanced medical treatments require expensive, high-tech equipment that can be made available only by a technologically progressive, economically rich society. Clearly you can't have much progress in medicine without the whole technological system and everything that goes with it.

122. Even if medical progress could be maintained without the rest of the technological system, it would by itself bring certain evils. Suppose for example that a cure for diabetes is discovered. People with a genetic tendency to diabetes will then be able to survive and reproduce as well as anyone else. Natural selection against genes for diabetes will cease and such genes will spread throughout the population. (This may be occurring to some extent already, since diabetes, while not curable, can be controlled through the use of insulin.) The same thing will happen with many other diseases susceptibility to which is affected by genetic factors (e.g., childhood cancer), resulting in massive genetic degradation of the population. The only solution will be some sort of eugenics program or extensive genetic engineering of human beings, so that man in the future will no longer be a creation of nature, or of chance, or of God (depending on your religious or philosophical opinions), but a manufactured product.

123. If you think that big government interferes in your life too much NOW, just wait till the government starts regulating the genetic constitution of your children. Such regulation will inevitably follow the introduction of genetic engineering of human beings, because the consequences of unregulated genetic engineering would be disastrous.[19]

124. The usual response to such concerns is to talk about "medical ethics." But a code of ethics would not serve to protect freedom in the face of medical progress; it would only make matters worse. A code of ethics

applicable to genetic engineering would be in effect a means of regulating the genetic constitution of human beings. Somebody (probably the upper middle class, mostly) would decide that such and such applications of genetic engineering were "ethical" and others were not, so that in effect they would be imposing their own values on the genetic constitution of the population at large. Even if a code of ethics were chosen on a completely democratic basis, the majority would be imposing their own values on any minorities who might have a different idea of what constituted an "ethical" use of genetic engineering. The only code of ethics that would truly protect freedom would be one that prohibited ANY genetic engineering of human beings, and you can be sure that no such code will ever be applied in a technological society. No code that reduced genetic engineering to a minor role could stand up for long, because the temptation presented by the immense power of biotechnology would be irresistible, especially since to the majority of people many of its applications will seem obviously and unequivocally good (eliminating physical and mental diseases, giving people the abilities they need to get along in today's world). Inevitably, genetic engineering will be used extensively, but only in ways consistent with the needs of the industrial-technological system.[20]

Technology is a More Powerful Social Force than the Aspiration for Freedom

125. It is not possible to make a LASTING compromise between technology and freedom, because technology is by far the more powerful social force and continually encroaches on freedom through REPEATED compromises. Imagine the case of two neighbors, each of whom at the outset owns the same amount of land, but one of whom is more powerful than the other. The powerful one demands a piece of the other's land. The weak one refuses. The powerful one says, "Okay, let's compromise. Give me half of what I asked." The weak one has little choice but to give in. Some time later the powerful neighbor demands another piece of land, again there is a compromise, and so forth. By forcing a long series of compromises on the weaker man, the powerful one eventually gets all of his land. So it goes in the conflict between technology and freedom.

126. Let us explain why technology is a more powerful social force than the aspiration for freedom.

127. A technological advance that appears not to threaten freedom often turns out to threaten it very seriously later on. For example, consider motorized transport. A walking man formerly could go where he pleased, go at his own pace without observing any traffic regulations, and was independent of technological support systems. When motor vehicles were introduced they appeared to increase man's freedom. They took no freedom away from the walking man, no one had to have an automobile if he didn't want one, and anyone who did choose to buy an automobile could travel much faster and farther than a walking man. But the introduction of motorized transport soon changed society in such a way as to restrict greatly man's freedom of locomotion. When automobiles became numerous, it became necessary to regulate their use extensively. In a car, especially in densely populated areas, one cannot just go where one likes at one's own pace; one's movement is governed by the flow of traffic and by various traffic laws. One is tied down by various obligations: license requirements, driver test, renewing registration, insurance, maintenance required for safety, monthly payments on purchase price. Moreover, the use of motorized transport is no longer optional. Since the introduction of motorized transport the arrangement of our cities has changed in such a way that the majority of people no longer live within walking distance of their place of employment, shopping areas and recreational opportunities, so that they HAVE TO depend on the automobile for transportation. Or else they must use public transportation, in which case they have even less control over their own movement than when driving a car. Even the walker's freedom is now greatly restricted. In the city he continually has to stop to wait for traffic lights that are designed mainly to serve auto traffic. In the country, motor traffic makes it dangerous and unpleasant to walk along the highway. (Note this important point that we have just illustrated with the case of motorized transport: When a new item of technology is introduced as an option that an individual can accept or not as he chooses, it does not necessarily REMAIN optional. In many cases the new technology changes society in such a way that people eventually find themselves FORCED to use it.)

128. While technological progress AS A WHOLE continually narrows our sphere of freedom, each new technical advance CONSIDERED BY ITSELF appears to be desirable. Electricity, indoor plumbing, rapid long-

distance communications...how could one argue against any of these things, or against any other of the innumerable technical advances that have made modern society? It would have been absurd to resist the introduction of the telephone, for example. It offered many advantages and no disadvantages. Yet, as we explained in paragraphs 59-76, all these technical advances taken together have created a world in which the average man's fate is no longer in his own hands or in the hands of his neighbors and friends, but in those of politicians, corporation executives and remote, anonymous technicians and bureaucrats whom he as an individual has no power to influence.[21] The same process will continue in the future. Take genetic engineering, for example. Few people will resist the introduction of a genetic technique that eliminates a hereditary disease. It does no apparent harm and prevents much suffering. Yet a large number of genetic improvements taken together will make the human being into an engineered product rather than a free creation of chance (or of God, or whatever, depending on your religious beliefs).

129. Another reason why technology is such a powerful social force is that, within the context of a given society, technological progress marches in only one direction; it can never be reversed. Once a technical innovation has been introduced, people usually become dependent on it, so that they can never again do without it, unless it is replaced by some still more advanced innovation. Not only do people become dependent as individuals on a new item of technology, but, even more, the system as a whole becomes dependent on it. (Imagine what would happen to the system today if computers, for example, were eliminated.) Thus the system can move in only one direction, toward greater technologization. Technology repeatedly forces freedom to take a step back but technology can never take a step back—short of the overthrow of the whole technological system.

130. Technology advances with great rapidity and threatens freedom at many different points at the same time (crowding, rules and regulations, increasing dependence of individuals on large organizations, propaganda and other psychological techniques, genetic engineering, invasion of privacy through surveillance devices and computers, etc.). To hold back any ONE of the threats to freedom would require a long and difficult social struggle. Those who want to protect freedom are overwhelmed by the sheer number of new attacks and the rapidity with which they develop, hence they become apathetic and no longer resist. To fight each of the threats separately would

be futile. Success can be hoped for only by fighting the technological system as a whole; but that is revolution, not reform.

131. Technicians (we use this term in its broad sense to describe all those who perform a specialized task that requires training) tend to be so involved in their work (their surrogate activity) that when a conflict arises between their technical work and freedom, they almost always decide in favor of their technical work. This is obvious in the case of scientists, but it also appears elsewhere: Educators, humanitarian groups, conservation organizations do not hesitate to use propaganda[14] or other psychological techniques to help them achieve their laudable ends. Corporations and government agencies, when they find it useful, do not hesitate to collect information about individuals without regard to their privacy. Law enforcement agencies are frequently inconvenienced by the constitutional rights of suspects and often of completely innocent persons, and they do whatever they can do legally (or sometimes illegally) to restrict or circumvent those rights. Most of these educators, government officials and law officers believe in freedom, privacy and constitutional rights, but when these conflict with their work, they usually feel that their work is more important.

132. It is well known that people generally work better and more persistently when striving for a reward than when attempting to avoid a punishment or negative outcome. Scientists and other technicians are motivated mainly by the rewards they get through their work. But those who oppose technological invasions of freedom are working to avoid a negative outcome, consequently there are few who work persistently and well at this discouraging task. If reformers ever achieved a signal victory that seemed to set up a solid barrier against further erosion of freedom through technical progress, most would tend to relax and turn their attention to more agreeable pursuits. But the scientists would remain busy in their laboratories, and technology as it progressed would find ways, in spite of any barriers, to exert more and more control over individuals and make them always more dependent on the system.

133. No social arrangements, whether laws, institutions, customs or ethical codes, can provide permanent protection against technology. History shows that all social arrangements are transitory; they all change or break down eventually. But technological advances are permanent within the context of a given civilization. Suppose for example that it were possible to arrive at some social arrangement that would prevent genetic engineering

from being applied to human beings, or prevent it from being applied in such a way as to threaten freedom and dignity. Still, the technology would remain, waiting. Sooner or later the social arrangement would break down. Probably sooner, given the pace of change in our society. Then genetic engineering would begin to invade our sphere of freedom, and this invasion would be irreversible (short of a breakdown of technological civilization itself). Any illusions about achieving anything permanent through social arrangements should be dispelled by what is currently happening with environmental legislation. A few years ago it seemed that there were secure legal barriers preventing at least SOME of the worst forms of environmental degradation. A change in the political wind, and those barriers begin to crumble.

134. For all of the foregoing reasons, technology is a more powerful social force than the aspiration for freedom. But this statement requires an important qualification. It appears that during the next several decades the industrial-technological system will be undergoing severe stresses due to economic and environmental problems, and especially due to problems of human behavior (alienation, rebellion, hostility, a variety of social and psychological difficulties). We hope that the stresses through which the system is likely to pass will cause it to break down, or at least will weaken it sufficiently so that a revolution against it becomes possible. If such a revolution occurs and is successful, then at that particular moment the aspiration for freedom will have proved more powerful than technology.

135. In paragraph 125 we used an analogy of a weak neighbor who is left destitute by a strong neighbor who takes all his land by forcing on him a series of compromises. But suppose now that the strong neighbor gets sick, so that he is unable to defend himself. The weak neighbor can force the strong one to give him his land back, or he can kill him. If he lets the strong man survive and only forces him to give the land back, he is a fool, because when the strong man gets well he will again take all the land for himself. The only sensible alternative for the weaker man is to kill the strong one while he has the chance. In the same way, while the industrial system is sick we must destroy it. If we compromise with it and let it recover from its sickness, it will eventually wipe out all of our freedom.

Simpler Social Problems Have Proved Intractable

136. If anyone still imagines that it would be possible to reform the system in such a way as to protect freedom from technology, let him consider how clumsily and for the most part unsuccessfully our society has dealt with other social problems that are far more simple and straightforward. Among other things, the system has failed to stop environmental degradation, political corruption, drug trafficking or domestic abuse.

137. Take our environmental problems, for example. Here the conflict of values is straightforward: economic expedience now versus saving some of our natural resources for our grandchildren.[22] But on this subject we get only a lot of blather and obfuscation from the people who have power, and nothing like a clear, consistent line of action, and we keep on piling up environmental problems that our grandchildren will have to live with. Attempts to resolve the environmental issue consist of struggles and compromises between different factions, some of which are ascendant at one moment, others at another moment. The line of struggle changes with the shifting currents of public opinion. This is not a rational process, nor is it one that is likely to lead to a timely and successful solution to the problem. Major social problems, if they get "solved" at all, are rarely or never solved through any rational, comprehensive plan. They just work themselves out through a process in which various competing groups pursuing their own (usually short-term) self-interest[23] arrive (mainly by luck) at some more or less stable modus vivendi. In fact, the principles we formulated in paragraphs 100-106 make it seem doubtful that rational, long-term social planning can EVER be successful.

138. Thus it is clear that the human race has at best a very limited capacity for solving even relatively straightforward social problems. How then is it going to solve the far more difficult and subtle problem of reconciling freedom with technology? Technology presents clear-cut material advantages, whereas freedom is an abstraction that means different things to different people, and its loss is easily obscured by propaganda and fancy talk.

139. And note this important difference: It is conceivable that our environmental problems (for example) may some day be settled through a rational, comprehensive plan, but if this happens it will be only because

it is in the long-term interest of the system to solve these problems. But it is NOT in the interest of the system to preserve freedom or small-group autonomy. On the contrary, it is in the interest of the system to bring human behavior under control to the greatest possible extent.[24] Thus, while practical considerations may eventually force the system to take a rational, prudent approach to environmental problems, equally practical considerations will force the system to regulate human behavior ever more closely (preferably by indirect means that will disguise the encroachment on freedom). This isn't just our opinion. Eminent social scientists (e.g., James Q. Wilson) have stressed the importance of "socializing" people more effectively.

Revolution is Easier than Reform

140. We hope we have convinced the reader that the system cannot be reformed in such a way as to reconcile freedom with technology. The only way out is to dispense with the industrial-technological system altogether. This implies revolution, not necessarily an armed uprising, but certainly a radical and fundamental change in the nature of society.

141. People tend to assume that because a revolution involves a much greater change than reform does, it is more difficult to bring about than reform is. Actually, under certain circumstances revolution is much easier than reform. The reason is that a revolutionary movement can inspire an intensity of commitment that a reform movement cannot inspire. A reform movement merely offers to solve a particular social problem. A revolutionary movement offers to solve all problems at one stroke and create a whole new world; it provides the kind of ideal for which people will take great risks and make great sacrifices. For this reason it would be much easier to overthrow the whole technological system than to put effective, permanent restraints on the development or application of any one segment of technology, such as genetic engineering, for example. Not many people will devote themselves with single-minded passion to imposing and maintaining restraints on genetic engineering, but under suitable conditions large numbers of people may devote themselves passionately to a revolution against the industrial-technological system. As we noted in paragraph 132, reformers seeking to limit certain aspects of technology would be working to avoid a negative

outcome. But revolutionaries work to gain a powerful reward—fulfillment of their revolutionary vision—and therefore work harder and more persistently than reformers do.

142. Reform is always restrained by the fear of painful consequences if changes go too far. But once a revolutionary fever has taken hold of a society, people are willing to undergo unlimited hardships for the sake of their revolution. This was clearly shown in the French and Russian Revolutions. It may be that in such cases only a minority of the population is really committed to the revolution, but this minority is sufficiently large and active so that it becomes the dominant force in society. We will have more to say about revolution in paragraphs 180-205.

Control of Human Behavior

143. Since the beginning of civilization, organized societies have had to put pressures on human beings for the sake of the functioning of the social organism. The kinds of pressures vary greatly from one society to another. Some of the pressures are physical (poor diet, excessive labor, environmental pollution), some are psychological (noise, crowding, forcing human behavior into the mold that society requires). In the past, human nature has been approximately constant, or at any rate has varied only within certain bounds. Consequently, societies have been able to push people only up to certain limits. When the limit of human endurance has been passed, things start going wrong: rebellion, or crime, or corruption, or evasion of work, or depression and other mental problems, or an elevated death rate, or a declining birth rate or something else, so that either the society breaks down, or its functioning becomes too inefficient and it is (quickly or gradually, through conquest, attrition or evolution) replaced by some more efficient form of society.[25]

144. Thus human nature has in the past put certain limits on the development of societies. People could be pushed only so far and no farther. But today this may be changing, because modern technology is developing ways of modifying human beings.

145. Imagine a society that subjects people to conditions that make them terribly unhappy, then gives them drugs to take away their unhappiness. Science fiction? It is already happening to some extent in our own society. It

is well known that the rate of clinical depression has been greatly increasing in recent decades. We believe that this is due to disruption of the power process, as explained in paragraphs 59-76. But even if we are wrong, the increasing rate of depression is certainly the result of SOME conditions that exist in today's society. Instead of removing the conditions that make people depressed, modern society gives them antidepressant drugs. In effect, antidepressants are a means of modifying an individual's internal state in such a way as to enable him to tolerate social conditions that he would otherwise find intolerable. (Yes, we know that depression is often of purely genetic origin. We are referring here to those cases in which environment plays the predominant role.)

146. Drugs that affect the mind are only one example of the methods of controlling human behavior that modern society is developing. Let us look at some of the other methods.

147. To start with, there are the techniques of surveillance. Hidden video cameras are now used in most stores and in many other places, computers are used to collect and process vast amounts of information about individuals. Information so obtained greatly increases the effectiveness of physical coercion (i.e., law enforcement).[26] Then there are the methods of propaganda, for which the mass communications media provide effective vehicles. Efficient techniques have been developed for winning elections, selling products, influencing public opinion. The entertainment industry serves as an important psychological tool of the system, possibly even when it is dishing out large amounts of sex and violence. Entertainment provides modern man with an essential means of escape. While absorbed in television, videos, etc., he can forget stress, anxiety, frustration, dissatisfaction. Many primitive peoples, when they don't have any work to do, are quite content to sit for hours at a time doing nothing at all, because they are at peace with themselves and their world. But most modern people must be constantly occupied or entertained, otherwise they get "bored," i.e., they get fidgety, uneasy, irritable.

148. Other techniques strike deeper that the foregoing. Education is no longer a simple affair of paddling a kid's behind when he doesn't know his lessons and patting him on the head when he does know them. It is becoming a scientific technique for controlling the child's development. Sylvan Learning Centers, for example, have had great success in motivating children to study, and psychological techniques are also used with more or

less success in many conventional schools. "Parenting" techniques that are taught to parents are designed to make children accept the fundamental values of the system and behave in ways that the system finds desirable. "Mental health" programs, "intervention" techniques, psychotherapy and so forth are ostensibly designed to benefit individuals, but in practice they usually serve as methods for inducing individuals to think and behave as the system requires. (There is no contradiction here; an individual whose attitudes or behavior bring him into conflict with the system is up against a force that is too powerful for him to conquer or escape from, hence he is likely to suffer from stress, frustration, defeat. His path will be much easier if he thinks and behaves as the system requires. In that sense the system is acting for the benefit of the individual when it brainwashes him into conformity.) Child abuse in its gross and obvious forms is disapproved in most if not all cultures. Tormenting a child for a trivial reason or no reason at all is something that appalls almost everyone. But many psychologists interpret the concept of abuse much more broadly. Is spanking, when used as part of a rational and consistent system of discipline, a form of abuse? The question will ultimately be decided by whether or not spanking tends to produce behavior that makes a person fit in well with the existing system of society. In practice, the word "abuse" tends to be interpreted to include any method of child-rearing that produces behavior inconvenient for the system. Thus, when they go beyond the prevention of obvious, senseless cruelty, programs for preventing "child abuse" are directed toward the control of human behavior on behalf of the system.

149. Presumably, research will continue to increase the effectiveness of psychological techniques for controlling human behavior. But we think it is unlikely that psychological techniques alone will be sufficient to adjust human beings to the kind of society that technology is creating. Biological methods probably will have to be used. We have already mentioned the use of drugs in this connection. Neurology may provide other avenues for modifying the human mind. Genetic engineering of human beings is already beginning to occur in the form of "gene therapy," and there is no reason to assume that such methods will not eventually be used to modify those aspects of the body that affect mental functioning.

150. As we mentioned in paragraph 134, industrial society seems likely to be entering a period of severe stress, due in part to problems of human behavior and in part to economic and environmental problems.

And a considerable proportion of the system's economic and environmental problems result from the way human beings behave. Alienation, low self-esteem, depression, hostility, rebellion; children who won't study, youth gangs, illegal drug use, rape, child abuse, other crimes, unsafe sex, teen pregnancy, population growth, political corruption, race hatred, ethnic rivalry, bitter ideological conflict (e.g., pro-choice vs. pro-life), political extremism, terrorism, sabotage, anti-government groups, hate groups. All these threaten the very survival of the system. The system will therefore be FORCED to use every practical means of controlling human behavior.

151. The social disruption that we see today is certainly not the result of mere chance. It can only be a result of the conditions of life that the system imposes on people. (We have argued that the most important of these conditions is disruption of the power process.) If the systems succeeds in imposing sufficient control over human behavior to assure its own survival, a new watershed in human history will have been passed. Whereas formerly the limits of human endurance have imposed limits on the development of societies (as we explained in paragraphs 143, 144), industrial-technological society will be able to pass those limits by modifying human beings, whether by psychological methods or biological methods or both. In the future, social systems will not be adjusted to suit the needs of human beings. Instead, human beings will be adjusted to suit the needs of the system.[27]

152. Generally speaking, technological control over human behavior will probably not be introduced with a totalitarian intention or even through a conscious desire to restrict human freedom.[28] Each new step in the assertion of control over the human mind will be taken as a rational response to a problem that faces society, such as curing alcoholism, reducing the crime rate or inducing young people to study science and engineering. In many cases, there will be a humanitarian justification. For example, when a psychiatrist prescribes an antidepressant for a depressed patient, he is clearly doing that individual a favor. It would be inhumane to withhold the drug from someone who needs it. When parents send their children to Sylvan Learning Centers to have them manipulated into becoming enthusiastic about their studies, they do so from concern for their children's welfare. It may be that some of these parents wish that one didn't have to have specialized training to get a job and that their kid didn't have to be brainwashed into becoming a computer nerd. But what can they

do? They can't change society, and their child may be unemployable if he doesn't have certain skills. So they send him to Sylvan.

153. Thus control over human behavior will be introduced not by a calculated decision of the authorities but through a process of social evolution (RAPID evolution, however). The process will be impossible to resist, because each advance, considered by itself, will appear to be beneficial, or at least the evil involved in making the advance will seem to be less than that which would result from not making it. (See paragraph 127.) Propaganda for example is used for many good purposes, such as discouraging child abuse or race hatred.[14] Sex education is obviously useful, yet the effect of sex education (to the extent that it is successful) is to take the shaping of sexual attitudes away from the family and put it into the hands of the state as represented by the public school system.

154. Suppose a biological trait is discovered that increases the likelihood that a child will grow up to be a criminal, and suppose some sort of gene therapy can remove this trait.[29] Of course most parents whose children possess the trait will have them undergo the therapy. It would be inhumane to do otherwise, since the child would probably have a miserable life if he grew up to be a criminal. But many or most primitive societies have a low crime rate in comparison with that of our society, even though they have neither high-tech methods of child-rearing nor harsh systems of punishment. Since there is no reason to suppose that more modern men than primitive men have innate predatory tendencies, the high crime rate of our society must be due to the pressures that modern conditions put on people, to which many cannot or will not adjust. Thus a treatment designed to remove potential criminal tendencies is at least in part a way of re-engineering people so that they suit the requirements of the system.

155. Our society tends to regard as a "sickness" any mode of thought or behavior that is inconvenient for the system, and this is plausible, because when an individual doesn't fit into the system it causes pain to the individual as well as problems for the system. Thus the manipulation of an individual to adjust him to the system is seen as a "cure" for a "sickness" and therefore as good.

156. In paragraph 127 we pointed out that if the use of a new item of technology is INITIALLY optional, it does not necessarily REMAIN optional, because the new technology tends to change society in such a way that it becomes difficult or impossible for an individual to function without

using that technology. This applies also to the technology of human behavior. In a world in which most children are put through a program to make them enthusiastic about studying, a parent will almost be forced to put his kid through such a program, because if he does not, then the kid will grow up to be, comparatively speaking, an ignoramus and therefore unemployable. Or suppose a biological treatment is discovered that, without undesirable side-effects, will greatly reduce the psychological stress from which so many people suffer in our society. If large numbers of people choose to undergo the treatment, then the general level of stress in society will be reduced, so that it will be possible for the system to increase the stress-producing pressures. This will lead more people to undergo the treatment; and so forth, so that eventually the pressures may become so heavy that few people will be able to survive without undergoing the stress-reducing treatment. In fact, something like this seems to have happened already with one of our society's most important psychological tools for enabling people to reduce (or at least temporarily escape from) stress, namely, mass entertainment (see paragraph 147). Our use of mass entertainment is "optional": No law requires us to watch television, listen to the radio, read magazines. Yet mass entertainment is a means of escape and stress-reduction on which most of us have become dependent. Everyone complains about the trashiness of television, but almost everyone watches it. A few have kicked the TV habit, but it would be a rare person who could get along today without using ANY form of mass entertainment. (Yet until quite recently in human history most people got along very nicely with no other entertainment than that which each local community created for itself.) Without the entertainment industry the system probably would not have been able to get away with putting as much stress-producing pressure on us as it does.

157. Assuming that industrial society survives, it is likely that technology will eventually acquire something approaching complete control over human behavior. It has been established beyond any rational doubt that human thought and behavior have a largely biological basis. As experimenters have demonstrated, feelings such as hunger, pleasure, anger and fear can be turned on and off by electrical stimulation of appropriate parts of the brain. Memories can be destroyed by damaging parts of the brain or they can be brought to the surface by electrical stimulation. Hallucinations can be induced or moods changed by drugs. There may or may not be an immaterial human soul, but if there is one it clearly is less powerful than the

biological mechanisms of human behavior. For if that were not the case then researchers would not be able so easily to manipulate human feelings and behavior with drugs and electrical currents.

158. It presumably would be impractical for all people to have electrodes inserted in their heads so that they could be controlled by the authorities. But the fact that human thoughts and feelings are so open to biological intervention shows that the problem of controlling human behavior is mainly a technical problem; a problem of neurons, hormones and complex molecules; the kind of problem that is accessible to scientific attack. Given the outstanding record of our society in solving technical problems, it is overwhelmingly probable that great advances will be made in the control of human behavior.

159. Will public resistance prevent the introduction of technological control of human behavior? It certainly would if an attempt were made to introduce such control all at once. But since technological control will be introduced through a long sequence of small advances, there will be no rational and effective public resistance. (See paragraphs 127,132, 153.)

160. To those who think that all this sounds like science fiction, we point out that yesterday's science fiction is today's fact. The Industrial Revolution has radically altered man's environment and way of life, and it is only to be expected that as technology is increasingly applied to the human body and mind, man himself will be altered as radically as his environment and way of life have been.

Human Race at a Crossroads

161. But we have gotten ahead of our story. It is one thing to develop in the laboratory a series of psychological or biological techniques for manipulating human behavior and quite another to integrate these techniques into a functioning social system. The latter problem is the more difficult of the two. For example, while the techniques of educational psychology doubtless work quite well in the "lab schools" where they are developed, it is not necessarily easy to apply them effectively throughout our educational system. We all know what many of our schools are like. The teachers are too busy taking knives and guns away from the kids to subject them to the latest techniques for making them into computer nerds.

Thus, in spite of all its technical advances relating to human behavior, the system to date has not been impressively successful in controlling human beings. The people whose behavior is fairly well under the control of the system are those of the type that might be called "bourgeois." But there are growing numbers of people who in one way or another are rebels against the system: welfare leeches, youth gangs, cultists, satanists, Nazis, radical environmentalists, militiamen, etc.

162. The system is currently engaged in a desperate struggle to overcome certain problems that threaten its survival, among which the problems of human behavior are the most important. If the system succeeds in acquiring sufficient control over human behavior quickly enough, it will probably survive. Otherwise it will break down. We think the issue will most likely be resolved within the next several decades, say 40 to 100 years.

163. Suppose the system survives the crisis of the next several decades. By that time it will have to have solved, or at least brought under control, the principal problems that confront it, in particular that of "socializing" human beings; that is, making people sufficiently docile so that their behavior no longer threatens the system. That being accomplished, it does not appear that there would be any further obstacle to the development of technology, and it would presumably advance toward its logical conclusion, which is complete control over everything on Earth, including human beings and all other important organisms. The system may become a unitary, monolithic organization, or it may be more or less fragmented and consist of a number of organizations coexisting in a relationship that includes elements of both cooperation and competition, just as today the government, the corporations and other large organizations both cooperate and compete with one another. Human freedom mostly will have vanished, because individuals and small groups will be impotent vis-à-vis large organizations armed with supertechnology and an arsenal of advanced psychological and biological tools for manipulating human beings, besides instruments of surveillance and physical coercion. Only a small number of people will have any real power, and even these probably will have only very limited freedom, because their behavior too will be regulated; just as today our politicians and corporation executives can retain their positions of power only as long as their behavior remains within certain fairly narrow limits.

164. Don't imagine that the system will stop developing further techniques for controlling human beings and nature once the crisis of the

next few decades is over and increasing control is no longer necessary for the system's survival. On the contrary, once the hard times are over the system will increase its control over people and nature more rapidly, because it will no longer be hampered by difficulties of the kind that it is currently experiencing. Survival is not the principal motive for extending control. As we explained in paragraphs 87-90, technicians and scientists carry on their work largely as a surrogate activity; that is, they satisfy their need for power by solving technical problems. They will continue to do this with unabated enthusiasm, and among the most interesting and challenging problems for them to solve will be those of understanding the human body and mind and intervening in their development. For the "good of humanity," of course.

165. But suppose on the other hand that the stresses of the coming decades prove to be too much for the system. If the system breaks down there may be a period of chaos, a "time of troubles" such as those that history has recorded at various epochs in the past. It is impossible to predict what would emerge from such a time of troubles, but at any rate the human race would be given a new chance. The greatest danger is that industrial society may begin to reconstitute itself within the first few years after the breakdown. Certainly there will be many people (power-hungry types especially) who will be anxious to get the factories running again.

166. Therefore two tasks confront those who hate the servitude to which the industrial system is reducing the human race. First, we must work to heighten the social stresses within the system so as to increase the likelihood that it will break down or be weakened sufficiently so that a revolution against it becomes possible. Second, it is necessary to develop and propagate an ideology that opposes technology and the industrial system. Such an ideology can become the basis for a revolution against industrial society if and when the system becomes sufficiently weakened. And such an ideology will help to assure that, if and when industrial society breaks down, its remnants will be smashed beyond repair, so that the system cannot be reconstituted. The factories should be destroyed, technical books burned, etc.

Human Suffering

167. The industrial system will not break down purely as a result of revolutionary action. It will not be vulnerable to revolutionary attack

unless its own internal problems of development lead it into very serious difficulties. So if the system breaks down it will do so either spontaneously, or through a process that is in part spontaneous but helped along by revolutionaries. If the breakdown is sudden, many people will die, since the world's population has become so overblown that it cannot even feed itself any longer without advanced technology. Even if the breakdown is gradual enough so that reduction of the population can occur more through lowering of the birth rate than through elevation of the death rate, the process of de-industrialization probably will be very chaotic and involve much suffering. It is naive to think it likely that technology can be phased out in a smoothly managed, orderly way, especially since the technophiles will fight stubbornly at every step. Is it therefore cruel to work for the breakdown of the system? Maybe, but maybe not. In the first place, revolutionaries will not be able to break the system down unless it is already in enough trouble so that there would be a good chance of its eventually breaking down by itself anyway; and the bigger the system grows, the more disastrous the consequences of its breakdown will be; so it may be that revolutionaries, by hastening the onset of the breakdown, will be reducing the extent of the disaster.

168. In the second place, one has to balance struggle and death against the loss of freedom and dignity. To many of us, freedom and dignity are more important than a long life or avoidance of physical pain. Besides, we all have to die sometime, and it may be better to die fighting for survival, or for a cause, than to live a long but empty and purposeless life.

169. In the third place, it is not at all certain that survival of the system will lead to less suffering than the breakdown of the system would. The system has already caused, and is continuing to cause, immense suffering all over the world. Ancient cultures, that for hundreds or thousands of years gave people a satisfactory relationship with each other and with their environment, have been shattered by contact with industrial society, and the result has been a whole catalog of economic, environmental, social and psychological problems. One of the effects of the intrusion of industrial society has been that over much of the world traditional controls on population have been thrown out of balance. Hence the population explosion, with all that that implies. Then there is the psychological suffering that is widespread throughout the supposedly fortunate countries of the West (see paragraphs 44, 45). No one knows what will happen as a result of ozone depletion, the greenhouse effect and other environmental problems

that cannot yet be foreseen. And, as nuclear proliferation has shown, new technology cannot be kept out of the hands of dictators and irresponsible Third World nations. Would you like to speculate about what Iraq or North Korea will do with genetic engineering?

170. "Oh!" say the technophiles, "Science is going to fix all that! We will conquer famine, eliminate psychological suffering, make everybody healthy and happy!" Yeah, sure. That's what they said 200 years ago. The Industrial Revolution was supposed to eliminate poverty, make everybody happy, etc. The actual result has been quite different. The technophiles are hopelessly naive (or self-deceiving) in their understanding of social problems. They are unaware of (or choose to ignore) the fact that when large changes, even seemingly beneficial ones, are introduced into a society, they lead to a long sequence of other changes, most of which are impossible to predict (paragraph 103). The result is disruption of the society. So it is very probable that in their attempts to end poverty and disease, engineer docile, happy personalities and so forth, the technophiles will create social systems that are terribly troubled, even more so than the present one. For example, the scientists boast that they will end famine by creating new, genetically engineered food plants. But this will allow the human population to keep expanding indefinitely, and it is well known that crowding leads to increased stress and aggression. This is merely one example of the PREDICTABLE problems that will arise. We emphasize that, as past experience has shown, technical progress will lead to other new problems that CANNOT be predicted in advance (paragraph 103). In fact, ever since the Industrial Revolution technology has been creating new problems for society far more rapidly that it has been solving old ones. Thus it will take a long and difficult period of trial and error for the technophiles to work the bugs out of their Brave New World (if they ever do). In the mean time there will be great suffering. So it is not at all clear that the survival of industrial society would involve less suffering than the breakdown of that society would. Technology has gotten the human race into a fix from which there is not likely to be any easy escape.

The Future

171. But suppose now that industrial society does survive the next several decades and that the bugs do eventually get worked out of the system, so that it functions smoothly. What kind of system will it be? We will consider several possibilities.

172. First let us postulate that the computer scientists succeed in developing intelligent machines that can do all things better than human beings can do them. In that case presumably all work will be done by vast, highly organized systems of machines and no human effort will be necessary. Either of two cases might occur. The machines might be permitted to make all of their own decisions without human oversight, or else human control over the machines might be retained.

173. If the machines are permitted to make all their own decisions we can't make any conjecture as to the results, because it is impossible to guess how such machines might behave. We only point out that the fate of the human race would be at the mercy of the machines. It might be argued that the human race would never be foolish enough to hand over all power to the machines. But we are suggesting neither that the human race would voluntarily turn power over to the machines nor that the machines would willfully seize power. What we do suggest is that the human race might easily permit itself to drift into a position of such dependence on the machines that it would have no practical choice but to accept all of the machines' decisions. As society and the problems that face it become more and more complex and as machines become more and more intelligent, people will let machines make more and more of their decisions for them, simply because machine-made decisions will bring better results than man-made ones. Eventually a stage may be reached at which the decisions necessary to keep the system running will be so complex that human beings will be incapable of making them intelligently. At that stage the machines will be in effective control. People won't be able to just turn the machines off, because they will be so dependent on them that turning them off would amount to suicide.

174. On the other hand it is possible that human control over the machines may be retained. In that case the average man may have control over certain private machines of his own, such as his car or his personal

computer, but control over large systems of machines will be in the hands of a tiny elite—just as it is today, but with two differences. Due to improved techniques the elite will have greater control over the masses; and because human work will no longer be necessary the masses will be superfluous, a useless burden on the system. If the elite is ruthless they may simply decide to exterminate the mass of humanity. If they are humane they may use propaganda or other psychological or biological techniques to reduce the birth rate until the mass of humanity becomes extinct, leaving the world to the elite. Or, if the elite consist of soft-hearted liberals, they may decide to play the role of good shepherds to the rest of the human race. They will see to it that everyone's physical needs are satisfied, that all children are raised under psychologically hygienic conditions, that everyone has a wholesome hobby to keep him busy, and that anyone who may become dissatisfied undergoes "treatment" to cure his "problem." Of course, life will be so purposeless that people will have to be biologically or psychologically engineered either to remove their need for the power process or to make them "sublimate" their drive for power into some harmless hobby. These engineered human beings may be happy in such a society, but they most certainly will not be free. They will have been reduced to the status of domestic animals.

175. But suppose now that the computer scientists do not succeed in developing artificial intelligence, so that human work remains necessary. Even so, machines will take care of more and more of the simpler tasks so that there will be an increasing surplus of human workers at the lower levels of ability. (We see this happening already. There are many people who find it difficult or impossible to get work, because for intellectual or psychological reasons they cannot acquire the level of training necessary to make themselves useful in the present system.) On those who are employed, ever-increasing demands will be placed: They will need more and more training, more and more ability, and will have to be ever more reliable, conforming and docile, because they will be more and more like cells of a giant organism. Their tasks will be increasingly specialized so that their work will be, in a sense, out of touch with the real world, being concentrated on one tiny slice of reality. The system will have to use any means that it can, whether psychological or biological, to engineer people to be docile, to have the abilities that the system requires and to "sublimate" their drive for power into some specialized task. But the statement that the people of

such a society will have to be docile may require qualification. The society may find competitiveness useful, provided that ways are found of directing competitiveness into channels that serve the needs of the system. We can imagine a future society in which there is endless competition for positions of prestige and power. But no more than a very few people will ever reach the top, where the only real power is (see end of paragraph 163). Very repellent is a society in which a person can satisfy his need for power only by pushing large numbers of other people out of the way and depriving them of THEIR opportunity for power.

176. One can envision scenarios that incorporate aspects of more than one of the possibilities that we have just discussed. For instance, it may be that machines will take over most of the work that is of real, practical importance, but that human beings will be kept busy by being given relatively unimportant work. It has been suggested, for example, that a great development of the service industries might provide work for human beings. Thus people would spend their time shining each other's shoes, driving each other around in taxicabs, making handicrafts for one another, waiting on each other's tables, etc. This seems to us a thoroughly contemptible way for the human race to end up, and we doubt that many people would find fulfilling lives in such pointless busy-work. They would seek other, dangerous outlets (drugs, crime, "cults," hate groups) unless they were biologically or psychologically engineered to adapt them to such a way of life.

177. Needless to say, the scenarios outlined above do not exhaust all the possibilities. They only indicate the kinds of outcomes that seem to us most likely. But we can envision no plausible scenarios that are any more palatable than the ones we've just described. It is overwhelmingly probable that if the industrial-technological system survives the next 40 to 100 years, it will by that time have developed certain general characteristics: Individuals (at least those of the "bourgeois" type, who are integrated into the system and make it run, and who therefore have all the power) will be more dependent than ever on large organizations; they will be more "socialized" than ever and their physical and mental qualities to a significant extent (possibly to a very great extent) will be those that are engineered into them rather than being the results of chance (or of God's will, or whatever); and whatever may be left of wild nature will be reduced to remnants preserved for scientific study and kept under the supervision and management of scientists (hence it will no longer be truly wild). In the long run (say a few centuries from now) it is

likely that neither the human race nor any other important organisms will exist as we know them today, because once you start modifying organisms through genetic engineering there is no reason to stop at any particular point, so that the modifications will probably continue until man and other organisms have been utterly transformed.

178. Whatever else may be the case, it is certain that technology is creating for human beings a new physical and social environment radically different from the spectrum of environments to which natural selection has adapted the human race physically and psychologically. If man is not adjusted to this new environment by being artificially re-engineered, then he will be adapted to it through a long and painful process of natural selection. The former is far more likely than the latter.

179. It would be better to dump the whole stinking system and take the consequences.

Strategy

180. The technophiles are taking us all on an utterly reckless ride into the unknown. Many people understand something of what technological progress is doing to us, yet take a passive attitude toward it because they think it is inevitable. But we (FC) don't think it is inevitable. We think it can be stopped, and we will give here some indications of how to go about stopping it.

181. As we stated in paragraph 166, the two main tasks for the present are to promote social stress and instability in industrial society and to develop and propagate an ideology that opposes technology and the industrial system. When the system becomes sufficiently stressed and unstable, a revolution against technology may be possible. The pattern would be similar to that of the French and Russian Revolutions. French society and Russian society, for several decades prior to their respective revolutions, showed increasing signs of stress and weakness. Meanwhile, ideologies were being developed that offered a new world-view that was quite different from the old one. In the Russian case revolutionaries were actively working to undermine the old order. Then, when the old system was put under sufficient additional stress (by financial crisis in France,

by military defeat in Russia) it was swept away by revolution. What we propose is something along the same lines.

182. It will be objected that the French and Russian Revolutions were failures. But most revolutions have two goals. One is to destroy an old form of society and the other is to set up the new form of society envisioned by the revolutionaries. The French and Russian revolutionaries failed (fortunately!) to create the new kind of society of which they dreamed, but they were quite successful in destroying the old society. We have no illusions about the feasibility of creating a new, ideal form of society. Our goal is only to destroy the existing form of society.

183. But an ideology, in order to gain enthusiastic support, must have a positive ideal as well as a negative one; it must be FOR something as well as AGAINST something. The positive ideal that we propose is Nature. That is, WILD nature: Those aspects of the functioning of the Earth and its living things that are independent of human management and free of human interference and control. And with wild nature we include human nature, by which we mean those aspects of the functioning of the human individual that are not subject to regulation by organized society but are products of chance, or free will, or God (depending on your religious or philosophical opinions).

184. Nature makes a perfect counter-ideal to technology for several reasons. Nature (that which is outside the power of the system) is the opposite of technology (which seeks to expand indefinitely the power of the system). Most people will agree that nature is beautiful; certainly it has tremendous popular appeal. The radical environmentalists ALREADY hold an ideology that exalts nature and opposes technology.[30] It is not necessary for the sake of nature to set up some chimerical utopia or any new kind of social order. Nature takes care of itself: It was a spontaneous creation that existed long before any human society, and for countless centuries many different kinds of human societies coexisted with nature without doing it an excessive amount of damage. Only with the Industrial Revolution did the effect of human society on nature become really devastating. To relieve the pressure on nature it is not necessary to create a special kind of social system, it is only necessary to get rid of industrial society. Granted, this will not solve all problems. Industrial society has already done tremendous damage to nature and it will take a very long time for the scars to heal. Besides, even preindustrial societies can do significant damage to nature. Nevertheless,

getting rid of industrial society will accomplish a great deal. It will relieve the worst of the pressure on nature so that the scars can begin to heal. It will remove the capacity of organized society to keep increasing its control over nature (including human nature). Whatever kind of society may exist after the demise of the industrial system, it is certain that most people will live close to nature, because in the absence of advanced technology there is no other way that people CAN live. To feed themselves they must be peasants, or herdsmen, or fishermen, or hunters, etc. And, generally speaking, local autonomy should tend to increase, because lack of advanced technology and rapid communications will limit the capacity of governments or other large organizations to control local communities.

185. As for the negative consequences of eliminating industrial society—well, you can't eat your cake and have it too. To gain one thing you have to sacrifice another.

186. Most people hate psychological conflict. For this reason they avoid doing any serious thinking about difficult social issues, and they like to have such issues presented to them in simple, black-and-white terms: THIS is all good and THAT is all bad. The revolutionary ideology should therefore be developed on two levels.

187. On the more sophisticated level the ideology should address itself to people who are intelligent, thoughtful and rational. The object should be to create a core of people who will be opposed to the industrial system on a rational, thought-out basis, with full appreciation of the problems and ambiguities involved, and of the price that has to be paid for getting rid of the system. It is particularly important to attract people of this type, as they are capable people and will be instrumental in influencing others. These people should be addressed on as rational a level as possible. Facts should never intentionally be distorted and intemperate language should be avoided. This does not mean that no appeal can be made to the emotions, but in making such appeal, care should be taken to avoid misrepresenting the truth or doing anything else that would destroy the intellectual respectability of the ideology.

188. On a second level, the ideology should be propagated in a simplified form that will enable the unthinking majority to see the conflict of technology vs. nature in unambiguous terms. But even on this second level the ideology should not be expressed in language that is so cheap, intemperate or irrational that it alienates people of the thoughtful and

rational type. Cheap, intemperate propaganda sometimes achieves impressive short-term gains, but it will be more advantageous in the long run to keep the loyalty of a small number of intelligently committed people than to arouse the passions of an unthinking, fickle mob who will change their attitude as soon as someone comes along with a better propaganda gimmick. However, propaganda of the rabble-rousing type may be necessary when the system is nearing the point of collapse and there is a final struggle between rival ideologies to determine which will become dominant when the old world-view goes under.

189. Prior to that final struggle, the revolutionaries should not expect to have a majority of people on their side. History is made by active, determined minorities, not by the majority, which seldom has a clear and consistent idea of what it really wants. Until the time comes for the final push toward revolution,[31] the task of revolutionaries will be less to win the shallow support of the majority than to build a small core of deeply committed people. As for the majority, it will be enough to make them aware of the existence of the new ideology and remind them of it frequently; though of course it will be desirable to get majority support to the extent that this can be done without weakening the core of seriously committed people.

190. Any kind of social conflict helps to destabilize the system, but one should be careful about what kind of conflict one encourages. The line of conflict should be drawn between the mass of the people and the power-holding elite of industrial society (politicians, scientists, upper-level business executives, government officials, etc.). It should NOT be drawn between the revolutionaries and the mass of the people. For example, it would be bad strategy for the revolutionaries to condemn Americans for their habits of consumption. Instead, the average American should be portrayed as a victim of the advertising and marketing industry, which has suckered him into buying a lot of junk that he doesn't need and that is very poor compensation for his lost freedom. Either approach is consistent with the facts. It is merely a matter of attitude whether you blame the advertising industry for manipulating the public or blame the public for allowing itself to be manipulated. As a matter of strategy one should generally avoid blaming the public.

191. One should think twice before encouraging any other social conflict than that between the power-holding elite (which wields technology) and the general public (over which technology exerts its power). For one thing,

other conflicts tend to distract attention from the important conflicts (between power-elite and ordinary people, between technology and nature); for another thing, other conflicts may actually tend to encourage technologization, because each side in such a conflict wants to use technological power to gain advantages over its adversary. This is clearly seen in rivalries between nations. It also appears in ethnic conflicts within nations. For example, in America many black leaders are anxious to gain power for African-Americans by placing black individuals in the technological power-elite. They want there to be many black government officials, scientists, corporation executives and so forth. In this way they are helping to absorb the African-American subculture into the technological system. Generally speaking, one should encourage only those social conflicts that can be fitted into the framework of the conflicts of power-elite vs. ordinary people, technology vs. nature.

192. But the way to discourage ethnic conflict is NOT through militant advocacy of minority rights (see paragraphs 21, 29). Instead, the revolutionaries should emphasize that although minorities do suffer more or less disadvantage, this disadvantage is of peripheral significance. Our real enemy is the industrial-technological system, and in the struggle against the system, ethnic distinctions are of no importance.

193. The kind of revolution we have in mind will not necessarily involve an armed uprising against any government. It may or may not involve physical violence, but it will not be a POLITICAL revolution. Its focus will be on technology and economics, not politics.[32]

194. Probably the revolutionaries should even AVOID assuming political power, whether by legal or illegal means, until the industrial system is stressed to the danger point and has proved itself to be a failure in the eyes of most people. Suppose for example that some "green" party should win control of the United States Congress in an election. In order to avoid betraying or watering down their own ideology they would have to take vigorous measures to turn economic growth into economic shrinkage. To the average man the results would appear disastrous: There would be massive unemployment, shortages of commodities, etc. Even if the grosser ill effects could be avoided through superhumanly skillful management, still people would have to begin giving up the luxuries to which they have become addicted. Dissatisfaction would grow, the "green" party would be voted out of office and the revolutionaries would have suffered a severe setback. For

this reason the revolutionaries should not try to acquire political power until the system has gotten itself into such a mess that any hardships will be seen as resulting from the failures of the industrial system itself and not from the policies of the revolutionaries. The revolution against technology will probably have to be a revolution by outsiders, a revolution from below and not from above.

195. The revolution must be international and worldwide. It cannot be carried out on a nation-by-nation basis. Whenever it is suggested that the United States, for example, should cut back on technological progress or economic growth, people get hysterical and start screaming that if we fall behind in technology the Japanese will get ahead of us. Holy robots! The world will fly off its orbit if the Japanese ever sell more cars than we do! (Nationalism is a great promoter of technology.) More reasonably, it is argued that if the relatively democratic nations of the world fall behind in technology while nasty, dictatorial nations like China, Vietnam and North Korea continue to progress, eventually the dictators may come to dominate the world. That is why the industrial system should be attacked in all nations simultaneously, to the extent that this may be possible. True, there is no assurance that the industrial system can be destroyed at approximately the same time all over the world, and it is even conceivable that the attempt to overthrow the system could lead instead to the domination of the system by dictators. That is a risk that has to be taken. And it is worth taking, since the difference between a "democratic" industrial system and one controlled by dictators is small compared with the difference between an industrial system and a non-industrial one.[33] It might even be argued that an industrial system controlled by dictators would be preferable, because dictator-controlled systems usually have proved inefficient, hence they are presumably more likely to break down. Look at Cuba.

196. Revolutionaries might consider favoring measures that tend to bind the world economy into a unified whole. Free trade agreements like NAFTA and GATT are probably harmful to the environment in the short run, but in the long run they may perhaps be advantageous because they foster economic interdependence between nations. It will be easier to destroy the industrial system on a worldwide basis if the world economy is so unified that its breakdown in any one major nation will lead to its breakdown in all industrialized nations.

197. Some people take the line that modern man has too much power, too much control over nature; they argue for a more passive attitude on the part of the human race. At best these people are expressing themselves unclearly, because they fail to distinguish between power for LARGE ORGANIZATIONS and power for INDIVIDUALS and SMALL GROUPS. It is a mistake to argue for powerlessness and passivity, because people NEED power. Modern man as a collective entity—that is, the industrial system—has immense power over nature, and we (FC) regard this as evil. But modern INDIVIDUALS and SMALL GROUPS OF INDIVIDUALS have far less power than primitive man ever did. Generally speaking, the vast power of "modern man" over nature is exercised not by individuals or small groups but by large organizations. To the extent that the average modern INDIVIDUAL can wield the power of technology, he is permitted to do so only within narrow limits and only under the supervision and control of the system. (You need a license for everything and with the license come rules and regulations.) The individual has only those technological powers with which the system chooses to provide him. His PERSONAL power over nature is slight.

198. Primitive INDIVIDUALS and SMALL GROUPS actually had considerable power over nature; or maybe it would be better to say power WITHIN nature. When primitive man needed food he knew how to find and prepare edible roots, how to track game and take it with homemade weapons. He knew how to protect himself from heat, cold, rain, dangerous animals, etc. But primitive man did relatively little damage to nature because the COLLECTIVE power of primitive society was negligible compared to the COLLECTIVE power of industrial society.

199. Instead of arguing for powerlessness and passivity, one should argue that the power of the INDUSTRIAL SYSTEM should be broken, and that this will greatly INCREASE the power and freedom of INDIVIDUALS and SMALL GROUPS.

200. Until the industrial system has been thoroughly wrecked, the destruction of that system must be the revolutionaries' ONLY goal. Other goals would distract attention and energy from the main goal. More importantly, if the revolutionaries permit themselves to have any other goal than the destruction of technology, they will be tempted to use technology as a tool for reaching that other goal. If they give in to that temptation, they will fall right back into the technological trap, because modern technology

is a unified, tightly organized system, so that, in order to retain SOME technology, one finds oneself obliged to retain MOST technology, hence one ends up sacrificing only token amounts of technology.

201. Suppose for example that the revolutionaries took "social justice" as a goal. Human nature being what it is, social justice would not come about spontaneously; it would have to be enforced. In order to enforce it the revolutionaries would have to retain central organization and control. For that they would need rapid long-distance transportation and communication, and therefore all the technology needed to support the transportation and communication systems. To feed and clothe poor people they would have to use agricultural and manufacturing technology. And so forth. So that the attempt to ensure social justice would force them to retain most parts of the technological system. Not that we have anything against social justice, but it must not be allowed to interfere with the effort to get rid of the technological system.

202. It would be hopeless for revolutionaries to try to attack the system without using SOME modern technology. If nothing else they must use the communications media to spread their message. But they should use modern technology for only ONE purpose: to attack the technological system.

203. Imagine an alcoholic sitting with a barrel of wine in front of him. Suppose he starts saying to himself, "Wine isn't bad for you if used in moderation. Why, they say small amounts of wine, are even good for you! It won't do me any harm if I take just one little drink...." Well, you know what is going to happen. Never forget that the human race with technology is just like an alcoholic with a barrel of wine.

204. Revolutionaries should have as many children as they can. There is strong scientific evidence that social attitudes are to a significant extent inherited. No one suggests that a social attitude is a direct outcome of a person's genetic constitution, but it appears that personality traits are partly inherited and that certain personality traits tend, within the context of our society, to make a person more likely to hold this or that social attitude. Objections to these findings have been raised, but the objections are feeble and seem to be ideologically motivated. In any event, no one denies that children tend on the average to hold social attitudes similar to those of their parents. From our point of view it doesn't matter all that much whether the attitudes are passed on genetically or through childhood training. In either case they ARE passed on.

205. The trouble is that many of the people who are inclined to rebel against the industrial system are also concerned about the population problem, hence they are apt to have few or no children. In this way they may be handing the world over to the sort of people who support or at least accept the industrial system. To ensure the strength of the next generation of revolutionaries the present generation should reproduce itself abundantly. In doing so they will be worsening the population problem only slightly. And the most important problem is to get rid of the industrial system, because once the industrial system is gone the world's population necessarily will decrease (see paragraph 167); whereas, if the industrial system survives, it will continue developing new techniques of food production that may enable the world's population to keep increasing almost indefinitely.

206. With regard to revolutionary strategy, the only points on which we absolutely insist are that the single, overriding goal must be the elimination of modern technology, and that no other goal can be allowed to compete with this one. For the rest, revolutionaries should take an empirical approach. If experience indicates that some of the recommendations made in the foregoing paragraphs are not going to give good results, then those recommendations should be discarded.

Two Kinds of Technology

207. An argument likely to be raised against our proposed revolution is that it is bound to fail, because (it is claimed) throughout history technology has always progressed, never regressed, hence technological regression is impossible. But this claim is false.

208. We distinguish between two kinds of technology, which we will call small-scale technology and organization-dependent technology. Small-scale technology is technology that can be used by small-scale communities without outside assistance. Organization-dependent technology is technology that depends on large-scale social organization. We are aware of no significant cases of regression in small-scale technology. But organization-dependent technology DOES regress when the social organization on which it depends breaks down. Example: When the Roman Empire fell apart the Romans' small-scale technology survived because any clever village craftsman could build, for instance, a water wheel, any skilled smith could make steel

by Roman methods, and so forth. But the Romans' organization-dependent technology DID regress. Their aqueducts fell into disrepair and were never rebuilt. Their techniques of road construction were lost. The Roman system of urban sanitation was forgotten, so that not until rather recent times did the sanitation of European cities equal that of ancient Rome.

209. The reason why technology has seemed always to progress is that, until perhaps a century or two before the Industrial Revolution, most technology was small-scale technology. But most of the technology developed since the Industrial Revolution is organization-dependent technology. Take the refrigerator for example. Without factory-made parts or the facilities of a post-industrial machine shop it would be virtually impossible for a handful of local craftsmen to build a refrigerator. If by some miracle they did succeed in building one it would be useless to them without a reliable source of electric power. So they would have to dam a stream and build a generator. Generators require large amounts of copper wire. Imagine trying to make that wire without modern machinery. And where would they get a gas suitable for refrigeration? It would be much easier to build an icehouse or preserve food by drying or pickling, as was done before the invention of the refrigerator.

210. So it is clear that if the industrial system were once thoroughly broken down, refrigeration technology would quickly be lost. The same is true of other organization-dependent technology. And once this technology had been lost for a generation or so it would take centuries to rebuild it, just as it took centuries to build it the first time around. Surviving technical books would be few and scattered. An industrial society, if built from scratch without outside help, can only be built in a series of stages: You need tools to make tools to make tools to make tools.... A long process of economic development and progress in social organization is required. And, even in the absence of an ideology opposed to technology, there is no reason to believe that anyone would be interested in rebuilding industrial society. The enthusiasm for "progress" is a phenomenon peculiar to the modern form of society, and it seems not to have existed prior to the 17th century or thereabouts.

211. In the late Middle Ages there were four main civilizations that were about equally "advanced": Europe, the Islamic world, India, and the Far East (China, Japan, Korea). Three of these civilizations remained more or less stable, and only Europe became dynamic. No one knows why Europe

became dynamic at that time; historians have their theories but these are only speculation. At any rate it is clear that rapid development toward a technological form of society occurs only under special conditions. So there is no reason to assume that a long-lasting technological regression cannot be brought about.

212. Would society EVENTUALLY develop again toward an industrial-technological form? Maybe, but there is no use in worrying about it, since we can't predict or control events 500 or 1,000 years in the future. Those problems must be dealt with by the people who will live at that time.

The Danger of Leftism

213. Because of their need for rebellion and for membership in a movement, leftists or persons of similar psychological type often are attracted to a rebellious or activist movement whose goals and membership are not initially leftist. The resulting influx of leftish types can easily turn a non-leftist movement into a leftist one, so that leftist goals replace or distort the original goals of the movement.

214. To avoid this, a movement that exalts nature and opposes technology must take a resolutely anti-leftist stance and must avoid all collaboration with leftists. Leftism is in the long run inconsistent with wild nature, with human freedom and with the elimination of modern technology. Leftism is collectivist; it seeks to bind together the entire world (both nature and the human race) into a unified whole. But this implies management of nature and of human life by organized society, and it requires advanced technology. You can't have a united world without rapid long-distance transportation and communication, you can't make all people love one another without sophisticated psychological techniques, you can't have a "planned society" without the necessary technological base. Above all, leftism is driven by the need for power, and the leftist seeks power on a collective basis, through identification with a mass movement or an organization. Leftism is unlikely ever to give up technology, because technology is too valuable a source of collective power.

215. The anarchist[34] too seeks power, but he seeks it on an individual or small-group basis; he wants individuals and small groups to be able to

control the circumstances of their own lives. He opposes technology because it makes small groups dependent on large organizations.

216. Some leftists may seem to oppose technology, but they will oppose it only so long as they are outsiders and the technological system is controlled by non-leftists. If leftism ever becomes dominant in society, so that the technological system becomes a tool in the hands of leftists, they will enthusiastically use it and promote its growth. In doing this they will be repeating a pattern that leftism has shown again and again in the past. When the Bolsheviks in Russia were outsiders, they vigorously opposed censorship and the secret police, they advocated self-determination for ethnic minorities, and so forth; but as soon as they came into power themselves, they imposed a tighter censorship and created a more ruthless secret police than any that had existed under the tsars, and they oppressed ethnic minorities at least as much as the tsars had done. In the United States, a couple of decades ago when leftists were a minority in our universities, leftist professors were vigorous proponents of academic freedom, but today, in those of our universities where leftists have become dominant, they have shown themselves ready to take away everyone else's academic freedom. (This is "political correctness.") The same will happen with leftists and technology: They will use it to oppress everyone else if they ever get it under their own control.

217. In earlier revolutions, leftists of the most power-hungry type, repeatedly, have first cooperated with non-leftist revolutionaries, as well as with leftists of a more libertarian inclination, and later have double-crossed them to seize power for themselves. Robespierre did this in the French Revolution, the Bolsheviks did it in the Russian Revolution, the communists did it in Spain in 1938 and Castro and his followers did it in Cuba. Given the past history of leftism, it would be utterly foolish for non-leftist revolutionaries today to collaborate with leftists.

218. Various thinkers have pointed out that leftism is a kind of religion. Leftism is not a religion in the strict sense because leftist doctrine does not postulate the existence of any supernatural being. But for the leftist, leftism plays a psychological role much like that which religion plays for some people. The leftist NEEDS to believe in leftism; it plays a vital role in his psychological economy. His beliefs are not easily modified by logic or facts. He has a deep conviction that leftism is morally Right with a capital R, and that he has not only a right but a duty to impose leftist morality on everyone. (However, many of the people we are referring to as "leftists" do not think

of themselves as leftists and would not describe their system of beliefs as leftism. We use the term "leftism" because we don't know of any better word to designate the spectrum of related creeds that includes the feminist, gay rights, political correctness, etc., movements, and because these movements have a strong affinity with the old left. See paragraphs 227-230.)

219. Leftism is totalitarian force. Wherever leftism is in a position of power it tends to invade every private corner and force every thought into a leftist mold. In part this is because of the quasi-religious character of leftism: Everything contrary to leftist beliefs represents Sin. More importantly, leftism is a totalitarian force because of the leftists' drive for power. The leftist seeks to satisfy his need for power through identification with a social movement, and he tries to go through the power process by helping to pursue and attain the goals of the movement (see paragraph 83). But no matter how far the movement has gone in attaining its goals the leftist is never satisfied, because his activism is a surrogate activity (see paragraph 41). That is, the leftist's real motive is not to attain the ostensible goals of leftism; in reality he is motivated by the sense of power he gets from struggling for and then reaching a social goal.[35] Consequently the leftist is never satisfied with the goals he has already attained; his need for the power process leads him always to pursue some new goal. The leftist wants equal opportunities for minorities. When that is attained he insists on statistical equality of achievement by minorities. And as long as anyone harbors in some corner of his mind a negative attitude toward some minority, the leftist has to re-educate him. And ethnic minorities are not enough; no one can be allowed to have a negative attitude toward homosexuals, disabled people, fat people, old people, ugly people, and on and on and on. It's not enough that the public should be informed about the hazards of smoking; a warning has to be stamped on every package of cigarettes. Then cigarette advertising has to be restricted if not banned. The activists will never be satisfied until tobacco is outlawed, and after that it will be alcohol, then junk food, etc. Activists have fought gross child abuse, which is reasonable. But now they want to stop all spanking. When they have done that they will want to ban something else they consider unwholesome, then another thing and then another. They will never be satisfied until they have complete control over all child-rearing practices. And then they will move on to another cause.

220. Suppose you asked leftists to make a list of ALL the things that were wrong with society, and then suppose you instituted EVERY social

change that they demanded. It is safe to say that within a couple of years the majority of leftists would find something new to complain about, some new social "evil" to correct; because, once again, the leftist is motivated less by distress at society's ills than by the need to satisfy his drive for power by imposing his solutions on society.

221. Because of the restrictions placed on their thought and behavior by their high level of socialization, many leftists of the oversocialized type cannot pursue power in the ways that other people do. For them the drive for power has only one morally acceptable outlet, and that is in the struggle to impose their morality on everyone.

222. Leftists, especially those of the oversocialized type, are True Believers in the sense of Eric Hoffer's book, *The True Believer*. But not all True Believers are of the same psychological type as leftists. Presumably a true-believing Nazi, for instance, is very different psychologically from a true-believing leftist. Because of their capacity for single-minded devotion to a cause, True Believers are a useful, perhaps a necessary, ingredient of any revolutionary movement. This presents a problem with which we must admit we don't know how to deal. We aren't sure how to harness the energies of the True Believer to a revolution against technology. At present all we can say is that no True Believer will make a safe recruit to the revolution unless his commitment is exclusively to the destruction of technology. If he is committed also to another ideal, he may want to use technology as a tool for pursuing that other ideal. (See paragraphs 200, 201.)

223. Some readers may say, "This stuff about leftism is a lot of crap. I know John and Jane who are leftish types and they don't have all these totalitarian tendencies." It's quite true that many leftists, possibly even a numerical majority, are decent people who sincerely believe in tolerating others' values (up to a point) and wouldn't want to use high-handed methods to reach their social goals. Our remarks about leftism are not meant to apply to every individual leftist but to describe the general character of leftism as a movement. And the general character of a movement is not necessarily determined by the numerical proportions of the various kinds of people involved in the movement.

224. The people who rise to positions of power in leftist movements tend to be leftists of the most power-hungry type, because power-hungry people are those who strive hardest to get into positions of power. Once the power-hungry types have captured control of the movement, there are

many leftists of a gentler breed who inwardly disapprove of many of the actions of the leaders, but cannot bring themselves to oppose them. They NEED their faith in the movement, and because they cannot give up this faith they go along with the leaders. True, SOME leftists do have the guts to oppose the totalitarian tendencies that emerge, but they generally lose, because the power-hungry types are better organized, are more ruthless and Machiavellian and have taken care to build themselves a strong power-base.

225. These phenomena appeared clearly in Russia and other countries that were taken over by leftists. Similarly, before the breakdown of communism in the USSR, leftish types in the West would seldom criticize that country. If prodded they would admit that the USSR did many wrong things, but then they would try to find excuses for the communists and begin talking about the faults of the West. They always opposed Western military resistance to communist aggression. Leftish types all over the world vigorously protested the U.S. military action in Vietnam, but when the USSR invaded Afghanistan they did nothing. Not that they approved of the Soviet actions; but, because of their leftist faith, they just couldn't bear to put themselves in opposition to communism. Today, in those of our universities where "political correctness" has become dominant, there are probably many leftish types who privately disapprove of the suppression of academic freedom, but they go along with it anyway.

226. Thus the fact that many individual leftists are personally mild and fairly tolerant people by no means prevents leftism as a whole from having a totalitarian tendency.

227. Our discussion of leftism has a serious weakness. It is still far from clear what we mean by the word "leftist." There doesn't seem to be much we can do about this. Today leftism is fragmented into a whole spectrum of activist movements. Yet not all activist movements are leftist, and some activist movements (e.g., radical environmentalism) seem to include both personalities of the leftist type and personalities of thoroughly un-leftist types who ought to know better than to collaborate with leftists. Varieties of leftists fade out gradually into varieties of non-leftists and we ourselves would often be hard-pressed to decide whether a given individual is or is not a leftist. To the extent that it is defined at all, our conception of leftism is defined by the discussion of it that we have given in this article, and we can only advise the reader to use his own judgment in deciding who is a leftist.

228. But it will be helpful to list some criteria for diagnosing leftism. These criteria cannot be applied in a cut and dried manner. Some individuals may meet some of the criteria without being leftists, some leftists may not meet any of the criteria. Again, you just have to use your judgment.

229. The leftist is oriented toward large-scale collectivism. He emphasizes the duty of the individual to serve society and the duty of society to take care of the individual. He has a negative attitude toward individualism. He often takes a moralistic tone. He tends to be for gun control, for sex education and other psychologically "enlightened" educational methods, for social planning, for affirmative action, for multiculturalism. He tends to identify with victims. He tends to be against competition and against violence, but he often finds excuses for those leftists who do commit violence. He is fond of using the common catch-phrases of the left, like "racism," "sexism," "homophobia," "capitalism," "imperialism," "neocolonialism," "genocide," "social change," "social justice," "social responsibility." Maybe the best diagnostic trait of the leftist is his tendency to sympathize with the following movements: feminism, gay rights, ethnic rights, disability rights, animal rights political correctness. Anyone who strongly sympathizes with ALL of these movements is almost certainly a leftist.[36]

230. The more dangerous leftists, that is, those who are most power-hungry, are often characterized by arrogance or by a dogmatic approach to ideology. However, the most dangerous leftists of all may be certain oversocialized types who avoid irritating displays of aggressiveness and refrain from advertising their leftism, but work quietly and unobtrusively to promote collectivist values, "enlightened" psychological techniques for socializing children, dependence of the individual on the system, and so forth. These crypto-leftists (as we may call them) approximate certain bourgeois types as far as practical action is concerned, but differ from them in psychology, ideology and motivation. The ordinary bourgeois tries to bring people under control of the system in order to protect his way of life, or he does so simply because his attitudes are conventional. The crypto-leftist tries to bring people under control of the system because he is a True Believer in a collectivistic ideology. The crypto-leftist is differentiated from the average leftist of the oversocialized type by the fact that his rebellious impulse is weaker and he is more securely socialized. He is differentiated from the ordinary well-socialized bourgeois by the fact that there is some deep lack

within him that makes it necessary for him to devote himself to a cause and immerse himself in a collectivity. And maybe his (well-sublimated) drive for power is stronger than that of the average bourgeois.

Final Note

231. Throughout this article we've made imprecise statements and statements that ought to have had all sorts of qualifications and reservations attached to them; and some of our statements may be flatly false. Lack of sufficient information and the need for brevity made it impossible for us to formulate our assertions more precisely or add all the necessary qualifications. And of course in a discussion of this kind one must rely heavily on intuitive judgment, and that can sometimes be wrong. So we don't claim that this article expresses more than a crude approximation to the truth.

232. All the same, we are reasonably confident that the general outlines of the picture we have painted here are roughly correct. Just one possible weak point needs to be mentioned. We have portrayed leftism in its modern form as a phenomenon peculiar to our time and as a symptom of the disruption of the power process. But we might possibly be wrong about this. Oversocialized types who try to satisfy their drive for power by imposing their morality on everyone have certainly been around for a long time. But we THINK that the decisive role played by feelings of inferiority, low self-esteem, powerlessness, identification with victims by people who are not themselves victims, is a peculiarity of modern leftism. Identification with victims by people not themselves victims can be seen to some extent in 19th-century leftism and early Christianity, but as far as we can make out, symptoms of low self-esteem, etc., were not nearly so evident in these movements, or in any other movements, as they are in modern leftism. But we are not in a position to assert confidently that no such movements have existed prior to modern leftism. This is a significant question to which historians ought to give their attention. •

ENDNOTES

[1] We are not asserting that all, or even most, bullies and ruthless competitors suffer from feelings of inferiority.

[2] During the Victorian period many oversocialized people suffered from serious psychological problems as a result of repressing or trying to repress their sexual feelings. Freud apparently based his theories on people of this type. Today the focus of socialization has shifted from sex to aggression.

[3] Not necessarily including specialists in engineering or the "hard" sciences.

[4] There are many individuals of the middle and upper classes who resist some of these values, but usually their resistance is more or less covert. Such resistance appears in the mass media only to a very limited extent. The main thrust of propaganda in our society is in favor of the stated values. The main reason why these values have become, so to speak, the official values of our society is that they are useful to the industrial system. Violence is discouraged because it disrupts the functioning of the system. Racism is discouraged because ethnic conflicts also disrupt the system, and discrimination wastes the talents of minority-group members who could be useful to the system. Poverty must be "cured" because the underclass causes problems for the system and contact with the underclass lowers the morale of the other classes. Women are encouraged to have careers because their talents are useful to the system and, more importantly, because by having regular jobs women become integrated into the system and tied directly to it rather than to their families. This helps to weaken family solidarity. (The leaders of the system say they want to strengthen the family, but what they really mean is that they want the family to serve as an effective tool for socializing children in accord with the needs of the system. We argue in paragraphs 51,52 that the system cannot afford to let the family or other small-scale social groups be strong or autonomous.)

[5] It may be argued that the majority of people don't want to make their own decisions but want leaders to do their thinking for them. There is an element of truth in this. People like to make their own decisions in small matters, but making decisions on difficult, fundamental questions requires facing up to psychological conflict, and most people hate psychological conflict. Hence they tend to lean on others in making difficult decisions. But it does not follow that they like to have decisions imposed on them without having any opportunity to influence those decisions. The majority of people are natural followers, not leaders, but they like to have direct personal access to their leaders, they want to be able to influence the

leaders and participate to some extent in making even the difficult decisions. At least to that degree they need autonomy.

[6] Some of the symptoms listed are similar to those shown by caged animals. To explain how these symptoms arise from deprivation with respect to the power process: common-sense understanding of human nature tells one that lack of goals whose attainment requires effort leads to boredom and that boredom, long continued, often leads eventually to depression. Failure to attain goals leads to frustration and lowering of self-esteem. Frustration leads to anger, anger to aggression, often in the form of spouse or child abuse. It has been shown that long-continued frustration commonly leads to depression and that depression tends to cause anxiety, guilt, sleep disorders, eating disorders and bad feelings about oneself. Those who are tending toward depression seek pleasure as an antidote; hence insatiable hedonism and excessive sex, with perversions as a means of getting new kicks. Boredom too tends to cause excessive pleasure-seeking since, lacking other goals, people often use pleasure as a goal. The foregoing is a simplification. Reality is more complex, and of course deprivation with respect to the power process is not the ONLY cause of the symptoms described. By the way, when we mention depression we do not necessarily mean depression that is severe enough to be treated by a psychiatrist. Often only mild forms of depression are involved. And when we speak of goals we do not necessarily mean long-term, thought-out goals. For many or most people through much of human history, the goals of a hand-to-mouth existence (merely providing oneself and one's family with food from day to day) have been quite sufficient.

[7] A partial exception may be made for a few passive, inward-looking groups, such as the Amish, which have little effect on the wider society. Apart from these, some genuine small-scale communities do exist in America today. For instance, youth gangs and "cults." Everyone regards them as dangerous, and so they are, because the members of these groups are loyal primarily to one another rather than to the system, hence the system cannot control them. Or take the gypsies. The gypsies commonly get away with theft and fraud because their loyalties are such that they can always get other gypsies to give testimony that "proves" their innocence. Obviously the system would be in serious trouble if too many people belonged to such groups. Some of the early-20th-century Chinese thinkers who were concerned with modernizing China recognized the necessity of breaking down small-scale social groups such as the family: "[According to Sun Yat-Sen] the Chinese people needed a new surge of patriotism, which would lead to a transfer of loyalty from the family to the state. ... [according to Li Huang] traditional attachments, particularly to the family, had to be abandoned if nationalism were to develop in China" (Chester C. Tan, *Chinese Political Thought in the Twentieth century*, page 125, page 297).

[8] Yes, we know that 19th-century America had its problems, and serious ones, but for the sake of brevity we have to express ourselves in simplified terms.

[9] We leave aside the "underclass." We are speaking of the mainstream.

[10] Some social scientists, educators, "mental health" professionals and the like are doing their best to push the social drives into group 1 by trying to see to it that everyone has a satisfactory social life.

[11] Is the drive for endless material acquisition really an artificial creation of the advertising and marketing industry? Certainly there is no innate human drive for material acquisition. There have been many cultures in which people have desired little material wealth beyond what was necessary to satisfy their basic physical needs (Australian aborigines, traditional Mexican peasant culture, some African cultures). On the other hand there have also been many preindustrial cultures in which material acquisition has played an important role. So we can't claim that today's acquisition-oriented culture is exclusively a creation of the advertising and marketing industry. But it is clear that the advertising and marketing industry has had an important part in creating that culture. The big corporations that spend millions on advertising wouldn't be spending that kind of money without solid proof that they were getting it back in increased sales. One member of FC met a sales manager a couple of years ago who was frank enough to tell him, "Our job is to make people buy things they don't want and don't need." He then described how an untrained novice could present people with the facts about a product and make no sales at all, while a trained and experienced professional salesman would make lots of sales to the same people. This shows that people are manipulated into buying things they don't really want.

[12] The problem of purposelessness seems to have become less serious during the last 15 years or so [this refers to the 15 years preceding 1995], because people now feel less secure physically and economically than they did earlier, and the need for security provides them with a goal. But purposelessness has been replaced by frustration over the difficulty of attaining security. We emphasize the problem of purposelessness because the liberals and leftists would wish to solve our social problems by having society guarantee everyone's security; but if that could be done it would only bring back the problem of purposelessness. The real issue is not whether society provides well or poorly for people's security; the trouble is that people are dependent on the system for their security rather than having it in their own hands. This, by the way, is part of the reason why some people get worked up about the right to bear arms; possession of a gun puts that aspect of their security in their own hands.

[13] Conservatives' efforts to decrease the amount of government regulation are of little benefit to the average man. For one thing, only a fraction of the regulations can

be eliminated because most regulations are necessary. For another thing, most of the deregulation affects business rather than the average individual, so that its main effect is to take power from the government and give it to private corporations. What this means for the average man is that government interference in his life is replaced by interference from big corporations, which may be permitted, for example, to dump more chemicals that get into his water supply and give him cancer. The conservatives are just taking the average man for a sucker, exploiting his resentment of Big Government to promote the power of Big Business.

[14] When someone approves of the purpose for which propaganda is being used in a given case, he generally calls it "education" or applies to it some similar euphemism. But propaganda is propaganda regardless of the purpose for which it is used.

[15] We are not expressing approval or disapproval of the Panama invasion. We only use it to illustrate a point.

[16] When the American colonies were under British rule there were fewer and less effective legal guarantees of freedom than there were after the American Constitution went into effect, yet there was more personal freedom in preindustrial America, both before and after the War of Independence, than there was after the Industrial Revolution took hold in this country. We quote from *Violence in America: Historical and Comparative Perspectives*, edited by Hugh Davis Graham and Ted Robert Gurr, chapter 12 by Roger Lane, pages 476-478: "The progressive heightening of standards of propriety, and with it the increasing reliance on official law enforcement [in 19th-century America]...were common to the whole society... [T]he change in social behavior is so long term and so wide-spread as to suggest a connection with the most fundamental of contemporary social processes; that of industrial urbanization itself. ...Massachusetts in 1835 had a population of some 660,940, 81 percent rural, overwhelmingly preindustrial and native born. Its citizens were used to considerable personal freedom. Whether teamsters, farmers or artisans, they were all accustomed to setting their own schedules, and the nature of their work made them physically independent of each other. ...Individual problems, sins or even crimes, were not generally cause for wider social concern. ...But the impact of the twin movements to the city and to the factory, both just gathering force in 1835, had a progressive effect on personal behavior throughout the 19th century and into the 20th. The factory demanded regularity of behavior, a life governed by obedience to the rhythms of clock and calendar, the demands of foreman and supervisor. In the city or town, the needs of living in closely packed neighborhoods inhibited many actions previously unobjectionable. Both blue- and white-collar employees in larger establishments were mutually dependent on their fellows; as one man's work fit into another's, so one man's business was no longer his own. The results of the new organization of life and work were apparent by 1900, when some 76 percent of the

2,805,346 inhabitants of Massachusetts were classified as urbanites. Much violent or irregular behavior which had been tolerable in a casual, independent society was no longer acceptable in the more formalized, cooperative atmosphere of the later period. ...The move to the cities had, in short, produced a more tractable, more socialized, more 'civilized' generation than its predecessors."

[17] Apologists for the system are fond of citing cases in which elections have been decided by one or two votes, but such cases are rare.

[18] "Today, in technologically advanced lands, men live very similar lives in spite of geographical, religious, and political differences. The daily lives of a Christian bank clerk in Chicago, a Buddhist bank clerk in Tokyo, and a Communist bank clerk in Moscow are far more alike than the life of any one of them is like that of any single man who lived a thousand years ago. These similarities are the result of a common technology...." L. Sprague de Camp, *The Ancient Engineers*, Ballantine edition, page 17. The lives of the three bank clerks are not IDENTICAL. Ideology does have SOME effect. But all technological societies, in order to survive, must evolve along APPROXIMATELY the same trajectory.

[19] Just think, an irresponsible genetic engineer might create a lot of terrorists.

[20] For a further example of undesirable consequences of medical progress, suppose a reliable cure for cancer is discovered. Even if the treatment is too expensive to be available to any but the elite, it will greatly reduce their incentive to stop the escape of carcinogens into the environment.

[21] Since many people may find paradoxical the notion that a large number of good things can add up to a bad thing, we illustrate with an analogy. Suppose Mr. A is playing chess with Mr. B. Mr. C, a grand master, is looking over Mr. A's shoulder. Mr. A of course wants to win his game, so if Mr. C points out a good move for him to make, he is doing Mr. A a favor. But suppose now that Mr. C tells Mr. A how to make ALL of his moves. In each particular instance he does Mr. A a favor by showing him his best move, but by making ALL of his moves for him he spoils his game, since there is no point in Mr. A's playing the game at all if someone else makes all his moves. The situation of modern man is analogous to that of Mr. A. The system makes an individual's life easier for him in innumerable ways, but in doing so it deprives him of control over his own fate.

[22] Here we are considering only the conflict of values within the mainstream. For the sake of simplicity we leave out of the picture "outsider" values like the idea that wild nature is more important than human economic welfare.

[23] Self-interest is not necessarily MATERIAL self-interest. It can consist in fulfillment of some psychological need, for example, by promoting one's own ideology or religion.

[24] A qualification: It is in the interest of the system to permit a certain prescribed degree of freedom in some areas. For example, economic freedom (with suitable limitations and restraints) has proved effective in promoting economic growth. but only planned, circumscribed, limited freedom is in the interest of the system. The individual must always be kept on a leash, even if the leash is sometimes long. (See paragraphs 94, 97.)

[25] We don't mean to suggest that the efficiency or the potential for survival of a society has always been inversely proportional to the amount of pressure or discomfort to which the society subjects people. That certainly is not the case. There is good reason to believe that many primitive societies subjected people to less pressure than European society did, but European society proved far more efficient than any primitive society and always won out in conflicts with such societies because of the advantages conferred by technology.

[26] If you think that more effective law enforcement is unequivocally good because it suppresses crime, then remember that crime as defined by the system is not necessarily what YOU would call crime. Today, smoking marijuana is a "crime," and, in some places in the U.S., so is possession of an unregistered handgun. Tomorrow, possession of ANY firearm, registered or not, may be made a crime, and the same thing may happen with disapproved methods of child-rearing, such as spanking. In some countries, expression of dissident political opinions is a crime, and there is no certainty that this will never happen in the U.S., since no constitution or political system lasts forever. If a society needs a large, powerful law enforcement establishment, then there is something gravely wrong with that society; it must be subjecting people to severe pressures if so many refuse to follow the rules, or follow them only because forced. Many societies in the past have gotten by with little or no formal law-enforcement.

[27] To be sure, past societies have had means of influencing human behavior, but these have been primitive and of low effectiveness compared with the technological means that are now being developed.

[28] However, some psychologists have publicly expressed opinions indicating their contempt for human freedom. And the mathematician Claude Shannon was quoted in *Omni* (August 1987) as saying, "I visualize a time when we will be to robots what dogs are to humans, and I'm rooting for the machines."

[29] This is no science fiction! After writing paragraph 154 we came across an article in *Scientific American* according to which scientists are actively developing techniques for identifying possible future criminals and for treating them by a combination of biological and psychological means. Some scientists advocate compulsory application of the treatment, which may be available in the near future. (See "Seeking the

Criminal Element," by W. Wayt Gibbs, *Scientific American*, March 1995.) Maybe you think this is okay because the treatment would be applied to those who might become violent criminals. But of course it won't stop there. Next, a treatment will be applied to those who might become drunk drivers (they endanger human life too), then perhaps to people who spank their children, then to environmentalists who sabotage logging equipment, eventually to anyone whose behavior is inconvenient for the system.

[30] A further advantage of nature as a counter-ideal to technology is that, in many people, nature inspires the kind of reverence that is associated with religion, so that nature could perhaps be idealized on a religious basis. It is true that in many societies religion has served as a support and justification for the established order, but it is also true that religion has often provided a basis for rebellion. Thus it may be useful to introduce a religious element into the rebellion against technology, the more so because Western society today has no strong religious foundation. Religion nowadays either is used as cheap and transparent support for narrow, short-sighted selfishness (some conservatives use it this way), or even is cynically exploited to make easy money (by many evangelists), or has degenerated into crude irrationalism (fundamentalist protestant sects, "cults"), or is simply stagnant (Catholicism, mainline Protestantism). The nearest thing to a strong, widespread, dynamic religion that the West has seen in recent times has been the quasi-religion of leftism, but leftism today is fragmented and has no clear, unified, inspiring goal. Thus there is a religious vacuum in our society that could perhaps be filled by a religion focused on nature in opposition to technology. But it would be a mistake to try to concoct artificially a religion to fill this role. Such an invented religion would probably be a failure. Take the "Gaia" religion for example. Do its adherents REALLY believe in it or are they just play-acting? If they are just play-acting their religion will be a flop in the end. It is probably best not to try to introduce religion into the conflict of nature vs. technology unless you REALLY believe in that religion yourself and find that it arouses a deep, strong, genuine response in many other people.

[31] Assuming that such a final push occurs. Conceivably the industrial system might be eliminated in a somewhat gradual or piecemeal fashion. (See paragraphs 4, 167 and Note 32.)

[32] It is even conceivable (remotely) that the revolution might consist only of a massive change of attitudes toward technology resulting in a relatively gradual and painless disintegration of the industrial system. But if this happens we'll be very lucky. It's far more probable that the transition to a non-technological society will be very difficult and full of conflicts and disasters.

[33] The economic and technological structure of a society are far more important than its political structure in determining the way the average man lives. (See paragraphs 95, 119 and Notes 16, 18.)

[34] This statement refers to our particular brand of anarchism. A wide variety of social attitudes have been called "anarchist," and it may be that many who consider themselves anarchists would not accept our statement of paragraph 215. It should be noted, by the way, that there is a nonviolent anarchist movement whose members probably would not accept FC as anarchist and certainly would not approve of FC's violent methods.

[35] Many leftists are motivated also by hostility, but the hostility probably results in part from a frustrated need for power.

[36] It is important to understand that we mean someone who sympathizes with these movements as they exist today in our society. One who believes that women, homosexuals, etc., should have equal rights is not necessarily a leftist. The feminist, gay rights, etc., movements that exist in our society have the particular ideological tone that characterizes leftism, and if one believes, for example, that women should have equal rights it does not necessarily follow that one must sympathize with the feminist movement as it exists today. •

Postscript to the Manifesto

The Manifesto, *Industrial Society and its Future*, has been criticized as "unoriginal," but this misses the point. The Manifesto was never intended to be original. Its purpose was to set forth certain points about modern technology in clear and relatively brief form, so that those points could be read and understood by people who would never work their way through a difficult text such as Jacques Ellul's *Technological Society*.

The accusation of unoriginality is in any case irrelevant. Is it important for the future of the world to know whether Ted Kaczynski is original or unoriginal? Obviously not! But it is indeed important for the future of the world to know whether modern technology has us on the road to disaster, whether anything short of revolution can avert that disaster, and whether the political left is an obstacle to revolution. So why have critics, for the most part, ignored the substance of the arguments raised in the Manifesto and wasted words on matters of negligible importance, such as the author's putative lack of originality and the defects of his style? Clearly, the critics can't answer the substance of the Manifesto's reasoning, so they try to divert their own and others' attention from its arguments by attacking irrelevant aspects of the Manifesto.

One doesn't need to be original to recognize that technological progress is taking us down the road to disaster, and that nothing short of the overthrow of the entire technological system will get us off that road. In other words, only by accepting a massive disaster now can we avoid a far worse disaster later. But most of our intellectuals—and here I use that term in a broad sense— prefer not to face up to this frightening dilemma because, after all, they are not very brave, and they find it more comfortable to spend their time perfecting society's solutions to problems left over from the 19th century, such as those of social inequality, colonialism, cruelty to animals, and the like.

I haven't read everything that's been written on the technology problem, and it's possible that the Manifesto may have been preceded by some other text that expounded the problem in equally brief and accessible form. But even so it would not follow that the Manifesto was superfluous. However familiar its points may be to social scientists, those points still have not come to the attention of many other people who ought to be aware of them. More importantly, the available knowledge on this subject is not being *applied*. I don't think many of our intellectuals nowadays would deny that there *is* a technology problem, but nearly all of them decline to address it. At best they discuss particular problems created by technological progress, such as global warming or the spread of nuclear weapons. The technology problem as a whole is simply ignored.

It follows that the facts about technological progress and its consequences for society cannot be repeated too often. Even the most intelligent people may refuse to face up to a painful truth until it has been drummed into their heads again and again.

I should add that, as with the Manifesto, no claim of originality is made for this book as a whole. The fact that I've cited authority for many of the ideas about human society that are presented here shows that those ideas are not new, and probably most of the other ideas too have previously appeared somewhere in print.

If there is anything new in my approach, it is that I've taken revolution seriously as a practical proposition. Many radical environmentalists and "green" anarchists talk of revolution, but as far as I am aware none of them have shown any understanding of how real revolutions come about, nor do they seem to grasp the fact that the exclusive target of revolution must be technology itself, not racism, sexism, or homophobia. A very few serious thinkers have suggested revolution against the technological system; for example, Ellul, in his *Autopsy of Revolution*. But Ellul only dreams of a revolution that would result from a vaguely defined, spontaneous spiritual transformation of society, and he comes very close to admitting that the proposed spiritual transformation is impossible. I on the other hand think it plausible that the preconditions for revolution may be developing in modern society, and I mean a real revolution, not fundamentally different in character from other revolutions that have occurred in the past. But this revolution will not become a reality without a well-defined revolutionary movement guided by suitable leaders—leaders who have a rational understanding of what they are doing, not enraged adolescents acting solely on the basis of emotion. •

The Truth About Primitive Life:

A Critique of Anarcho-Primitivism

1. As the Industrial Revolution proceeded, modern society created for itself a self-congratulatory myth, the myth of "progress": From the time of our remote, ape-like ancestors, human history had been an unremitting march toward a better and brighter future, with everyone joyously welcoming each new technological advance: animal husbandry, agriculture, the wheel, the construction of cities, the invention of writing and of money, sailing ships, the compass, gunpowder, the printing press, the steam engine, and, at last, the crowning human achievement—modern industrial society! Prior to industrialization, nearly everyone was condemned to a miserable life of constant, backbreaking labor, malnutrition, disease, and an early death. Aren't we so lucky that we live in modern times and have lots of leisure and an array of technological conveniences to make our lives easy?

Today I think there are relatively few thoughtful, honest and well-informed people who still believe in this myth. To lose one's faith in "progress" one has only to look around and see the devastation of our environment, the spread of nuclear weapons, the excessive frequency of depression, anxiety disorders and psychological stress, the spiritual emptiness of a society that nourishes itself principally with television and computer games...one could go on and on.

The myth of progress may not yet be dead, but it is dying. In its place another myth has been growing up, a myth that has been promoted especially by the anarcho-primitivists, though it is widespread in other quarters as well. According to this myth, prior to the advent of civilization no one ever had to work, people just plucked their food from the trees and popped it into their mouths and spent the rest of their time playing ring-around-the-rosie with the flower children. Men and women were equal, there was no disease, no competition, no racism, sexism or homophobia, people lived in harmony with the animals and all was love, sharing, and cooperation.

Admittedly, the foregoing is a caricature of the anarcho-primitivists' vision. Most of them—I hope—are not quite as far out of touch with reality as that. They nevertheless are pretty far out of touch with it, and it's high time for someone to debunk their myth. Because that is the purpose of this article, I will say little here about the positive aspects of primitive societies. I do want to make clear, however, that one can truthfully say about such societies a great deal that is positive. In other words, the anarcho-primitivist myth is not one hundred percent myth; it does include some elements of reality.

2. Let's begin with the concept of "primitive affluence." It seems to be an article of faith among anarcho-primitivists that our hunting-and-gathering ancestors had to work an average of only two to three hours a day, or two to four hours a day...the figures given vary, but the maximum stated never exceeds four hours a day, or 28 hours a week (average).[1] People who give these figures usually do not state precisely what they mean by "work," but the reader is led to assume that it includes all of the activities necessary to meet the practical exigencies of the hunter-gatherers' way of life.

Characteristically, the anarcho-primitivists usually fail to cite their source for this supposed information, but it seems to be derived mainly from two essays, one by Marshall Sahlins ("The Original Affluent Society"),[2] and the other by Bob Black ("Primitive Affluence").[3] Sahlins claimed that for the Bushmen of the Dobe region of Southern Africa, the "work week was approximately 15 hours."[4] For this information he relied on the studies of Richard B. Lee. I do not have direct access to Lee's works, but I do have a copy of an article by Elizabeth Cashdan in which she summarizes Lee's results much more carefully and completely than Sahlins does.[5] Cashdan flatly contradicts Sahlins: According to her, Lee found that the Bushmen he studied worked more than 40 hours per week.[6]

In a part of his essay that many anarcho-primitivists have found convenient to overlook, Bob Black acknowledges the 40-hour workweek and explains the foregoing contradiction: Sahlins followed early work of Lee that considered only time spent in hunting and foraging. When all necessary work was considered, the workweek was more than doubled.[7]

The work omitted from consideration by Sahlins and the anarcho-primitivists was probably the most disagreeable part of the Bushmen's workweek, too, since it consisted largely of food-preparation and firewood collection.[8] I speak from extensive personal experience with wild foods: Preparing such foods for use is very often a pain in the neck. It is far more pleasant to *gather* nuts, *dig* roots, or *hunt* game than it is to *crack* nuts, *clean* roots, or *skin* and *butcher* game—or to collect firewood and cook over an open fire.

The anarcho-primitivists also err in assuming that Lee's findings can be applied to hunter-gatherers generally. It's not even clear that those findings are applicable on a year-round basis to the Bushmen studied by Lee. Cashdan cites evidence that Lee's research may have been done at the time of year when his Bushmen worked least.[9] She also mentions two other hunting-and-gathering peoples who have been shown quantitatively to spend far more time in hunting and foraging than Lee's Bushmen did,[10] and she points out that Lee may have seriously underestimated women's working time because he failed to include time spent on childcare.[11]

I'm not familiar with any other exact quantitative studies of hunter-gatherers' working time, but it is certain that at least some additional hunter-gatherers worked a great deal more than the 40-hour week of Lee's Bushmen. Gontran de Poncins stated that the Eskimos with whom he lived about 1939–1940 had "no significant degree of leisure," and that they "toiled and moiled fifteen hours a day merely in order to get food and stay alive."[12] He probably did not mean that they worked 15 hours *every* day, but it's clear from his account that his Eskimos worked plenty hard.

Among the Mbuti pygmies principally studied by Paul Schebesta, on days when the women did not fetch a supply of fruits and vegetables from the gardens of their village-dwelling neighbors, their gathering excursions in the forest lasted between five and six hours. Apart from their food-gathering, the women had considerable additional work to do. Each afternoon, for example, a woman had to go again into the forest and come back to camp panting and bowed under a huge load of firewood. The women worked far more than the men, but it seems clear from Schebesta's account that the men nevertheless worked much more than the three or four hours a day claimed by the anarcho-primitivists.[13] Colin Turnbull studied Mbuti pygmies who hunted with nets. Due to the advantage conferred by the nets, these Mbuti only needed to hunt about 20 hours per week. But for them: "Netmaking is virtually a full-time occupation...in which both men and women indulge whenever they have both the spare time and the inclination."[14]

The Siriono, who lived in a tropical forest in Bolivia, were not pure hunter-gatherers, since they did plant crops to a limited extent at certain times of the year. But they lived mostly by hunting and gathering.[15] According to the anthropologist Holmberg, Siriono men hunted, on average, every other day.[16] They started at daybreak and returned to camp typically between four and six o'clock in the afternoon.[17] This makes on average at least 11 hours of hunting, and at three and a half days a week it comes to 38 hours of hunting per week, at the least. Since the men also did a significant amount of work on days when they did not hunt,[18] their workweek, averaged over the year, had to be far more than 40 hours. And but little of this was agricultural work.[19] Actually, Holmberg estimated that the Siriono spent about half their waking time in hunting and foraging,[20] which would mean roughly 56 hours a week in these activities alone. With other work included, the workweek would have had to be far more than 60 hours. The Siriono woman "enjoys even less respite from labor than her husband," and "the obligation of bringing her children to maturity leaves little time for rest."[21] Holmberg's book contains many other indications of how hard the Siriono had to work.[22]

In "The Original Affluent Society," Sahlins gives, in addition to Lee's Bushmen, other examples of hunting-and-gathering peoples who supposedly worked little, but in most of these cases he either offers no quantitative estimate of working time, or he offers an estimate only of time spent in hunting and gathering. If Lee's Bushmen can be taken as a guide, this would be well under half the total working time.[23] However, for two groups of Australian Aborigines Sahlins does give quantitative estimates of time spent in "hunting, plant collecting, preparing foods and repairing weapons." In the first group the average weekly time each worker spent in these activities was about 26 1/2 hours; in the second group about 36 hours. But this does not include all work; it says nothing, for example, about time spent on child care, in collecting firewood, in moving camp, or in making and repairing implements other than weapons. If all necessary work were counted, the workweek of the second group would surely be over 40 hours. The workweek of the first group did not represent that of a normal hunting-and-gathering band, since the first group had no children to feed. Sahlins himself, moreover, questions the validity of inferences drawn from these data.[24] Of course, even if occasional examples could be found of hunting-and-gathering peoples whose total working time was as little as three hours a day, that would matter little for present purposes, since we are concerned here not with exceptional cases but with the typical working time of hunter-gatherers.

Whatever hunter-gatherers' working hours may have been, much of their work was physically very strenuous. Siriono men typically covered about 15 miles a day on their hunting excursions, and they sometimes covered as much as 40 miles.[25] Covering such a distance in trackless wilderness[26] requires far more effort than covering the same distance over a road or a groomed trail.

"In walking and running through swamp and jungle the naked hunter is exposed to thorns, to spines, and to insect pests.... [W]hile the food quest is differentially rewarding because food for survival is always eventually obtained, it is also always punishing because of the fatigue and pain inevitably associated with hunting, fishing and collecting food."[27]

"Men often dissipate their anger toward other men by hunting.... [E]ven if they do not kill anything they return home too tired to be angry."[28]

Even picking wild fruit could be dangerous[29] and could take considerable work[30] for the Siriono.[31] The Siriono made little use of wild roots,[32] but it is well known that many hunter-gatherers relied heavily on roots for food. Usually, gathering edible roots in the wilderness is not like pulling carrots out of the soft, cultivated soil of a garden. More typically the ground is hard, or covered with tough sod that you have to hack through in order to get at the roots. I wish I could take certain anarcho-primitivists out in the mountains, show them where the edible roots grow, and invite them to get their dinner by digging for it. By the time they had enough yampa roots or camas bulbs for a halfway square meal, their blistered hands would disabuse them of any idea that primitives didn't have to work for a living.

Hunter-gatherers' work was often monotonous, too. This is true for example of root-digging when the roots are small, as is the case with many of the roots that were used by the Indians of western North America, such as bitterroot and the aforementioned yampa and camas. Picking berries is monotonous if you spend many hours at it.

Or try tanning a deerskin. A raw, dry deerskin is stiff, like cardboard, and if you bend it, it will crack, just as cardboard will. In order to become usable as clothing or blankets, animal skins must be tanned. Assuming you want to leave the hair on the skin, as for winter clothing, there are only three indispensable steps to tanning a deerskin. First, you must carefully remove every bit of flesh from the skin. Fat in particular must be removed with scrupulous care, because any bit of fat left on the skin will rot it. Next, the skin must be softened. Finally, it must be smoked. If not smoked it will dry stiff and hard after a wetting and will have to be softened all over again. By far the most time-consuming step is the softening. It takes many hours of kneading the skin in your hands, or drawing it back and forth over the head of a spike driven into a block of wood, and the work is very monotonous indeed. I speak from personal experience.

An argument sometimes offered is that hunter-gatherers who survived into recent times lived in tough environments, since all of the more hospitable lands had been taken over by agricultural peoples. Supposedly, prehistoric hunter-gatherers who occupied fertile country must have worked far less than recent hunter-gatherers living in deserts or other unproductive environments.[33] This *may* be true, but the argument is speculative, and I'm skeptical of it.

I'm a bit rusty now, but I used to have considerable familiarity with the edible wild plants of the eastern United States, which is one of the most fertile regions in the world, and I would be surprised if one could live and raise a family there by hunting and gathering with less than a forty-hour workweek. The region contains a wide variety of edible plants, but living off them would not be as easy as you might think. Take nuts, for example. Black walnuts, white walnuts (butternuts), and hickory nuts are extremely nutritious and often abundant. The Indians used to collect huge piles of them.[34] If you found a few good trees in October, you could probably gather enough nuts in an hour or less to feed yourself for a whole day. Sounds great, doesn't it?

Yes, it does sound great—if you've never tried to crack a black walnut. Maybe Arnold Schwarzenegger could crack a black walnut with an ordinary nutcracker—if the nutcracker didn't break first—but a person of average physique couldn't do it. You have to whack the nut with a hammer; and the inside of the nut is divided up by partitions that are as thick and hard as the outer shell, so you have to break the nut into several fragments and then tediously pick out the bits of meat. The process is time-consuming. In order to get enough food for a day, you might have to spend most of the day just cracking nuts and picking out the bits of meat. Wild white walnuts (not to be confused with the domesticated English walnuts that you buy in the store) are much like black ones. Hickory nuts are not as difficult to crack, but they still have the hard internal partitions and they are usually much smaller than black walnuts.

The Indians got around these problems by putting the nuts into a mortar and pounding them into tiny bits, shells, meats, and all. Then they would boil the mixture and put it aside to cool. The fragments of shell would settle to the bottom of the pot while the pulverized meats would settle in a layer above the shells; thus the meats could be separated from the shells.[35] This was certainly more efficient than cracking the nuts individually, but as you can see it still required considerable work.

The Indians of the eastern U.S. utilized other wild foods that required more-or-less laborious preparation to make them edible.[36] It is hardly likely that they would have used such foods if foods that were more easily prepared had been readily available in sufficient quantity.

Euell Gibbons, an expert on edible wild plants, reported an episode of living off the country in the eastern United States.[37] It's difficult to say what his experience tells us about primitive people's working hours, since he did not give a quantitative accounting of the time he spent in foraging. In any case, he and his partners only foraged for food and processed it; they did not have to tan skins or make their own clothing, tools, utensils, or shelter; they had no children to feed; and they supplemented their diet with high-calorie store-bought foods: cooking-oil, sugar, and flour. On at least one occasion they used an automobile for transportation.

But let's assume for the sake of argument that in the fertile regions of the world wild foods were once so abundant that it was possible to live off the country year round with an average of only, say, three hours of work per day. With such abundant resources it would not be necessary for hunter-gatherers to travel in search of food. One would expect them to become sedentary, and in that case they would be able to accumulate wealth and form well-developed social hierarchies. Hence they would lose at least some of the qualities that anarcho-primitivists value in *nomadic* hunter-gatherers. Even the anarcho-primitivists do not deny that the Indians of the Northwest Coast of North America were sedentary hunter-gatherers who accumulated wealth and had well-developed social hierarchies.[38] The evidence suggests the existence of similar hunting-and-gathering societies elsewhere where the abundance of natural resources permitted it, for example, along the major rivers of Europe.[39] Thus the anarcho-primitivists are caught in a bind: Where natural resources were abundant enough to minimize work, they also maximized the likelihood of the social hierarchies that anarcho-primitivists abhor.

However, I have not been trying to prove that primitive man was less fortunate in his working life than modern man is. In my opinion the contrary was true. Probably at least some nomadic hunter-gatherers had more leisure time than modern employed Americans do. It's true that the roughly 40-hour workweek of Richard Lee's Bushmen was about equal to the standard American workweek. But modern Americans are burdened with many demands on their time outside their hours of employment. I myself, when working at a 40-hour job, have generally felt busy: I've had to shop for groceries, go to the bank, do the laundry, fill out income-tax forms, take the car in for maintenance, get a haircut, go to the dentist...there was always something that needed to be done. Many of the people I now correspond with likewise complain of being busy. In contrast, the male Bushman's time was genuinely his own outside of his working hours; he could spend his non-working time as he pleased. Bushman women of reproductive age may have had much less leisure time because, like women of all societies, they were burdened with the care of small children.

But leisure is a modern concept, and the emphasis that anarcho-primitivists put on it is evidence of their servitude to the values of the civilization that they claim to reject. The amount of time expended in work is not what matters. Many authors have discussed what is wrong with work in modern society, and I see no reason to go over that ground again. What does matter is that, apart from monotony, what is wrong with work in modern society is *not* wrong with the work of nomadic hunter-gatherers.

The hunter-gatherer's work is challenging, both in terms of physical effort and in terms of the level of skill required.[40] The hunter-gatherer's work is purposeful, and its purpose is not abstract, remote, or artificial but concrete, very real, and directly important to the worker: He works to satisfy the physical needs of himself, his family, and other people to whom he is personally close. Above all, the nomadic hunter-gatherer is a *free* worker: He is not exploited, he is subservient to no boss, no one gives him orders;[41] he designs his own workday, if not as an individual then as a member of a group that is small enough so that every individual can participate meaningfully in the decisions that are made.[42] Modern jobs tend to be psychologically stressful, but there are reasons to believe that primitive people's work typically involved little psychological stress.[43] Hunter-gatherers' work is often monotonous, but it is my view that monotony generally causes primitive people relatively little discomfort. Boredom, I think, is largely a civilized phenomenon and is a product of psychological stresses that are characteristic of civilized life. This admittedly is a matter of personal opinion, I can't prove it, and a discussion of it would take us beyond the scope of this article. Here I will only say that my opinion is based largely on my own experience of living outside the technoindustrial system.

How hunter-gatherers felt about their own work is difficult to say, since anthropologists and others who visited primitive peoples (at least those whose reports I've read) usually do not seem to have asked such questions. But the following from Holmberg's account of the Siriono is worth noting: "They are relatively apathetic to work (*tába tába*), which includes such distasteful tasks as housebuilding, gathering firewood, clearing, planting, and tilling of fields. In quite a different class, however, are such pleasant occupations as hunting (*gwáta gwáta*) and collecting (*déka déka*, 'to look for'), which are regarded more as diversions than as work."[44]

This despite the fact that, as we saw earlier, the Siriono's hunting and collecting activities were exceedingly time-consuming, fatiguing, strenuous, and physically demanding.

3. Another element of the anarcho-primitivist myth is the belief that hunter-gatherers, at least the nomadic ones, had gender equality. John Zerzan, for example, has asserted this in *Future Primitive*[45] and elsewhere.[46] Probably some hunter-gatherer societies did have full gender equality, though I don't know of a single unarguable example. I do know of hunting-and-gathering cultures that had a relatively high degree of gender equality but fell short of full equality. In other nomadic hunter-gatherer societies male dominance was unmistakable, and in some such societies it reached the level of out-and-out brutality toward women.

Probably the most touted example of gender equality among hunter-gatherers is that of Richard Lee's Bushmen, whom we mentioned earlier in our discussion of the hunter-gatherer's working life. It should be noted at the outset that it would be very risky to assume that Lee's conclusions concerning the Dobe Bushmen could be applied to the Bushmen of the Kalahari region generally. Different groups of Bushmen differed culturally;[47] they didn't even all speak the same language.[48]

At any rate, relying largely on Richard Lee's studies, Nancy Bonvillain states that among the Dobe Bushmen (whom she calls "Ju/'hoansi"), "social norms clearly support the notion of equality of women and men,"[49] and that their "society overtly validates equality of women and men."[50] So the Dobe Bushmen had gender equality, right?

Well, maybe not. Look at some of the facts that Bonvillain herself offers in the same book: "[M]ost leaders and camp spokespersons are men. Although women and men participate in group discussions and decision making, … men's talk in discussions involving both genders amounts to about two-thirds of the total."[51]

Much worse are the forced marriages of girls in their early teens to men much older than themselves.[52] It's true that practices that seem cruel to us may not be experienced as cruel by people of other cultures on whom they are imposed. But Bonvillain quotes words of a Bushman woman that show that at least some girls did experience their forced marriages as cruel: "I cried and cried."[53] "I ran away again and again. A part of my heart kept thinking, 'How come I'm a child and have taken a husband?'"[54] Moreover, "because seniority confers prestige…the greater age, experience, and maturity of husbands may make wives socially, if not personally, subordinate."[55] Thus, while the Dobe Bushmen no doubt had some of the elements of gender equality, one would have to stretch a point pretty far to claim that they had full gender equality.

On the basis of his personal experience, Colin Turnbull stated that among the Mbuti pygmies of Africa, a "woman is in no way the social inferior of a man,"[56] and that "the woman is not discriminated against."[57] That sounds like gender equality … until you look at the concrete facts that Turnbull himself offers in the very same books: "A certain amount of wife-beating is considered good, and the wife is expected to fight back."[58] "He said that he was very content with his wife, and he had not found it necessary to beat her at all often."[59] Man throws wife to the ground and slaps her.[60] Husband beats wife.[61] Man beats sister.[62] Kenge beats his sister.[63] "Perhaps he should have beaten her harder, Tungana [an old man] said, for some girls like being beaten."[64] "Amabosu countered by smacking her firmly across the face. Normally Ekianga would have approved of such manly assertion of authority over a disloyal wife…."[65] Turnbull mentions two instances of men giving orders to their wives.[66] I have not found any instance in Turnbull's books of wives giving orders to their husbands. Pipestem obtained by *wife* is referred to as husband's property.[67] "[A boy] has to have [a girl's] permission before intercourse can take place. The men say that once they lie down with a girl, however, if they want her they take her by surprise, when petting her, and force her to their will."[68] Nowadays we would call that "date rape," and the young man involved would risk a long prison sentence.

For the sake of balance, let's note that Turnbull found among the Mbuti no instance of what we would call "street rape" as opposed to "date rape";[69] husbands were not supposed to hit their wives on the head or in the face;[70] and in at least one case in which a man took to beating his wife too frequently and severely, his campmates eventually found means to end the abuse without the use of force and without overt interference.[71] It should also be borne in mind that the significance of a beating depends on the cultural context. In our society it is a great humiliation to be struck by another person, especially by one who is bigger and stronger than oneself. But since blows were commonplace among the Mbuti,[72] it is probably safe to assume that they were not felt as particularly humiliating.

Nevertheless, it is quite clear that some degree of male dominance was present among the Mbuti.

Among the Siriono: "A woman is subservient to her husband";[73] "The extended family is generally dominated by the oldest active male."[74] "[Women] are dominated by the men."[75] "If a man is out in the forest alone with a woman...he may throw her to the ground roughly and take his prize [sex] without so much as saying a word."[76] Parents definitely preferred to have male children.[77] "Although the title *ererékwa* is reserved by the men for a chief, if one asks a woman, 'Who is your *ererékwa*?' she will invariably reply, 'My husband.'"[78] On the other hand, the Siriono never beat their wives,[79] and "[w]omen enjoy about the same privileges as men. They get as much or more food to eat, and they enjoy the same sexual freedom."[80]

According to Bonvillain, Eskimo men "dominate their wives and daughters. Men's dominance is not total, however...."[81] She describes gender relations among the Eskimos in some detail,[82] which may or may not be slanted to reflect her feminist ideology.

Among the Eskimos with whom Gontran de Poncins lived, husbands clearly held overt authority over their wives[83] and sometimes beat them.[84] Yet, through their talent for persuasion, wives had great power over their husbands:

"It might seem...that the native woman lived altogether in a state of abject inferiority to the male Eskimo, but this is not the case. What she loses in authority, as compared to the white woman, she makes up, by superior cunning, in many other ways. Native women are very shrewd, and they almost never fail to get what they want....

"It was a perpetual joy to watch this comedy, this almost wordless struggle in which the wife...inevitably got the better of the husband. There does not exist an Eskimo woman untrained in the art of wheedling, not one unable to repeat with tireless and yet insinuating insistence the mention of what she wants, until the husband, worn down by her persistence, gives way....

"Women were behind everything in this Eskimo world."[85]

"[I]t is not necessary to be a feminist to ask: 'But what of the status of Eskimo women?' Their status...suits them well enough; and I have indicated here and there in these pages that they are not only the mistresses of their households but also, in most Eskimo families, the shrewd prompters of their husbands' decisions."[86]

However, Poncins may have overstated the extent of Eskimo women's power, since it was not sufficient to enable them to avoid unwanted sex: Wife-lending among these Eskimos was determined by the men, and the wives had to accept being lent whether they liked it or not.[87] At least in some cases, apparently, the women resented this rather strongly.[88]

The Australian Aborigines' treatment of their women was nothing short of abominable. Women had almost no power to choose their own husbands.[89] They are described as having been "owned" by the men, who chose their husbands for them.[90] Young women were often forced to marry old men, and then they had to work to provide their aged husbands with the necessities of life.[91] Not surprisingly, a young woman frequently resisted a forced marriage by running away. She was then beaten severely with a club and returned to her husband. If she persisted in running away, she might even have a spear driven into her thigh.[92] A woman trapped in a distasteful marriage might enjoy the consolation of having a lover on the side, but, while this was "semitolerated," it could lead to violence.[93] A woman might even go to the length of eloping with her lover. However:

"They would be followed, and if caught, as a punishment the girl became, for the time being, the common property of her pursuers. The couple were then brought back to the camp where, if they were of the right totem division to marry, the man would have to stand up to a trial by having spears thrown at him by the husband and his relations...and the girl was given a beating by her relatives.... If [the couple] were not of the right totem division to marry, they would both be speared when found, as their sin was unforgivable."[94]

Although there was "real harmony and mutual understanding in most Aboriginal families," wife-beating was practiced.[95] According to A. P. Elkin, under some circumstances—for example, on certain ceremonial occasions—women had to submit to compulsory sex, which "implies that woman is but an object to be used in certain socially established ways."[96] The women, says Elkin, "may often not object,"[97] but: "They sometimes live in terror of the use which is made of them at some ceremonial times."[98]

Of course, no claim is made here that all of the foregoing conditions prevailed in all parts of aboriginal Australia. Culture was not uniform across the continent.

Coon says that the Australians were nomadic, but he also states that in parts of southeastern Australia, namely, "the better-watered parts... particularly Victoria and the Murray River country," the aborigines were "relatively sedentary."[99] According to Massola, in the drier parts of southeastern Australia the aborigines had to cover long distances between fast-drying wells in times of drought.[100] This corresponds with the high degree of nomadism described for other arid parts of Australia, where "Aborigines moved from waterhole to waterhole along well-defined tracks in small family groups. The whole camp moved and rarely established bases."[101] In stating that in "the better-watered parts" the aborigines were "relatively sedentary," Coon doubtless means that "[i]n fertile regions there were well-established camping areas, close to water...where people always camped at certain times of year. Camps were bases from which people made forays into the surrounding bush for food, returning in the late afternoon or spending a few days away."[102]

Coon says that in part of the well-watered Murray River country each territorial clan had a headman and a council consisting mainly of men, though in a few cases women were also elected to the council; whereas, farther to the north and west, there was little formal leadership and "control over the women and younger males was shared between" the men aged from 30 to 50.[103] Thus Australian women had very little overt political power. Yet, as among Poncins's Eskimos, certainly in our society, and probably in every society, the women often exercised great influence over their menfolk.[104]

The Tasmanians also were nomadic hunter-gatherers (though some were "relatively sedentary"),[105] and it's not clear that they treated women any better than the Australians did. "In one account we are told that a band living near Hobart Town before the colonists' arrival was raided by neighbors who killed the men who tried to stop them and took away their women. And there are other accounts of individual cases of marriage by capture. Sometimes when a man from a neighboring band had the right to marry a girl, but neither she nor her parents liked him, it is said that they killed the girl rather than give her up."[106] "The other tribes considered [a certain tribe] cowards, and...raided them to steal their women."[107] "Woorrady...raped and killed a sister-in-law."[108]

Here I should make clear that it is not my intention to argue against gender equality. I myself am enough a product of modern industrial society to feel that women and men should have equal status. My purpose at this point is simply to exhibit the facts concerning the relations between the sexes in hunting-and-gathering societies.

4. There is a problem involved in any attempt to draw conclusions about original, "pure" hunter-gatherer cultures from reported observations of living hunter-gatherer societies. If we have a description of a primitive culture, it ordinarily will have been written by some civilized person. If the description is detailed, then, by the time it was written, the primitive people described very likely will have had significant contact, direct or indirect, with civilization, and such contact can bring about dramatic changes in a primitive culture. Elizabeth Marshall Thomas, in the epilogue to the 1989 edition of her book *The Harmless People*,[109] describes the catastrophically destructive effect of civilization on the Bushmen she knew. Harold B. Barclay has pointed out that (for example) modern Eskimos "are quite pleased with their high powered rifles, motorboats and so forth."[110] "So forth" would include snowmobiles. Hence, Barclay says, "hunter gatherers today...are in no sense identical to hunter gatherers of a thousand or ten thousand years ago."[111] According to Cashdan, writing in 1989, "all hunter-gatherers in the world today are in contact, directly or indirectly, with the world economy. This fact should caution us against viewing today's hunter-gatherers as 'snapshots' of the past."[112]

Of course, in seeking evidence of the way human beings lived prior to the advent of civilization, no one in his right mind would turn to peoples who used motorboats, snowmobiles, and high-powered rifles,[113] or to peoples whose cultures had obviously been grossly disrupted by the intrusion of civilized societies. We look for accounts of hunter-gatherers written (at least) several decades ago and at a time when—as far as we can tell—their cultures had not been seriously altered by contact with civilization. But it's not always easy to tell whether contact with civilization has altered a primitive culture. Coon is clearly aware of this problem, and in his excellent survey of hunter-gatherer cultures he gives the following example of how seemingly slight interference from civilization can have a dramatic effect on a primitive culture: When "well-meaning missionaries... hand[ed] out steel axes" to the Yir Yoront aborigines of Australia, the "Yir Yoront world almost came to an end. The men lost their authority over their wives, a generation gap appeared," and a system of trade stretching over hundreds of miles was disrupted.[114]

Richard Lee's Bushmen are perhaps the favorite example for anarcho-primitivists and leftish anthropologists who want to present a politically correct image of hunter-gatherers, and Lee's Bushmen were among the least "pure" of the hunter-gatherers we've mentioned here. They may not even have always been hunter-gatherers.[115] In any case they had probably been trading with agricultural and pastoral peoples for a couple of thousand years.[116] The Kung Bushmen whom Mrs. Thomas knew had metal acquired through trade,[117] and the same apparently was true of Lee's Bushmen.[118] Mrs. Thomas writes: "In the ten to twenty years after we started our work, many academics [this presumably includes Richard Lee] developed an enormous interest in the Bushmen. Many of them went to Botswana to visit groups of Kung Bushmen, and for a time in Botswana, the anthropologist/Bushman ratio seemed almost one to one."[119] Obviously, the presence of so many anthropologists may itself have affected the behavior of the Bushmen.

In the 1950s,[120] when Turnbull studied them, still more in the 1920s and 1930s[121] when Schebesta studied them, the Mbuti apparently had not had much direct contact with civilization, so that Schebesta went so far as to claim that "the Mbuti not only racially, but also psychologically and in terms of cultural history, are a primeval phenomenon (Urphänomen) among the races and peoples of the Earth."[122] Yet the Mbuti had already begun to be somewhat affected by civilization a few years before Schebesta's first visit to them.[123] And for centuries before that, the Mbuti had lived in close contact (which included extensive trade relations) with non-civilized, village-dwelling cultivators of crops.[124] As Schebesta wrote, "The belief that the Mbuti have been hermetically sealed off from the outer world has been laid to rest once and for all."[125] Turnbull goes farther: "This is in no way to say that the [social] structure to be found among the Mbuti is representative of an original pygmy hunting and gathering structure; in fact probably far from it, for the repercussions of the invasion of the forest by the village cultivators have been enormous."[126]

Though some of Gontran de Poncins's Eskimos were "purer" than others,[127] it appears that all of them had at least some trade goods from the whites. If any reader cares to take the trouble to track down the earliest primary sources—perhaps some of Vilhjalmur Stefansson's work—so as to approach as closely as possible to an original and "pure" Eskimo culture, I would be interested to hear of his or her findings. But it is possible that even long before European contact the Eskimos' culture may have been affected by something that they received from a non-hunting society; for their sled dogs may not have originated with hunter-gatherers.[128]

With the Siriono we come closer to purity than we do with the Bushmen, the Mbuti, or Poncins's Eskimos. The Siriono did not even have dogs,[129] and even though they cultivated crops to a limited extent anthropologists regarded their culture as Paleolithic (Old Stone Age).[130] Some of the Siriono studied by Holmberg had had little or no contact with whites prior to Holmberg's arrival[131] and, among those Siriono, European tools were rarely encountered[132] until Holmberg himself introduced them.[133] Instead, the Siriono made their tools of naturally-occurring local materials.[134] The Siriono moreover were so primitive that they could not count beyond three.[135] Nevertheless, Siriono culture might have been affected by contact with more "advanced" societies, since Holmberg thought the Siriono were "probably a remnant of an ancient population that was exterminated, absorbed, or engulfed by more civilized invaders."[136]

Lauriston Sharp even suggested that the Siriono might have "degenerated" [sic] "from a more advanced technical condition," though Holmberg rejected this view and Sharp himself considered it "irrelevant."[137] In addition, the Siriono might have been affected indirectly by European civilization, since probably at least some of the diseases from which they suffered, e.g., malaria, had been brought to the Americas by Europeans.[138]

It's not surprising that most of the hunter-gatherers I've mentioned here—like those cited by the anarcho-primitivists and the politically correct anthropologists—were affected by direct or indirect contact with agricultural or pastoral peoples even long before their first contact with Europeans, because outside of Australia, Tasmania, and the far west and north of North America "populations which remained faithful to the old hunter-gatherer way of life were small and scattered."[139] Consequently, with the possible exception of some who lived on small islands, they necessarily had some form of contact with surrounding non–hunter-gatherer populations.

Probably the Australian Aborigines and the Tasmanians were the hunter-gatherers who were purest when Europeans first found them. Australia was the only continent that was inhabited exclusively by hunter-gatherers until the white man's arrival, and Tasmania, an island just to the south of Australia, was even more isolated. But Tasmania may have been visited by Polynesians, and in the north of Australia there was some limited contact with people from Indonesia and New Guinea prior to the arrival of Europeans.[140] Still earlier contact with outsiders, who may or may not have been hunter-gatherers, is probable.[141]

Thus, we have no conclusive proof that hunter-gatherer cultures that survived into recent times had not been seriously affected by contact with non-hunter-gatherers by the time the first descriptions of them were written. Consequently, more or less uncertainty is involved in using reports on recent hunter-gatherer societies to draw conclusions about gender relations among prehistoric hunter-gatherers. And any conclusions drawn from archaeological remains about the social relationships between men and women can only be highly speculative.

So, if you like, you can reject all evidence from descriptions of recent hunter-gatherer cultures, and in that case we know almost nothing about the gender relations of prehistoric hunter-gatherers. Or (with the necessary reservations) you can accept the evidence from recent hunter-gatherer societies, and in that case the evidence clearly points to a significant degree of male dominance. In either case, there is no evidence to support the anarcho-primitivists' belief that all or most human societies had full gender equality prior to the advent of agriculture and animal husbandry some 10 thousand years ago.

5. Our review of the facts concerning gender relations in recent hunter-gatherers societies helps to reveal something of the psychology of the anarcho-primitivists and that of their cousins, the politically correct anthropologists.

The anarcho-primitivists, and many politically correct anthropologists, cite any evidence they can find that hunter-gatherers had gender equality, while systematically ignoring the abundant evidence of gender inequality found in eyewitness reports of hunter-gatherer cultures. For example, the anthropologist Haviland, in his textbook *Cultural Anthropology*, states that an "important characteristic of the food-foraging [hunter-gatherer] society is its egalitarianism."[142] He acknowledges that the two sexes may have had different status in such societies, but claims that "status differences by themselves do not imply any necessary inequality," and that in "traditional food-foraging societies, nothing necessitated special deference of women to men."[143] If you check the pages listed in Haviland's index for the entries "Bushmen," "Ju/'hoansi" (another name for the Dobe Bushmen), "Eskimo," "Inuit" (another name for Eskimos), "Mbuti," "Tasmanian," "Australian," and "Aborigine" (the Siriono are not listed in the index), you will find no mention of wife-beating, forced marriage, forced sexual intercourse, or any of the other indications of male dominance that I've cited above.

Haviland does not deny that these things occurred. He does not claim, for example, that Turnbull merely invented his stories of wife-beating among the Mbuti, or that such-and-such evidence shows that Australian Aboriginal women were not subjected to involuntary sex before the arrival of Europeans. He simply ignores these issues, as if they didn't exist. And it's not that Haviland isn't aware of the issues. For example, he quotes from A. P. Elkin's book, *The Australian Aborigines*,[144] an indication that he not only is familiar with the book but considers it a reliable source of information. Yet Elkin's book, which I cited earlier, provides ample evidence of Australian Aboriginal men's tyranny over their women[145]—evidence that Haviland fails to mention.

It's pretty clear what is going on: Equality of the sexes is a fundamental tenet of the mainstream ideology of modern society. As highly socialized members of that society, politically correct anthropologists believe in the principle of gender equality with something akin to religious conviction, and they feel a need to give us little moral lessons by holding up for our admiration examples of the gender equality that supposedly prevailed when the human race was in a pristine and unspoiled state. This portrayal of primitive cultures is driven by the anthropologists' own need to reaffirm their faith, and has nothing to do with an honest search for truth.

To take another example, I've written to John Zerzan four times inviting him to back up his claims about gender equality among hunter-gatherers.[146] The answers he gave me were vague and evasive.[147] I would gladly publish here Zerzan's letters to me on this subject so that the reader could judge them for himself. However, I wrote to Zerzan requesting permission to publish his letters, and he denied me that permission.[148] With his letters he sent me photocopies of pages from a few books that contained vague, general statements ostensibly supporting his claims about gender equality; for instance, this statement by John E. Pfeiffer, who is neither a specialist nor an eyewitness of primitive behavior, but a popularizer: "For reasons unknown sexism arrived with settling and farming, with the emergence of complex society."[149]

Zerzan also sent me a photocopy of a page from Bonvillain's book containing the following statement: "In foraging band [hunter-gatherer] societies, the potential for gender equality is perhaps the greatest...."[150] But Zerzan did not include copies of the pages on which Bonvillain said that male dominance was evident in some hunter-gatherer societies such as that of the Eskimos, or the pages on which she gave information that cast grave doubt on her own claim of gender equality among the Dobe Bushmen, as I discussed above.

Zerzan himself acknowledged that the material he sent me was "obviously not definitive," though he asserted that it was "completely representative in general."[151] When I pressed him for further backing for his claims,[152] he sent me a copy of his essay *Future Primitive*, from the book of the same name.[153] In this essay he cites most of his sources by giving only the authors' last names and their publications' dates; the reader presumably is expected to look up further information in a table of references provided elsewhere in the book. Since Zerzan did not send me a copy of the table of references, I had no way of checking his sources. I pointed this out to him,[154] but he still failed to send me a copy of his table of references. In any case, there is good reason to suspect that Zerzan was uncritical in selecting his sources. For example, he quotes the late Laurens van der Post;[155] but in his book *Teller of Many Tales*, J. D. F. Jones, a former admirer of Laurens van der Post, has exposed the latter as a liar and a fraud.

Even if taken at face value, the information in *Future Primitive* gives us nothing solid on the subject of gender relations. Vague, general statements are of little use. As I pointed out earlier, Bonvillain and Turnbull made general assertions about gender equality among the Bushmen and the Mbuti respectively, and those assertions were contradicted by concrete facts that Bonvillain and Turnbull themselves reported in the same books. On subjects other than gender equality, some of the statements in *Future Primitive* are demonstrably false. To take a couple of examples:

(i) Zerzan, relying on one "De Vries," claims that among hunter-gatherers childbirth is "without difficulty or pain."[156] Oh, really? Here's Mrs. Thomas, writing from her personal experience among the Bushmen: "Bushmen women give birth alone...unless a girl is bearing her first child, in which case her mother may help her, or unless the birth is extremely difficult, in which case a woman may ask the help of her mother or another woman.... [A] woman in labor may clench her teeth, may let her tears come or bite her hands until blood flows, but she may never cry out to show her agony."[157]

Since natural selection eliminates the weak and the defective among hunter-gatherers and since primitive women's work keeps them in good physical condition, it is probably true that childbirth, on average, was not as difficult among hunter-gatherers as it is for modern women. For Mbuti women, according to Schebesta, delivery was usually easy (though this does not imply that it was free of pain). On the other hand, breech deliveries were much feared and usually ended fatally both for the mother and the for child.[158]

(ii) Relying on one "Duffy," Zerzan claims that the Mbuti "look on any form of violence between one person and another with great abhorrence and distaste, and never represent it in their dancing or their playacting."[159] But Hutereau and Turnbull independently have provided eyewitness accounts according to which the Mbuti did indeed playact violence between human beings.[160] More important, there was plenty of *real-life* violence among the Mbuti. Accounts of physical fights and beatings are scattered throughout Turnbull's books, *The Forest People* and *Wayward Servants*. To cite just one of the numerous examples, Turnbull mentions a woman who lost three teeth in fighting with another woman over a man.[161] I've already mentioned Turnbull's statements about wife-beating among the Mbuti.

It's worth noting that Zerzan apparently believes that our ancestors were capable of mental telepathy.[162] But particularly revealing is Zerzan's quotation of "Shanks and Tilley": "The point of archaeology is not merely to interpret the past but to change the manner in which the past is interpreted in the service of social reconstruction in the present."[163] This is virtually open advocacy of the proposition that archaeologists should slant their findings for political purposes. What better evidence could there be of the massive politicization that has taken place in American anthropology over the last 35 or 40 years? In view of this politicization, anything in recent anthropological literature that portrays primitive peoples' behavior as politically correct must be viewed with the utmost skepticism.

After citing to Zerzan some of the examples of gender inequality that I've discussed above, I questioned his honesty on the ground that he had "systematically excluded nearly all of the evidence...that undercuts the idealized picture of hunter-gatherer societies" that he wanted to present.[164] Zerzan answered that he "did not find many credible sources" that contradicted his outlook.[165] This statement strains credulity. Some of the examples that I cited to Zerzan (and have discussed above) were from books on which he himself had relied—those of Bonvillain and Turnbull.[166] Yet he somehow managed to overlook all of the evidence in those books that contradicted his claims. Since Zerzan has read widely about hunter-gatherer societies, and since the Australian Aborigines are among the best-known hunter-gatherers, I find it very difficult to believe that he has never come across any accounts of the Australians' mistreatment of women. Yet he never mentions such accounts—not even for the purpose of refuting them.

One does not necessarily have to assume any conscious dishonesty on Zerzan's part. As Nietzsche said, "The most common lie is the lie one tells to oneself; lying to others is relatively the exception."[167] In other words, self-deception often precedes deception of others. An important factor here may be one that is well known to professional propagandists: People tend to block out—to fail to perceive or to remember—information that they find uncongenial.[168] Since information that discredits one's ideology is highly uncongenial, it follows that people will tend to block out such information. A young anarcho-primitivist with whom I've corresponded has provided me with an amazing example of this phenomenon. He wrote to me: "there is no question about the persistance [sic] of patriarchy in all other oceanic societies, but none seems apparent in the [Australian] Aborigines—According to A. P. Elkin's *The Australian Aborigines* wives were not held in a restrictive marriage at all...."[169] It was apparent that my anarcho-primitivist friend had read Elkin's discussion of women's position in Australian Aboriginal society. I've cited above some of the relevant pages of Elkin's book, such as those on which he states that Australian Aboriginal women sometimes lived in terror of the compulsory sex to which they were subjected at some ceremonial times. Any reasonably rational person who will take the trouble to read those pages[170] will find himself hard-pressed to explain how my anarcho-primitivist friend could have read that material and then claimed in all seriousness that no patriarchy seemed apparent in Australian Aboriginal society—unless my friend simply blocked out of his mind the information that he found ideologically unacceptable. My friend did not question the accuracy of Elkin's information; in fact, he was relying on Elkin as an authority. He simply remained oblivious to the information that indicated patriarchy among the Australian Aborigines.

By this time it should be sufficiently clear to the reader that what the anarcho-primitivists (and many anthropologists) are up to has nothing to do with a rational search for the truth about primitive cultures. Instead, they have been developing a myth.

6. I've already had occasion at several points to mention violence among nomadic hunter-gatherers. Examples of violence, including deadly violence, among hunter-gatherers are abundant. To mention only a few such examples:

"One account has been published of a mortal battle between an inland band of Tasmanians having access to ochre, and a coastal band who had agreed to exchange seashells for the other's product. The inland people brought their ochre, but the coastal people arrived empty handed. Men were killed because of a breach of faith over the two materials, neither of which was edible or of any other practical use. In other words, the Tasmanians were just as 'human' as the rest of us."[171]

The Tasmanians made their spears "in two lengths...the shorter ones were for hunting, the longer ones for fighting."[172]

Among the hunter-gatherers of the Andaman Islands, "grievances were remembered, and revenge might be taken later.... The raiders either crept through the jungle or approached in canoes. They leaped on their victims by surprise, quickly shot [with arrows] all the men and women unable to escape, and took away any uninjured children, to adopt them....

"If enough members of the group survived to reconstitute the band, they might eventually grow numerous enough to seek revenge, and a lengthy feud might arise.... [Peace efforts were] initiated by the women because it was they who had kept the hostilities alive, egging on their men..."[173]

Among at least some groups of Australian Aborigines, women at times would provoke their menfolk to deadly violence against other men.[174] Among the Eskimos with whom Gontran de Poncins lived, there was "a good deal of killing," and it was sometimes a woman who persuaded a man to kill another man.[175] Paintings made in rock shelters by prehistoric hunter-gatherers of eastern Spain show groups of men fighting each other with bows and arrows.[176]

One could go on and on. But I don't want to give the impression that all hunter-gatherer societies were violent. Turnbull refers to numerous nonlethal fights and beatings among the Mbuti, but in those of his books that I've read he mentions not a single case of homicide.[177] This suggests that *deadly* violence was rare among the Mbuti at the time when Turnbull knew them. Siriono women sometimes fought physically, striking each other with sticks, and there was a good deal of aggression among the children, even with sticks or burning brands used as weapons.[178] But men rarely fought each other with weapons,[179] and the Siriono were not warlike.[180] Under extreme provocation they did kill certain whites and missionized Indians,[181] but among the Siriono themselves intentional homicide was almost unknown.[182] Among the Bushmen whom Mrs. Thomas knew aggression of any kind was minimal, though she makes clear that this was not necessarily true of all Bushman groups.[183]

It is important, too, to realize that deadly violence among primitives is not even remotely comparable to modern warfare. When primitives fight, two little bands of men shoot arrows or swing war-clubs at one another because they *want* to fight; or because they are defending themselves, their families, or their territory. In the modern world soldiers fight because they are forced to do so, or, at best, because they have been brainwashed into believing in some kook ideology such as that of Nazism, socialism, or what American politicians choose to call "freedom." In any case the modern soldier is merely a pawn, a dupe who dies not for his family or his tribe but for the politicians who exploit him. If he's unlucky, maybe he does not die but comes home horribly crippled in a way that would never result from an arrow- or a spear-wound. Meanwhile, thousands of non-combatants are killed or mutilated. The environment is ravaged, not only in the war zone, but also back home, due to the accelerated consumption of natural resources needed to feed the war machine. In comparison, the violence of primitive man is relatively innocuous.

That, however, isn't good enough for the anarcho-primitivists or for today's politically correct anthropologists. They can't deny altogether the existence of violence among hunter-gatherers, since the evidence for it is incontrovertible. But they will stretch the truth as far as they think they can get away with in order to minimize the amount of violence in the human past. It's worthwhile to give an example that illustrates the silliness of some of the reasoning that they use. In reference to *Homo habilis*, a physically primitive ancestor of modern man, the anthropologist Haviland writes: "They obtained their meat not by killing live animals but by scavenging.... *Homo habilis* got meat by scavenging from carcasses of dead animals, rather than hunting live ones. We know this because the marks of stone tools on the bones of butchered animals commonly overlie marks the teeth of carnivores made. Clearly, *Homo habilis* did not get to the prey first."[184]

But, as Haviland certainly ought to know, many or most predatory animals engage *both* in hunting *and* in scavenging. For example, bears, African lions, martens, wolverines, wolves, coyotes, foxes, jackals, hyenas, the raccoon dog of Asia, the Komodo dragon, and some vultures both hunt and scavenge.[185] Thus, the fact that *Homo habilis* engaged in scavenging provides no evidence whatsoever that he did not also hunt.

I emphasize that I do not know or care whether *Homo habilis* hunted. I see no reason why it should be important for us to know whether our half-human ancestors two million years ago were bloodthirsty killers, peaceful vegetarians, or something in between. The point here is simply to show what kind of reasoning some anthropologists will resort to in their effort to make the human past look as politically correct as possible.

Since political correctness has warped the portrayal not only of the human past but of wild nature generally, it should be pointed out that deadly violence among wild animals is not confined to predation of one species upon another. Killing of one member of a species by another member of the same species does occur. For example, it is well known that wild chimpanzees often kill other chimpanzees.[186] Elephants sometimes kill one another in fights, and the same is true of wild pigs.[187] Among the sea birds called brown boobies, two eggs are laid in each nest. After the eggs are hatched, one of the chicks attacks the other and forces it out of the nest, so that it dies.[188] Komodo dragons sometimes eat one another,[189] and there is evidence that cannibalism occurred among some dinosaurs.[190] Último Reducto has pointed out to me that there is incontrovertible evidence of cannibalism among prehistoric humans.[191]

I do want to make clear that it is by no means my intention to exalt violence. I prefer to see people (and animals) get along smoothly with one another. My purpose is only to expose the irrationality of the politically correct image of primitive peoples and of wild nature.

7. An important element of the anarcho-primitivist myth is the belief that hunter-gatherer societies were free of competition and were characterized instead by sharing and cooperation.

Colin Turnbull's early writings on the Mbuti pygmies seem to be quite frank, but his work leaned increasingly toward political correctness as time went by.[192] Writing in 1983 (18 and 21 years, respectively, after he had published *Wayward Servants* and *The Forest People*), Turnbull noted that Mbuti children had no competitive games,[193] and after referring to the high value that he claimed modern society placed on "competition" and "economic independence,"[194] he contrasted these with "the well-tried primitive values of family-writ-large: interdependence, cooperation, and reliance on community…rather than on self…."[195]

But according to Turnbull's own earlier work, physical fighting was commonplace among the Mbuti.[196] If a physical fight isn't a form of competition, then what is? It's clear in fact that the Mbuti were a very quarrelsome people, and, in addition to physical fights, there were many verbal disputes among them.[197] Generally speaking, any dispute, whether it is settled physically or verbally, is a form of competition: the interests of one person conflict with those of another, and their quarreling is an effort by each to promote his own interests at the other's expense. The Mbuti's jealousies also were evidence of competitive impulses.[198]

Two things for which the Mbuti competed were mates and food. I've already mentioned a case of two women who fought over a man,[199] and quarreling over food apparently was common.[200]

It's worth noting that Turnbull, in his early work, described the Mbuti as "individualists."[201] There is abundant evidence of competitiveness and/or individualism among other primitive peoples. The Nuer (African pastoralists), the pagan Germanic tribes, the Carib Indians, the Siriono (who lived mainly by hunting and gathering), the Navajo, the Apaches, the Plains Indians, and North American Indians generally have all been described explicitly as "individualistic."[202] But "individualism" is a vague word that may mean different things to different people, so it's more helpful to look at definite facts that have been reported. Some of the works that I cite in Note 202 do back up with facts their application of the term "individualistic" to the peoples mentioned. Holmberg writes:

"[W]hen an Indian [Siriono] has reached adulthood he displays an individualism and apathy toward his fellows that is remarkable. The apparent unconcern of one individual for another—even within the family—never ceased to amaze me while I was living with the Siriono. Frequently men would depart for the hunt alone—without so much as a goodbye—and remain away from the band for weeks at a time without any concern on the part of their fellow tribesmen or even their wives....

"Unconcern with one's fellows is manifested on every hand. On one occasion Ekwataia...went hunting. On his return darkness overcame him about five hundred yards from camp. The night was black as ink, and Ekwataia lost his way. He began to call for help—for someone to bring him fire or to guide him into camp by calls. No one paid heed to his requests.... After about half an hour, his cries ceased, and his sister Seáci, said: 'A jaguar probably got him.' When Ekwataia returned the following morning, he told me that he had spent the night sitting on the branch of a tree to avoid being eaten by jaguars."[203]

Holmberg repeatedly remarks on the uncooperative character of the Siriono, and says that those of them who became disabled by age or sickness were simply abandoned by the others.[204]

Among other primitive peoples, individualism takes other forms. For example, among most of the North American Indians, warfare was a decidedly individualistic enterprise. "The Indians, being highly individualistic and often fighting more for personal glory than group advantage, never developed a science of warfare."[205] According to the Cheyenne Indian Wooden Leg:

"[W]hen any battle actually began it was a case of every man for himself. There were no ordered groupings, no systematic movements in concert, no compulsory goings and comings. Warriors...mingled indiscriminately... every one looked out for himself only, or each helped a friend if such help were needed and if the able one's personal inclination just then was toward friendly helpfulness.... The Sioux tribes...fought their battles as a band of individuals, the same as we fought ours, and the same as was the way of all Indians I ever knew."[206]

During the first half of the 20th century, Stanley Vestal interviewed many Plains Indians who still remembered the old days. According to him:

"It cannot be too often repeated that—except when defending his camp—the Indian was totally indifferent to the general result of a fight: all he cared about was his own coups. Time and again old men have said to me, in discussing a given battle, 'Nothing happened that day,' meaning simply that the speaker had been unable to count a coups."[207] "Plains Indians could not wage war by plan. They had...no discipline.... On the rare occasions when they did have a plan, some ambitious young man was sure to launch a premature attack...."[208]

Compare this with modern man's way of waging war: Troops move in obedience to carefully elaborated plans; every man has a specific task to perform in cooperation with other men, and he performs it not for personal glory but for the advantage of the army as a whole. Thus, in warfare, it is modern man who is cooperative and primitive man who is, generally speaking, an individualist.

Primitive individualism is not confined to warfare. Among the Indians of subarctic North America, who were hunter-gatherers, there was an "individualistic relationship to the supernatural," "self-reliance," and a "high value placed on personal autonomy."[209] Australian Aboriginal children were "taught to be self-reliant."[210] Among the Woodland Indians of the eastern United States, "great emphasis was placed on self-reliance and individual competence,"[211] and the Navajo "insist[ed] upon self-reliance."[212] The Nuer of Africa extolled the virtues of "stubbornness" and "independence"; "Their only test of character is whether one can stand up for oneself."[213]

Evidence of competition among primitives is ample. In addition to the Mbuti, at least some other hunter-gatherers competed for mates or for food. "One cannot remain long with the Siriono without noting that quarreling and wrangling are ubiquitous."[214] The majority of quarrels "arose directly over questions of food," but sexual jealousy also led to fights and quarrels among the Siriono.[215] The Australian Aborigines fought for the possession of women.[216] Poncins reports the case of one Eskimo who killed another in order to take his wife, and he states that any Eskimo would kill in order to prevent his wife from being taken from him.[217]

Notwithstanding Turnbull's remark that Mbuti children had no competitive games, some Mbuti adults did play tug-of-war, which clearly is a competitive game;[218] and certain other primitive peoples too had competitive games. Massola mentions war games among the Australian Aborigines, and a ball game in which "[t]he boy who caught the ball the greatest number of times was considered to be the winner."[219] The game of lacrosse originated among the Algonkin Indians.[220] Navaho children of both sexes had foot-races,[221] and among the Plains Indians almost all of the boys' games were competitive.[222] The Cheyenne Indian Wooden Leg described some of the competitive sports in which his people had engaged: "Horse races, foot races, wrestling matches, target shooting with guns or with arrows, tossing the arrows by hand, swimming, jumping and other like contests..."[223] The Cheyenne also competed in war, in hunting, and "in all worthy activities."[80]

Richard E. Leakey quotes Richard Lee thusly: "Sharing deeply pervades the behavior and values of !Kung [Bushmen] foragers...sharing [is] central to the conduct of life in foraging societies." Leakey adds: "This ethic is not confined to the !Kung: it is a feature of hunter-gatherers in general."[225]

Of course, we share too. We pay taxes. Our tax money is used to help poor or disabled people through public-assistance programs, and to carry on other public activities that are supposed to promote the general welfare. Employers share with their employees by paying them wages.

But, aha! you answer, we share only because we are forced to do so. If we tried to evade payment of taxes we would go to prison; if an employer offered insufficient wages and benefits, no one would work for him, or perhaps he would have trouble with the union or with the minimum-wage laws. The difference is that hunter-gatherers shared voluntarily, out of loving, open-hearted generosity ... right?

Well, not exactly. Just as our sharing is governed by tax laws, union contracts, and the like, sharing in hunter-gatherer societies was commonly governed by "rigid procedural rules" that "must be followed in order to keep the peace."[226] Many hunter-gatherers were just as grudging about sharing their food as we are about paying our taxes, and just as anxious to make sure that they got not a bit less than what the rules entitled them to.

Among Richard Lee's Bushmen: "Distribution [of meat] is done with great care, according to a set of rules...improper meat distributions can be the cause of bitter wrangling among close relatives."[227] Among the

Tikerarmiut Eskimos, even though the rules for distribution of whale meat "were scrupulously followed…there still might be vociferous arguments."[228] The Siriono had food taboos that might have served as rules for the distribution of meat, but the taboos were very often disregarded.[229] Though the Siriono did share food, they did so with extreme reluctance:[230] "People constantly complain and quarrel about the distribution of food.… Enía said to me one night: 'When someone comes near the house, women hide the meat.… Women even push meat up their vaginas to hide it.'"[231] "If, for instance, a person does share food with a kinsman, he has the right to expect some in return. Reciprocity, however, is almost always forced, and is sometimes even hostile. Indeed, sharing rarely occurs without a certain amount of mutual distrust and misunderstanding."[232] The Mbuti had rules for sharing meat,[233] but there was, "often as not, a great deal of squabbling over the division of the game."[234] "Once an animal is killed…it is taken…to be shared out on return to the camp.… This is not to say that sharing takes place without any dispute or acrimony. On the contrary, the arguments that ensue when the hunt returns to camp are frequently long and loud…."[235] "When the hunt returns to camp…men and women alike, but particularly women, may be seen furtively concealing some of their spoils under the leaves of their roofs, or in empty pots nearly";[236] "It would be a rare Mbuti woman who did not conceal a portion of the catch in case she was forced to share with others."[237]

The fact that some hunter-gatherers often quarreled over the distribution of food conflicts with the anarcho-primitivists' claims about "primitive affluence." If food was so easy to get, then why would people quarrel over it? It should also be noted that the general rule of sharing among hunter-gatherers applied mainly to meat. There was relatively little sharing of vegetable foods,[238] even though vegetable foods often constituted the greater part of the diet.[239]

But I don't want to give the impression that all primitive peoples or all hunter-gatherers were radical individualists who never cooperated and never shared except under compulsion. The Siriono, in terms of their selfishness, callousness, and uncooperativeness, were an extreme case. Among most of the primitive peoples about whom I've read there seems to have been a reasonable balance between cooperation and competition, sharing and selfishness, individualism and community spirit.

In stating that hunter-gatherers did not usually share vegetable foods, shellfish, or the like outside of the household, Coon also indicates that such foods might indeed be shared with other families if the latter were hungry.[240] Notwithstanding their individualistic traits, the Cheyenne (and probably other Plains Indians) placed a high value on generosity (i.e., voluntary sharing),[241] and the same was true of the Nuer.[242] The Eskimos with whom Gontran de Poncins lived were so generous in sharing their belongings that Poncins described their community as "quasi-communist" and stated that "all labored in common with no hint of selfishness."[243] (Poncins did note, however, that an Eskimo expected every gift to be repaid eventually with a return gift.)[244] The importance to the Mbuti of cooperation in hunting and in some other activities is described by Turnbull,[245] who also states that failure to share in time of need was a "crime,"[246] and that the Mbuti shared to some extent even when there was no necessity for sharing.[247]

In contrast to the callousness shown by the Siriono, the old or crippled among the Mbuti were treated with a care and respect that derived mainly from affection and a sense of responsibility.[248] Poncins's Eskimos would abandon helpless old people to die when it became too difficult to take care of them any longer, but they must have done this reluctantly, because as long as they had the old people with them, "they look after the aged on the trail, running back so often to the sled to see if the old people are warm enough, if they are comfortable, if they are not perhaps hungry and want a bit of fish."[249]

Just as one could go on and on citing examples of selfishness, competition, and aggression among hunter-gatherers, so one could go on and on citing examples of generosity, cooperation, and love among them. I've emphasized primarily examples showing selfishness, competition, and aggression only because of the need to debunk the anarcho-primitivist myth that portrays the life of hunter-gatherers as a kind of politically correct Garden of Eden.

In any case, when Colin Turnbull contrasts modern "competition," "independence," and reliance on "self" with "the well-tried primitive values of interdependence, cooperation, and reliance on community," he simply makes a fool of himself. As we've already seen, the latter values are not particularly characteristic of primitive societies. And a moment's thought shows that in modern society self-reliance has become practically impossible, while cooperation and interdependence are developed to an infinitely greater degree than could ever be the case in a primitive society.

A modern nation is a vast, highly-organized system in which every part is dependent on every other part. The factories and oil refineries could not function without the electricity provided by power plants, the power plants need replacement parts produced in the factories, the factories require materials that could not be transported without the fuel provided by oil refineries. The factories, refineries, and power plants could not function without the workers. The workers need food produced on farms, the farms require fuel and spare parts for tractors and machinery, hence cannot do without the refineries and factories…and so forth. And even a modern nation is no longer a self-sufficient unit. Increasingly, every country is dependent on the global economy. Since the modern individual could not survive without the goods and services provided by the worldwide technoindustrial machine, it is absurd today to speak of self-reliance.

To keep the whole machine running, a vast, elaborately choreographed system of cooperation is necessary. People have to arrive at their places of employment at precisely designated times, and do their work in accord with detailed rules and procedures in order to ensure that every individual's performance meshes with everyone else's. In order for traffic to flow smoothly and without accidents or congestion, people must cooperate by complying with numerous traffic regulations. Appointments must be kept, taxes paid, licenses procured, laws obeyed, etc., etc., etc. There has never existed a primitive society that has had such a far-reaching and elaborate system of cooperation, or one that has regulated the behavior of the individual in such detail. Under these circumstances, the claim that modern society is characterized by "independence" and "self-reliance," in opposition to primitive "interdependence" and "cooperation," appears bizarre.

It might be answered that modern people cooperate with the system only because they are forced to do so, whereas at least part of primitive man's cooperation is more or less voluntary. This of course is true, and the reason for it is clear. Precisely because our system of cooperation is so highly developed, it is exceedingly demanding and therefore so burdensome to the individual that few people would comply with it if they didn't fear losing their jobs, paying a fine, or going to jail. Primitive man's cooperation can be partly voluntary for the very reason that far less cooperation is required of primitive man than of modern man.

What gives modern society a superficial appearance of individualism, independence, and self-reliance is the vanishing of the ties that formerly linked individuals into *small-scale* communities. Today, nuclear families commonly have little connection to their next-door neighbors or even to their cousins. Most people have friends, but friends nowadays tend to use each other only for entertainment. They do not usually cooperate in economic or other serious, practical activities, nor do they offer each other much physical or economic security. If you become disabled, you don't expect your friends to support you. You depend on insurance or on the welfare department.

But the ties of cooperation and mutual assistance that once bound the hunter-gatherer to his band have not simply vanished into thin air. They have been replaced by ties that bind us to the technoindustrial system as a whole, and bind us much more tightly than the hunter-gatherer was bound to his band. It is absurd to say that a person is independent, self-reliant, or an individualist because he belongs to a collectivity of hundreds of millions of people rather than to one of 30 or 50 people.

As for competition, it is more firmly leashed in our society than it was in most primitive societies. As we've seen, two Mbuti women might compete for a man with their fists; they might compete for food by filching some or by having a shouting match over the division of meat. Australian Aboriginal men fought over women with deadly weapons.[250] But such direct and unrestrained competition cannot be tolerated in modern society because it would disrupt the elaborate and finely-tuned system of cooperation. So our society has developed outlets for the competitive impulse that are harmless, or even useful, to the system. Men today do not compete for women, or vice versa, by fighting. Men compete for women by earning money and driving prestigious cars; women compete for men by cultivating charm and appearance. Corporation executives compete by striving for promotions. In this context, *competition* among the executives is a device that encourages them to *cooperate* with the corporation, for the person who wins the promotion is the one who best serves the corporation. It could plausibly be argued that competitive sports in modern society function as an outlet for aggressive and competitive impulses that would have serious disruptive consequences if they were expressed in the way that many primitive peoples express such impulses.

Clearly, the system needs people who are cooperative, obedient, and willing to accept dependence. As the historian Von Laue puts it: "Industrial society, after all, requires an incredible docility at the base of its freedoms [sic]."[251] For this reason, community, cooperation, and helping others have become deeply-ingrained, fundamental values of modern society.

But what about the value supposedly placed on independence, individualism, and competition? Whereas the words "community," "cooperation," and "helping" in our society are unequivocally accepted as "good," the words "individualism" and "competition" are tense, two-edged words that must be used with some care if one wishes to avoid risk of a negative reaction. To illustrate with an anecdote, when I was in the seventh or eighth grade our teacher, who was apt to be somewhat rough with the kids, asked a girl to name the country that she lived in. The girl was not very bright and apparently did not know the full name of the United States of America, so she answered simply: "The United States." "The United States of what?," asked the teacher. The girl just sat there with a blank expression. The teacher kept badgering her for an answer until she ventured a guess: "The United States of Community?"

Why "community"? Because of course "community" was a goody-goody word, the kind of word that a kid would use to get brownie points with a teacher. Would any kid in a similar situation have answered "United States of Competition" or "United States of Individualism"? Not likely!

It is routinely taken for granted that words like "community," "cooperation," "helping," and "sharing" represent something positive, but "individualism" is seldom used in the mainstream media or in the educational system in an unequivocally positive sense. "Competition" is more often used in a positive sense, but typically it us used that way only in specific contexts in which competition is useful (or at least harmless) to the system. For example, competition is considered desirable in the business world because it weeds out inefficient companies, spurs other companies to become more efficient, and promotes economic and technological progress. But only leashed competition—that is, competition that abides by rules designed to make it harmless or useful—is commonly spoken of favorably. And, when treated in a positive sense, competition is always justified in terms of communitarian values. Thus, business competition is considered good because it promotes efficiency and progress, which supposedly are good for the community as a whole.

"Independence," too, is a "good" word only when used in certain ways. For example, when one speaks of making disabled people "independent" one never thinks of making them independent of the system. One means only that they are to be provided with gainful employment so that the community will not be burdened with the cost of supporting them. Once they have found a job they are every bit as dependent on the system as they were when they lived on welfare, and they have a great deal less freedom to decide how to spend their time.

So why do politically correct anthropologists and others like them contrast the supposedly primitive values of "community," "cooperation," "sharing," and "interdependence" with what they claim are the modern values of "competition," "individualism," and "independence"? Certainly an important part of the answer is that politically correct people have absorbed too well the values that the system's propaganda has taught them, including the values of "cooperation," "community," "helping," and so forth. Another value they have absorbed from propaganda is that of "tolerance," which in cross-cultural contexts tends to translate into condescending approval of non-Western cultures.

A well-socialized modern anthropologist is therefore faced with a conflict: Since he is supposed to be tolerant, he finds it difficult to say anything bad about primitive cultures. But primitive cultures provide abundant examples of behavior that is decidedly bad from the point of view of modern Western values. So the anthropologist has to censor much of the "bad" behavior out of his descriptions of primitive cultures in order to avoid showing them in a negative light. In addition, due to his own excessively thorough socialization, the politically correct anthropologist has a need to rebel.[252] He is too well socialized to discard the fundamental values of modern society, so he expresses his hostility toward that society by distorting facts to make it seem that modern society deviates from its own stated values to a much greater extent than it actually does. Thus the anthropologist ends by magnifying the competitive and individualistic aspects of modern society while grossly understating these aspects of primitive societies.

There's more to it than that, of course, and I can't claim to understand fully the psychology of these people. It seems obvious, for example, that the politically correct portrayal of hunter-gatherers is motivated in part by an impulse to construct an image of a pure and innocent world existing at the dawn of time, analogous to the Garden of Eden, but the basis of this impulse is not clear to me.

8. What about hunter-gatherers' relations with animals? Some anarcho-primitivists seem to think that animals and humans once "coexisted," and that although animals nowadays sometimes eat humans, "such attacks by animals are comparatively rare," and "these animals are short of food due to the encroachment of civilization and are acting more out of extreme hunger and desperation. It is also due to our ignorance of the animal's gestures and scents, despoiled foliage or other signals our ancestor's [sic] knew but our domestication has now denied us."[253]

It is certainly true that the hunter-gatherer's knowledge of animals' habits made him safer in the wilderness than a modern man would be. It is also true that attacks on humans by wild animals are and have been relatively infrequent, probably because animals have learned the hard way that it is risky to prey on humans. But to hunter-gatherers in many environments wild animals did represent a significant danger. The Siriono hunter was "occasionally exposed to attacks from jaguars, crocodiles, and poisonous snakes."[254] Leopards, forest buffalo, and crocodiles were a real threat to the Mbuti.[255] On the other hand, remarkably, the Kadar (hunter-gatherers of India) were said to have "a truce with tigers, which in the old days left them strictly alone."[256] This is the only case of the kind that I know of.

Hunter-gatherers represented a much greater danger to animals than vice versa, since of course they hunted animals for food. Even the Kadar, who had no hunting weapons and lived mainly on wild yams, occasionally used their digging sticks to kill small animals for food.[257] Hunting methods could be cruel. Mbuti pygmies would stab an elephant in the belly with a poisoned spear; the animal would then die of peritonitis (inflammation of the abdominal lining) during the next 24 hours.[258] The Bushmen shot game with poisoned arrows, and the animals died slowly over a period that could be as long as three days.[259] Prehistoric hunter-gatherers slaughtered animals on a mass basis by driving herds of them over cliffs or bluffs.[260] The process was fairly gruesome and presumably was painful to the animals, since some of them were not killed outright by their fall but only disabled. The Indian Wooden Leg said: "I have helped in the chasing of antelope bands over a cliff.... Many of them were killed or got broken legs. We clubbed to death the injured ones."[261] This is not exactly the kind of thing that appeals to animal-rights activists.

Anarcho-primitivists may want to claim that hunter-gatherers inflicted suffering on animals only to the extent that they had to do so in order to get meat. But this is not true. A good deal of hunter-gatherers' cruelty was gratuitous. In *The Forest People*, Turnbull reported:

"The youngster...had speared [the sindula] with his first thrust, pinning the animal to the ground through the fleshy part of the stomach. But the animal was still very much alive, fighting for freedom.... Maipe put another spear into its neck, but it still writhed and fought. Not until a third spear pierced its heart did it give up the struggle.

"...[T]he Pygmies...stood around in an excited group, pointing at the dying animal and laughing. One boy, about nine years old, threw himself on the ground and curled up in a grotesque heap and imitated the sindula's last convulsions....

"At other times I have seen Pygmies singeing feathers off birds that were still alive, explaining that the meat is more tender if death comes slowly. And the hunting dogs, valuable as they are, get kicked around mercilessly from the day they are born to the day they die."[262]

A few years later, in *Wayward Servants*, Turnbull wrote: "The moment of killing is best described as a moment of intense compassion and reverence. The fun that is sometimes subsequently made of the dead animal, particularly by the youths, appears to be almost a nervous reaction, and there is an element of fear in their behavior. On the other hand, a bird caught alive may deliberately be toyed with, its feathers singed off over the fire while it is still fluttering and squawking until it is finally burned or suffocated to death. This again is usually done by the youths who take the same nervous pleasure in the act; very rarely a young hunter may absent-mindedly [!?] do the same thing. Older hunters and elders generally disapprove, but do not interfere.

"The respect seems to be not for animal life but for the game as a gift of the forest...."[263]

This does not seem entirely consistent with what Turnbull reported earlier in *The Forest People*. Maybe Turnbull was already beginning to swing toward political correctness when he wrote *Wayward Servants*. But even if we take the statements of *Wayward Servants* at face value, the fact remains that the Mbuti did treat animals with unnecessary cruelty, whether or not they felt "compassion and reverence" for them.

If the Mbuti did have compassion for animals, they were probably exceptional in that regard. Hunter-gatherers seem typically to be callous toward animals. The Eskimos with whom Gontran de Poncins lived kicked and beat their dogs brutally.[264] The Siriono sometimes captured young animals alive and brought them back to camp, but they gave them nothing to eat, and the animals were treated so roughly by the children that they soon died.[265]

It should be noted that many hunting-and-gathering peoples did have a sense of reverence for or closeness to wild animals. I've already quoted Colin Turnbull's statement to that effect in the case of the Mbuti. Coon states that "it is virtually a standard rule among hunters that they should never mock or otherwise insult any wild creature whose life they have brought to an end."[266] (As the passages I've quoted from Turnbull show, there were exceptions to this "standard rule.") Venturing into speculation, Coon adds that "hunters sense the unity of nature and the combination of humility and responsibility of their role in it."[267] Wissler describes the closeness to and reverence toward nature (including wild animals) of the North American Indians.[268] Holmberg mentions the Siriono's "bonds" and "kinship" with the animal world.[269] But, as we've already seen, these "bonds" and this "kinship" did not prevent physical cruelty to animals.

Clearly, animal-rights activists would be horrified at the way hunter-gatherers often treated animals. For people who look to hunting-and-gathering cultures as their social ideal, it therefore makes no sense to maintain alliances with the animal-rights movement.

9. To mop up as it were, I'll mention briefly a few other elements of the anarcho-primitivist myth.

According to the myth, racism is an artifact of civilization. But it's not clear that this is actually true. Of course, most primitive peoples couldn't be racists, because they never came in contact with any member of a race different from their own. But where contacts between different races did occur, I'm not aware of any reason to believe that hunter-gatherers were less prone to racism than modern man is. The Mbuti pygmies were distinguishable from their village-dwelling neighbors not only by their shorter stature but also by their facial features and by the lighter color of their skin.[270] The Mbuti referred to the villagers as "black savages" and "animals," and did not consider them to be real people.[271] The villagers similarly referred to the Mbuti as "savages" and "animals," nor did they consider the Mbuti to be real people.[272] It's true that the villagers often took Mbuti wives, but this seems to have been only because their own women, in the forest environment, had very low fertility, whereas Mbuti women bore plenty of children.[273] First-generation offspring of mixed marriages were considered inferior.[274] (Worth noting is that while Mbuti women often married villagers and lived in the villages, villager women hardly ever married Mbuti men, because the women "shunned the hard Gypsy life of the forest nomads and preferred the settled village life."[275] Moreover, the mixed-blood offspring of Mbuti-villager unions usually remained in the villages and "only rarely found their way back to the forest, because they preferred the more comfortable village life to the tough life of the forest."[276] This is hardly consistent with the anarcho-primitivists' image of the hunter-gatherer's life as one of ease and plenty.)

In the foregoing case of mutual racial antagonism only one side—the Mbuti—consisted of hunter-gatherers, the villagers being cultivators of crops. For a possible example of racism in which both sides were hunter-gatherers, the Indians of the North American subarctic and the Eskimos hated and feared one another; they seldom met except to fight.[277]

How about homophobia? That wasn't unknown among hunter-gatherers either. According to Mrs. Thomas, homosexuality was not permitted among the Bushmen whom she knew[278] (though it does not necessarily follow that this was true of all Bushman groups). Among the Mbuti, according to Turnbull, "homosexuality is never alluded to except as a great insult, under the most dire provocation."[279]

The publisher of the anarcho-primitivist "zine" *Species Traitor* stated in a letter to me that in hunter-gatherer cultures "people had no property."[280] This is not true. Various forms of private property did exist among hunter-gatherers—and not only among sedentary ones like the Northwest Coast Indians. It is well known that most hunting-and-gathering peoples had *collective* property in land. That is, each band of 30 to 130 people owned the territory in which it lived. Coon provides an extended discussion of this.[281] It is less well known that hunter-gatherers, even nomadic ones, could also hold rights to natural resources as *individual* property, and in some cases such rights could even be inherited.[282] For example, among Mrs. Thomas's Bushmen: "[E]ach group...has a very specific territory which that group alone may use, and they respect their boundaries rigidly.... [I]f a person is born in a certain area he or she has a right to eat the melons that grow there and all the veld food.... [A] man may eat the melons wherever his wife can and wherever his father and mother could, so that every Bushman has in this way some kind of rights in many places. Gai, for example, ate melons at Ai a ha'o because his wife's mother was born there, as well as at his own birthplace, the Okwa Omaramba...."[283]

Among the Veddas (hunter-gatherers of Ceylon), "the band territory was subdivided for individual band members, who could pass their property on to their children."[284] Among certain Australian Aborigines there existed a system of inherited rights to goods obtained in trade for stones extracted from a quarry.[285] Among some other Australian Aborigines, certain fruit trees were privately owned.[286] The Mbuti used termites as food, and among them termite hills could be owned by individuals.[287]

Portable items such as tools, clothing, and ornaments usually were owned by individual hunter-gatherers.[288]

Turnbull mentions the argument of one W. Nippold to the effect that hunter-gatherers, including the Mbuti, had a highly developed sense of private property. Turnbull counters that this is "a debatable point, and largely a semantic problem."[289] Here there is no need for us to split hairs about what does and what does not constitute private property, or what would be a "highly developed sense" of it. Suffice it to say that the unqualified belief that hunter-gatherers did not have private property is only another element of the anarcho-primitivist myth.

It's important to note, however, that *nomadic* hunter-gatherers did not *accumulate* property to the extent of being able to use their wealth to dominate other people.[290] The hunter-gatherer ordinarily had to carry all of his property on his own back whenever he shifted camp, or at best he had to carry it in a canoe or on a dog-sled or travois.[291] By any of these means only a limited amount of property can be transported, hence an upper bound is imposed on the amount of property that a nomad can usefully accumulate.

Property in rights to natural resources does not need to be transported, so in theory even a nomadic hunter-gatherer could accumulate an unlimited amount of that kind of property. But in practice I am not aware of any instance in which anyone belonging to a nomadic hunting-and-gathering band accumulated enough property in rights to natural resources to enable him to dominate other people by means of it. Under the conditions of the nomadic hunting-and-gathering life, it would obviously be very difficult for any individual to enforce an exclusive right to more natural resources than he could utilize personally.

Given the absence of accumulated wealth among nomadic hunter-gatherers, it might be supposed that there would be no social hierarchies among the latter, but this is not quite true.

Clearly there is not much room for social hierarchy in a nomadic band that contains at most 130 people (including children), and typically well under half that number. Moreover, some hunting-and-gathering peoples made a conscious, consistent, and apparently quite successful effort to prevent anyone from setting himself or herself up above the level of the others. For example, among the Mbuti, there were "no chiefs or councils of elders,"[292] "[i]ndividual authority is unthinkable,"[293] and "[a]ny attempt at the assumption of individual authority, or even of excessive influence, is sharply countered by ridicule or ostracism."[294] In fact, Turnbull emphasizes throughout his books the Mbuti's zeal in opposing the assumption by anyone of an elevated status.[295]

The Indians of sub-arctic North America had no chiefs.[296] The Siriono did have chiefs, but: "The prerogatives of chieftainship are few.... [A chief] makes suggestions as to migrations, hunting trips, etc., but these are not always followed by his tribesmen. As a mark of status, however, a chief always possesses more than one wife.

"While chiefs complain a great deal that other members of the band do not satisfy their obligations to them, little heed is paid to their requests....

"In general, however, chiefs fare better than other members of the band. Their requests more frequently bear fruit than those of others...."[297]

The Bushmen whom Mrs. Thomas knew "have no chiefs or kings, only headmen who in function are virtually indistinguishable from the people they lead, and sometimes a band will not even have a headman."[298] Richard Lee's Kung Bushmen had no chiefs,[299] and like the Mbuti they made a conscious effort to prevent anyone from setting himself up above the others.[300]

However, some other Kung Bushmen did have chiefs or headmen, the headmanship was hereditary, and the headmen had real authority, for the "headman or chief...decides who shall go where and when on collecting expeditions, because the timing of the yearly round is critical to ensure the food supply."[301] This is what Coon says about the Bushmen in the area of the Gautscha water hole, and since Mrs. Thomas knew these Bushmen,[302] it's not clear how one would reconcile Coon's statement with her remark that "headmen…in function are virtually indistinguishable from the people they lead…." I don't have access to proper library facilities; I don't even have a complete copy of Mrs. Thomas's book, only photocopies of some pages, so I'll have to leave this problem to any reader who may be sufficiently interested to take it up.

Be that as it may, in some parts of Australia there were "powerful chiefs, whom the settlers called kings…. '[T]he king…wore a very elaborate turban crown and was always carried on the shoulders of the men.'"[303] In Tasmania too there were "territorial chiefs of considerable power, and…in some cases at least their office was hereditary."[304]

Thus, while social stratification was absent or slight in many or most nomadic hunting-and-gathering societies, the sweeping assumption that all hierarchy was absent in all such societies is not true.

It is commonly assumed, and not only by anarcho-primitivists, that hunter-gatherers were good conservationists. On this subject I don't have much information, but from what I do know it seems that hunter-gatherers had a mixed record as conservationists.

The Mbuti look very good. Schebesta believed that they had voluntarily limited their population in order to avoid overburdening their natural resources[305] (though, at least in the part of his work that I have read, he does not explain his grounds for this belief). According to Turnbull, "there is very definitely a strongly felt and stated urge to use every part of the animal, and never to kill more than is necessary for the band's needs for the day. This in fact may be one reason why the Mbuti are so reluctant to kill an excess of game and preserve it for exchange with the villagers."[306]

Turnbull also states that "in the view of mammalogists such as Van Gelder the [Mbuti] hunters are indeed the finest conservationists any conservation-minded government could wish for."[307]

On the other hand, when Turnbull took an Mbuti name Kenge to visit a game preserve out on the plains, Kenge was told "that he would see more game than he had ever seen in the forest, but he was not to try and hunt any. Kenge could not understand this, because to his mind game is meant to be hunted."[308]

According to Coon, the ethic of the Tikerarmiut Eskimos forbade them to trap more than four wolves, wolverines, foxes, or marmots on any one day. However, this ethic quickly broke down when white traders arrived and tempted the Tikerarmiut with trade goods that they could obtain in exchange for the pelts of the animals named.[309]

As soon as they acquired steel axes, the Siriono began destroying the wild fruit trees of their region because it was easier to harvest the fruit by cutting the tree down than by climbing it.[310]

It is well known that some hunter-gatherers intentionally set wildfires because they knew that burned-over land would produce more of the edible plants that they favored.[311] I consider this practice recklessly destructive. It is believed that prehistoric hunter-gatherers, through over-hunting, caused or at least contributed to the extinction of some species of large mammals,[312] though as far as I know this has never been definitely proved.

The foregoing doesn't even scratch the surface of the question of conservation versus environmental recklessness on the part of hunter-gatherers. It's a question that deserves thorough investigation.

10. I can't generalize broadly since I've communicated personally with only a few anarcho-primitivists, but it's clear that the beliefs of at least some anarcho-primitivists are impervious to any facts that conflict with them. One can point out to these people any number of facts of the kind I've presented here and quote the words of writers who actually visited hunter-gatherers at a time when the latter were still relatively unspoiled, yet the true-believing anarcho-primitivist will always find rationalizations, no matter how strained, to discount all inconvenient facts and maintain his belief in the myth.

One is reminded of the response of fundamentalist Christians to any rational attack on their beliefs. Whatever facts one may point out, the fundamentalist will always find some argument, however far-fetched, to explain them away and justify his belief in the literal, word-for-word truth of the Bible.

Actually, there is about anarcho-primitivism a distinct flavor of early Christianity. The anarcho-primitivists' hunting-and-gathering utopia corresponds to the Garden of Eden, where Adam and Eve lived in ease and without sin (Genesis 2). The invention of agriculture and civilization corresponds to the Fall: Adam and Eve ate fruit from the tree of knowledge (Genesis 3:6), were cast out of the Garden (Genesis 3:24), and thereafter had to earn their bread with the sweat of their brow by tilling the soil (Genesis 3:19, 23).

They moreover lost gender equality, since Eve became subordinate to her husband (Genesis 3:16). The revolution that anarcho-primitivists hope will overthrow civilization corresponds to the Day of Judgment, the day of destruction on which Babylon will fall (Revelation 18:2). The return to primitive utopia corresponds to the arrival of the Kingdom of God, wherein "there shall be no more death, neither sorrow, nor crying, neither shall there be any more pain…" (Revelation 21:4).

Today's activists who risk their bodies by engaging in masochistic resistance tactics, such as chaining themselves across roads to prevent the passage of logging trucks, correspond to the Christian martyrs—the true believers who "were beheaded for the witness of Jesus, and for the word of God" (Revelation 20:4). Veganism corresponds to the dietary restrictions of many religions, such as the Christian fast during Lent. Like anarcho-primitivists, the early Christians emphasized egalitarianism ("whosoever shall exalt himself shall be abased," Matthew 23:12) and sharing ("distribution was made unto every man according as he had need," Acts 4:35).

The psychological affinity between anarcho-primitivism and early Christianity does not augur well. As soon as the emperor Constantine gave the Christians an opportunity to become powerful they sold out, and ever since then Christianity, more often than not, has served as a prop for the established powers.

11. In the present article I've been mainly concerned to debunk the anarcho-primitivist myth, and for that reason I've emphasized certain aspects of primitive societies that will be seen as negative from the standpoint of modern values. But there is another side to this coin: Nomadic hunting-and-gathering societies showed many traits that were highly attractive. Among other things, there is reason to believe that such societies were relatively free of the psychological problems that bedevil modern man, such as chronic stress, anxiety or frustration, depression, eating and sleep disorders, and so forth; that people in such societies, in certain critically important respects (though not in all respects) had far more personal autonomy than modern man does; and that hunter-gatherers were better satisfied with their way of life than modern man is with his.

Why does this matter? Because it shows that chronic stress, anxiety and frustration, depression, and so forth, are not inevitable parts of the human condition, but are disorders brought on by modern civilization. Nor is servitude an inevitable part of the human condition: The example of at least some nomadic hunter-gatherer societies shows that true freedom is possible.

Even more important: Regardless of whether they were good conservationists or poor ones, primitive peoples were incapable of damaging their environment to anything remotely approaching the extent to which modern man is damaging his. Primitives simply didn't have the power to do that much damage. They may have used fire recklessly and they may have exterminated some species through overhunting, but they had no way to dam large rivers, to cover thousands of square miles of the Earth's surface with cities and pavement, or to produce the vast quantities of toxic chemicals and radioactive waste with which modern civilization threatens to ruin the world for good and all. Nor did primitives have any means of releasing the deadly-dangerous forces represented by genetic engineering and by the super-intelligent computers that may soon be developed. These are dangers that scare even the technophiles themselves.[313]

So I agree with the anarcho-primitivists that the advent of civilization was a great disaster and that the Industrial Revolution was an even greater one. I further agree that a revolution against modernity, and against civilization in general, is necessary. But you can't build an effective revolutionary movement out of soft-headed dreamers, lazies, and charlatans. You have to have tough-minded, realistic, practical people, and people of that kind don't need the anarcho-primitivists' mushy utopian myth.

CONCLUDING NOTE

When I wrote this article I had only begun to read II. Band, I. Teil of Schebesta's *Die Bambuti-Pygmäen vom Ituri*. Since reading the latter, and owing to the nature of the discrepancies that I found between Turnbull's account and that of Schebesta, I've been forced to entertain serious doubts about the reliability of Turnbull's work on the Mbuti pygmies. I now suspect that Turnbull consciously or unconsciously slanted his description of the Mbuti to make them appear more attractive to modern leftish intellectuals like himself. However, I do not consider it necessary now to rewrite this article in such a way as to eliminate the reliance on Turnbull, because I've cited Turnbull mainly for information that makes the Mbuti appear *un*attractive, e.g., for their wife-beating, fighting, and quarreling over food. Given the nature of Turnbull's bias, it seems safe to assume that, if anything, he would have *under*stated the amount of wife-beating, fighting, and quarreling that he observed. But I think it is only fair to warn the reader that where Turnbull ascribes attractive or politically correct traits to the Mbuti, a certain degree of skepticism may be in order.

I would like to thank a number of people who sent me books, articles, or other information pertaining to primitive societies, and without whose help the present article could not have been written: Facundo Bermudez, Chris J., Marjorie Kennedy, Alex Obledo, Patrick Scardo, Kevin Tucker, John Zerzan, and six other people who perhaps would not want their names to be mentioned publicly. But most of all I want to thank the woman I love (deceased as of December 31, 2006), who provided me with more useful information than anyone else did, including two volumes of Paul Schebesta's wonderful work on the Mbuti pygmies. •

NOTES

Because most of the works cited here are cited repeatedly, citations are given in abbreviated form. For bibliographical details, see the accompanying LIST OF WORKS CITED (p. [*186]).

"*Encycl. Brit.*" means "*The New Encyclopædia Britannica,*" Fifteenth Edition, 2003.

[1] Example: "What is 'Green Anarchy'?," by the Black and Green Network, *Green Anarchy* #9, September 2002, page 13 ("the hunter-gatherer workday usually did not exceed three hours").

[2] Sahlins, pages 1-39.

[3] Bob Black, *Primitive Affluence*; see List of Works Cited.

[4] Sahlins, page 21.

[5] Cashdan, *Hunters and Gatherers: Economic Behavior in Bands.*

[6] *Ibid.*, page 23.

[7] Bob Black, pages 12-13. Cashdan, page 23.

[8] Cashdan, pages 23-24.

[9] *Ibid.*, page 24.

[10] *Ibid.*, pages 24-25.

[11] *Ibid.*, page 26.

[12] Poncins, pages 111, 126.

[13] Schebesta, II. Band, I. Teil, pages 9, 17-20, 89, 93-96, 119, 159-160 (men make implements during their "leisure" hours), 170, Bildtafel X (photo of women with huge loads of firewood on their backs).

[14] Turnbull, *Change and Adaptation*, page 18; *Forest People*, page 131.

[15] Holmberg, pages 48-51, 63, 67, 76-77, 82-83, 223, 265.

[16] *Ibid.*, pages 75-76.

[17] *Ibid.*, pages 100, 101.

[18] *Ibid.*, pages 63, 76, 100.

[19] *Ibid.*, page 223.

[20] *Ibid.*, page 222.

[21] *Ibid.*, page 224.

[22] *Ibid.*, pages 87, 107, 157, 213, 220, 246, 248-49, 254, 268.

[23] Cashdan, page 23.

[24] Sahlins, pages 15-17, 38-39.

[25] Holmberg, pages 107, 222.

[26] The Siriono's wilderness was not strictly trackless, since they did develop paths by repeatedly using the same routes. Holmberg, page105. How little these paths resembled the groomed trails found in our national forests may be judged from the fact that they were "scarcely visible" (page 51), "never cleared" (page 105), and "impossible for the uninitiated to follow" (page 106).

[27] Holmberg, page 249.

[28] *Ibid.*, page 157.

[29] *Ibid.*, pages 65, 249.

[30] *Ibid.*, page 65.

[31] There was nothing exceptional about the strenuousness of the Siriono's hunting and foraging activities. E.g.: "The bushmen had followed the wildebeest's trail

through thorns and over the parching desert...." Thomas, page 198. "The men had followed the buffalo's track for three days...." *Ibid.*, page 190. The strenuousness of the Eskimos' life can be judged from a reading of Poncins, *Kabloona*. See the accounts of hunting excursions by Wooden Leg, a Northern Cheyenne Indian (fatigue, snow-blindness, frozen feet). Marquis, pages 8, 9.

[32] Holmberg, page 65.

[33] This argument is suggested, for example, by Haviland, page 167.

[34] Fernald and Kinsey, page 149.

[35] *Ibid.*, page 148. Gibbons, page 217.

[36] Examples are found in Fernald and Kinsey, passim.

[37] Gibbons, chapter titled "The Proof of the Pudding."

[38] Coon, pages 36, 179-180, 226, 228, 230, 262.

[39] Cashdan, page 22. Coon, pages 268-69, 390; see also page 253.

[40] For skill see, e.g., Poncins, pages 14-15, 38-39, 160, 209-210; Schebesta, II. Band, I. Teil, page 7; Holmberg, pages 120-21, 275; Coon, pages 14, 49, 75, 82-83.

[41] This is somewhat of an oversimplification, since compulsory authority and the giving of orders were not unknown among nomadic hunter-gatherers, but generally speaking a high level of personal autonomy in such societies is indicated by a reading of the works cited in this article. See, e.g., Turnbull, *Forest People*, page 83; Poncins, page 174.

[42] Nomadic hunter-gatherers ordinarily lived in bands that contained between 30 and 130 individuals, including children and babies, and in many cases these bands split up into still smaller groups. Coon, page 191. Cashdan, page 21. Siriono often hunted singly or in pairs; maximum size of hunting party was six or seven men. Holmberg, page 51. Efé pygmies commonly hunted in groups of two to four. Coon, page 88.

[43] I'll reserve the discussion of stress for some other occasion, but see, e.g., Poncins, pages 212-13, 273, 292. Schebesta, II. Band, I. Teil, page 18, writes: "The economic activity of the hunter-gatherer knows neither haste nor hurry, nor agonizing worry over the daily bread."

[44] Holmberg, page 101.

[45] "[L]ife before domestication/agriculture was in fact largely one of leisure,...sexual equality...." Zerzan, *Future Primitive*, page 16.

[46] "[U]ntil just 10,000 years ago...humans lived in keeping with an egalitarian ethos with ample leisure time, gender equality...." Zerzan, "Whose Future?," *Species Traitor* No 1. Pages in this publication are not numbered.

[47] Thomas, pages 11, 284-87.

[48] *Encycl. Brit.*, Vol. 22, article "Languages of the World," section "African Languages," subsection "Khoisan Languages," pages 757-760.

[49] Bonvillain, page 21.

[50] *Ibid.*, page 24.

[51] *Ibid.*, page 21.

[52] *Ibid.*, pages 21-22.

[53] *Ibid.*, page 22.

[54] *Ibid.*, page 23.

[55] *Ibid.*, pages 21-22.

[56] Turnbull, *Wayward Servants*, page 270.

[57] Turnbull, *Forest People*, page 154.

[58] Turnbull, *Wayward Servants*, page 287.

[59] Turnbull, *Forest People*, page 205.

[60] Turnbull, *Wayward Servants*, page 211.

[61] *Ibid.*, page 192.

[62] Turnbull, *Forest People*, page 204.

[63] *Ibid.*, pages 207-08.

[64] *Ibid.*, page 208.

[65] *Ibid.*, page 122.

[66] Turnbull, *Wayward Servants*, pages 288-89. *Forest People*, page 265.

[67] Turnbull, *Forest People*, pages 115-16.

[68] Turnbull, *Wayward Servants*, page 137.

[69] "I know of no cases of rape…." Turnbull, *Wayward Servants*, page 121. I can account for the apparent contradiction between this statement and the passage quoted a moment ago only by supposing that since Turnbull was writing before the concept of "date rape" had emerged, he did not consider that forced intercourse in the elima hut, under the circumstances he described, constituted rape. Hence, when he said he knew of no rape among the Mbuti, he was probably referring to something more or less equivalent to what we would call "street rape" as opposed to "date rape."

[70] Turnbull, *Wayward Servants*, page 189. However, Turnbull is perhaps inconsistent on this point. Note the passage I quoted a moment ago about Amabosu smacking his wife across the face and Ekianga's reaction.

[71] *Ibid.*, pages 287-89.

[72] Numerous examples are scattered through *Wayward Servants* and *Forest People.*

[73] Holmberg, page 125.

[74] *Ibid.*, page 129.

[75] *Ibid.*, page 147.

[76] *Ibid.*, page 163.

[77] *Ibid.*, page 202.

[78] *Ibid.*, page 148.

[79] *Ibid.*, page 128.

[80] *Ibid.*, page 147.

[81] Bonvillain, page 295.

[82] *Ibid.*, pages 38-45.

[83] Poncins, pages 113-14, 126.

[84] *Ibid.*, page 198. See also page 117.

[85] *Ibid.*, pages 114-15.

[86] *Ibid.*, page 126.

[87] *Ibid.*, page 113.

[88] *Ibid.*, pages 112-13. See also Coon, page 223 ("often the wives lent say that they do not enjoy this").

[89] Elkin, pages 132-33. Massola, page 73.

[90] Massola, pages 74, 76.

[91] *Ibid.*, page 75. Elkin, pages 133-34.

[92] Massola, page 76.

[93] Elkin, page 136. Massola, pages 73, 75. Coon, pages 260-61.

[94] Massola, pages 75-76.

[95] *Ibid.*, pages 76-77.

[96] Elkin, pages 135, 137-38.

[97] *Ibid.*, page 138.

[98] *Ibid.*, page 138 (footnote 12).

[99] Coon, pages 105, 217, 253.

[100] Massola, page 78.

[101] *Encycl. Brit.*, Vol. 14, article "Australia," page 437.

[102] *Ibid.*

[103] Coon, pages 253, 255.

[104] Massola, page 77.

[105] Coon, pages 105, 217.

[106] *Ibid.*, page 215.

[107] *Ibid.*, page 336.

[108] *Ibid.*, page 252.

[109] Thomas, pages 262-303.

[110] Harold B. Barclay, letter to editor, *Anarchy: A Journal of Desire Armed*, spring/summer 2002, pages 70-71.

[111] *Ibid.*

[112] Cashdan, page 21.

[113] The Eskimos described by poncins used rifles to some extent, but these apparently were not their main means of procuring food; and they had no motorboats or snowmobiles.

[114] Coon, page 276.

[115] Haviland, page 168 ("some of the Bushmen of Southern Africa, have at times been farmers and at others pastoral nomads").

[116] *Ibid.*, page 167. Cashdan, pages 43-44.

[117] Thomas, page 94.

[118] Pfeiffer, *Emergence of Man*, pages 345-46. Pfeiffer is not a reliable source of information, but anyone with access to good library facilities will be able to consult Richard Lee's own writings.

[119] Thomas, page 284.

[120] Turnbull, *Forest People*, pages 20, 21, 27 & unnumbered page of information at end of book.

[121] Schebesta, I. Band, pages 37, 46, 48.

[122] *Ibid.*, page 404.

[123] *Ibid.*, pages 141-42.

[124] *Ibid.*, passim. e.g., I. Band, page 87; II. Band, I. Teil, page 11.

[125] *Ibid.*, i. Band, page 92.

[126] Turnbull, *Wayward Servants*, page 16. See also pages 88-89.

[127] Poncins, pages 161-62.

[128] Coon, pages 58-59.

[129] Holmberg, page 69. Richard Lee's Bushmen did have dogs. Sahlins, "The Original Affluent Society," page 23. So did the Mbuti. Turnbull, *Forest People*, page 101. Schebesta, II. Band, I. Teil, pages 89-93.

[130] Lauriston Sharp, in Holmberg, page xii.

[131] Holmberg, pages xx-xxii, 1-3.

[132] *Ibid.*, page 26.

[133] *Ibid.*, page xxiii.

[134] *Ibid.*, pages 25-26.

[135] *Ibid.*, page 121.

[136] *Ibid.*, page 10.

[137] *Ibid.*, page xii.

[138] See *ibid.*, pages 207, 225-26, "The principal ailments of which the Siriono are victims are malaria, dysentery, hookworm, and skin diseases," page 226. Malaria, at least, was probably introduced to the Americas by Europeans. *Encycl. Brit.*, Vol. 7, article "Malaria," page 725.

[139] Leakey, page 201 (map caption).

[140] Coon, pages 25 (footnote), 67.

[141] *Encycl. Brit.*, Vol. 14, article "Australia," page 434.

[142] Haviland, page 173.

[143] *Ibid.*

[144] *Ibid.*, page 395.

[145] Elkin, pages 130-38.

[146] Letters from the author to John Zerzan: 2/13/03, page 2; 3/16/03; 5/2/03, pages 5-6; 4/18/04, page 1.

[147] Letters from John Zerzan to the author: 3/2/03; 3/18/03; 3/26/03; 5/12/03; 4/28/04; 5/22/04. The only thing Zerzan said in his letters that I consider worth answering at this point is his claim that the sources I had cited to him were "out of date" (letter to the author, 5/22/04, page 2). He offered no explanation of this statement. As a former student of history, Zerzan should be aware of the importance of going back to primary sources whenever possible. In the present context, that means going back to eyewitness accounts based on observation of hunter-gatherer societies at a time when these were still relatively unspoiled. But

for *at least* 30 years there have been no more unspoiled primitive peoples. Hence, any primary sources that are useful for present purposes must date back *at least* 30 years (i.e., to before 1975) and usually longer than that. It's true that here and in my letters to Zerzan I've relied not only on primary but also on secondary sources, due to the fact that my incarceration limits my access to primary sources. But Zerzan offered no evidence whatever to discredit the information that I cited to him from secondary sources (or from primary ones, either). Nor have any of the more "up to date" sources that I've seen offered anything to disprove the information in question. They mostly just ignore that information, as if it didn't exist. The whole issue gets shoved under the carpet.

[148] Letter from the author to John Zerzan, 5/11/04. Letter from John Zerzan to the author, 5/20/04.

[149] Pfeiffer, *Emergence of Society*, page 464? I can't give the page number with certainty, because it is "cut off" on the photocopy that Zerzan sent me.

[150] Bonvillain, page 294. The photocopy that Zerzan sent me was actually from the 1995 edition of the same book, in which the identical sentence appears on page 271.

[151] Letter from John Zerzan to the author, 3/2/03 (footnote).

[152] Letter from the author to John Zerzan, 5/2/03, pages 5-6.

[153] Zerzan, *Future Primitive and Others essays.*

[154] Letter from the author to John Zerzan, 4/18/04, page 1.

[155] Zerzan, "Future primitive," page 32.

[156] *Ibid.*, page 33.

[157] Thomas, pages 156-57.

[158] Schebesta, I. Band, page 203.

[159] Zerzan, "Future Primitive," page 36.

[160] Turnbull, *Wayward Servants*, page 138 & footnote 2. A. Hutereau, *Les Négrilles de l'Uelle et de l'Ubangi*, Congo, 1924, I, 4, page 709; quoted by Schebesta, II. Band, I. Teil, page 258 (see also page 15 [footnote 6]).

[161] Turnbull, *Wayward Servants*, page 206.

[162] Zerzan, "Future Primitive," page 26. In an interview with Julien Nitzberg, *Mean* magazine, April 2001, page 69, Zerzan said, "Freud...believed that before language, it's likely that people were pretty telepathic...." In my letter to him of 5/2/03, page 6, I asked Zerzan to refer me to the place in Freud's works where Freud had made such a statement, but Zerzan never answered that question.

[163] Zerzan, "Future Primitive," page 15.

[164] Letter from the author to John Zerzan, 4/18/04, page 6.

[165] Letter from John Zerzan to the author, 4/28/04.

[166] Zerzan sent me a photocopy of a page from Bonvillain's book with his letter of 3/2/03. In "Future Primitive," pages 34, 36, Zerzan cites "Turnbull (1962)" and "Turnbull (1965)." This presumably refers to *Forest People* and *Wayward Servants*. In "Future Primitive," page 33, Zerzan also cites Mrs. Thomas's book, yet he conveniently forgets Mrs. Thomas's statements about childbirth when he claims (on the same page of "Future Primitive") that childbirth is "without difficulty or pain" among hunter-gatherers.

[167] Nietzsche, page 186.

[168] *Encycl. Brit.* Vol. 26, article "Propaganda," page 176.

[169] Letter from the publisher of *Species Traitor* to the author, 4/7/03, page 6.

[170] Elkin, pages 130-38.

[171] Coon, page 172.

[172] *Ibid.*, page 75.

[173] *Ibid.*, pages 243-44.

[174] Massola, page 77.

[175] Poncins, pages 115-120, 125, 162-65, 237-38, 244.

[176] *Encycl. Brit.*, Vol. 28, article "Spain," page 18.

[177] Apart from infanticide. Schebesta and Turnbull agree that when twins were born only one member of the pair was allowed to live. Schebesta, I. Band, page 138. Turnbull, *Wayward Servants*, page 130. Schebesta further states (same page) that babies born crippled were done away with. Turnbull, however, mentions a girl who was born with a "diseased" hip but was allowed to live. Turnbull, *Forest People*, page 265. Schebesta, II. Band, I. Teil, pages 274, 277, indicates that trespassing and theft could lead to deadly violence, but Turnbull mentions no such thing.

[178] Holmberg, pages 126-27, 157, 209-210.

[179] *Ibid.*, page 157.

[180] *Ibid.*, pages 11, 158-59.

[181] *Ibid.*, pages 114, 159.

[182] *Ibid.*, page 152.

[183] Thomas, pages 284-87.

[184] Haviland, pages 77, 78.

[185] It's common knowledge that coyotes and at least some species of bears both hunt and scavenge. For lions, martens, foxes, jackals, hyenas, raccoon dogs, Komodo dragons, and vultures, see *Encycl. Brit.*, Vol. 4, page 910; Vol. 6, pages 196, 454, 945; Vol. 7, pages 383, 884; Vol. 9, page 876; Vol. 12, page 439; Vol. 17, page 449; Vol. 23, page 421. For wolves and wolverines, see *Encyclopedia Americana*, International Edition, 1998, Vol. 29, pages 94-95, 102.

[186] See, e.g., *Time* magazine, 8/19/02, page 56.

[187] *Encycl. Brit.*, Vol. 23, article "Mammals," pages 436, 449-450.

[188] "Sibling Desperado," *Science News*, Vol. 163, February 15, 2003.

[189] *Encycl. Brit.*, Vol. 6, article "Komodo Dragon," page 945.

[190] *Ibid.*, Vol. 17, article "Dinosaurs," page 319.

[191] E.g., Olalla Cernuda, "Hallada en Atapuerca la evidencia más antigua de canibalismo en la historia de la Humanidad," *El Mundo*, 22 julio de 2006.

[192] Here are a couple of examples that illustrate the politically correct tendency of Turnbull's later work: In 1983, Turnbull wrote that he objected to the word "pygmy" because "it invites the assumption that height is a significant factor, whereas, in the Ituri, it is of remarkable insignificance to both the Mbuti and their neighbors, the taller Africans who live around them." *Change and Adaptation*, first page of the Introduction. but 21 years earlier Turnbull had written: "The fact that they [the Mbuti] average less than four and a half feet in height is of no concern to them; their taller neighbors, who jeer at them for being so puny, are as clumsy as elephants...." *Forest People*, page 14. "They [a certain group of pygmies] pitied me for my height, which made me so clumsy...." *Ibid.*, page 239. Turnbull also claimed in 1983 that the Mbuti had never fought in resistance to the taller Africans' invasion of their forest, *Change and Adaptation*, page 20. But Schebesta, I. Band, pages 81-84, reported oral traditions according to which many of the Mbuti had indeed fought the villagers, and so effectively that they had driven them (for a time) entirely out of the eastern part of the forest at some point during the first half of the 19th century. Oral traditions are unreliable, but these stories were so widespread as to indicate a certain probability that some such fighting had occurred. Turnbull did not explain how he knew that these traditions were wrong and that the Mbuti had not fought. Turnbull was familiar with Schebesta's work. See, e.g., *Forest People*, page 20.

[193] Turnbull, *Change and Adaptation*, page 44.

[194] *Ibid.*, page 154.

[195] *Ibid.*, page 158.

[196] Turnbull mentions physical fighting in *Forest People*, pages 110, 122-23, and in *Wayward Servants*, pages 188, 191, 201, 205, 206, 212.

[197] Turnbull, *Forest People*, pages 33, 107, 110; *Wayward Servants*, pages 105, 106, 113, 157, 212, 216.

[198] Turnbull mentions jealousies in *Wayward Servants*, pages 103, 118, 157.

[199] Turnbull, *Wayward Servants*, page 206.

[200] Turnbull, *Forest People*, page 107; *Wayward Servants*, pages 157, 191, 198, 201.

[201] Turnbull, *Wayward Servants*, page 183.

[202] Evans-Pritchard, page 90. Davidson, pages 10, 205. Reichard, pages xviii, xxi, xxxvii. Debo, page 71. Wissler, page 287. Holmberg, pages 151, 259, 270 (footnote 5). *Encycl. Brit.*, Vol. 2, article "Carib," page 866; Vol. 13, article "American Peoples, Native," page 380.

[203] Holmberg, pages 259-260.

[204] *Ibid.*, pages 93, 102, 224-26, 228, 256-57, 259, 270 (footnote 5)).

[205] Leach, page 130.

[206] Marquis, pages 119-122.

[207] vestal, page 60.

[208] *Ibid.*, page 179.

[209] *Encycl. Brit.*, Vol. 13, article "American Peoples, Native," pages 351-52, 360.

[210] Massola, page 72.

[211] *Encycl. Brit.*, Vol. 13, article "American Peoples, Native," pages 384, 386.

[212] Reichard, page xxxix.

[213] Evans-Pritchard, pages 90, 181-83.

[214] Holmberg, page 153.

[215] *Ibid.*, pages 126-27, 141, 154.

[216] Coon, pages 260-61.

[217] Poncins, pages 125, 244.

[218] Schebesta, II. Band, I. Teil, page 241.

[219] Massola, pages 78-80.

[220] Wissler, pages 223, 304.

[221] Reichard, page 265.

[222] *Encycl. Brit.*, Vol. 13, article "American Peoples, Native," page 381.

[223] Marquis, page 39.

[224] *Ibid.*, pages 64, 66, 120, 277.

[225] Leakey, page 107.

[226] Coon, pages 176-77. Cashdan, pages 37-38, refers to "precise" or "formal" rules of meat-sharing among Australian Aborigines, Mbuti pygmies, and Kung Bushmen.

[227] Richard B. Lee, quoted by Bonvillain, page 20.

[228] Coon, page 125.

[229] Holmberg, pages 79-81.

[230] *Ibid.*, pages 87-89, 154-56.

[231] *Ibid.*, pages 154-55.

[232] *Ibid.*, page 151.

[233] Cashdan, page 37. Turnbull, *Forest People*, pages 96-97. Schebesta, II. Band, I. Teil, pages 96, 97.

[234] Turnbull, *Forest People*, page 107.

[235] Turnbull, *Wayward and Servants*, pages 157-58. Schebesta, II. Band, I. Teil, page 97, mentions a fierce quarrel over the distribution of meat that "almost led to bloodshed."

[236] Turnbull, *Wayward Servants*, page 120.

[237] *Ibid.*, page 198.

[238] Coon, page 176. Cashdan, page 38. Bonvillain, page 20. Turnbull, *Wayward Servants*, page 167. *Encycl. Brit.*, Vol. 14, article "Australia," page 438.

[239] Cashdan, page 28. Coon, pages 72-73. Bonvillain, page 20. *Encycl. Brit.*, Vol. 14, article "Australia," page 438. Turnbull, *Wayward Servants*, page 178, possibly underestimated the importance of vegetable foods in the Mbuti's diet ("hunting and gathering being equally important to the economy"). According to Schebesta, I. Band, pages 70-71, 198; II. Band, I. Teil, pages 11, 13-14, the Mbuti nourished themselves principally on vegetable products. At most 30% of their diet consisted of animal products, and of that 30% a considerable part consisted not of meat but of foods such as snails and caterpillars that were gathered like vegetables, not hunted.

[240] Coon, page 176.

[241] Marquis, page 159.

[242] Evans-Pritchard, page 90.

[243] Poncins, pages 78-79.

[244] *Ibid.*, page 121.

[245] Turnbull, *Wayward Servants*, e.g., page 105.

[246] *Ibid.*, pages 199-200 (footnote 5).

[247] *Ibid.*, page 113.

[248] *Ibid.*, page 153.

[249] Poncins, page 237.

[250] Coon, page 260.

[251] Von Laue, page 202.

[252] For discussion of this and some of the other psychological points made in this paragraph, see the Unabomber Manifesto, "Industrial Society and Its Future," paragraphs 6-32, 213-230.

[253] "The Forgotten Language Among Humans and Nature," *Species Traitor*, issue 2, Winter 2002. Pages in this publication are not numbered.

[254] Holmberg, page 249. see also pages 61, 117, 260.

[255] Turnbull, *Forest People*, pages 35, 58, 79, 179; *Wayward Servants*, pages 165, 168. Schebesta, I. Band, page 68. Coon, page 71.

[256] Coon, page 156.

[257] *Ibid.*, pages 156, 158, 196.

[258] Turnbull, *Change and Adaptation*, page 20; *Wayward Servants*, page 164. Schebesta, Ii. Band, I. Teil, pages 107-111, describes other cruel methods of killing elephants.

[259] Thomas, pages 94, 190.

[260] Wissler, pages 14, 270. Coon, page 88.

[261] Marquis, page 88.

[262] Turnbull, *Forest People*, page 101. Schebesta, II. Band, I. Teil, page 90, also states that the Mbuti kicked their hunting dogs.

[263] Turnbull, *Wayward Servants*, page 161.

[264] Poncins, pages 29, 30, 49, 189, 196, 198-99, 212, 216.

[265] Holmberg, pages 69-70, 208.

[266] Coon, page 119.

[267] *Ibid.*

[268] Wissler, pages 124, 304-06.

[269] Holmberg, pages 111, 195.

[270] Turnbull, *Forest People*, pages 14, 33. Schebesta, I. Band, passim, e.g., pages 107, 181-84, 355.

[271] Turnbull, *Forest People*, pages 47, 120, 167; *Wayward Servants*, pages 61, 82; *Change and Adaptation*, page 92.

[272] Turnbull, *Forest People*, pages 47, 234.

[273] Schebesta, I. Band, pages 106-07, 137.

[274] *Ibid.*, page 107.

[275] *Ibid.*, page 108.

[276] *Ibid.*, page 110.

[277] Wissler, page 221. See also Poncins, page 165 (Eskimo kills two Indians), and *Encycl. Brit.*, Vol. 13, article "American Peoples, Native," page 360 (subarctic Indians fight Eskimos).

[278] Thomas, page 87.

[279] Turnbull, *Wayward Servants*, page 122.

[280] Letter to the author from publisher of *Species Traitor*, 4/7/03, page 7.

[281] Coon, pages 191-95.

[282] *Ibid.*, page 194.

[283] Thomas, pages 10, 82-83. see also Cashdan, page 41.

[284] Cashdan, page 41. See also Coon, page 198.

[285] Coon, page 275.

[286] *Ibid.*, page 168.

[287] Schebesta, II. Band, I. Teil, pages 14, 21-22, 275-76.

[288] Cashdan, page 40. See also *ibid.*, page 37, and Schebesta, II. Band, I. Teil, pages 276-78.

[289] Turnbull, *Wayward Servants*, page 199 (footnote 5).

[290] See Coon, page 268. Schebesta, II. Band, I. Teil, pages 8, 18, remarks on the Mbuti's lack of interest in accumulating wealth.

[291] See Coon, pages 57-67.

[292] Turnbull, *Wayward Servants*, page 14.

[293] *Ibid.*, page 181.

[294] *Ibid.*, page 228.

[295] Turnbull, *Forest People*, pages 110, 125; *Wayward Servants*, pages 27, 28, 42, 178-181, 183, 187, 256, 274, 294, 300. Schebesta, II. Band, I. Teil, page 8, says that the Mbuti lacked any inclination to be domineering (Herrschsucht).

[296] *Encycl. Brit.*, Vol. 13, article "American Peoples, Native," page360.

[297] Holmberg, pages 148-49.

[298] Thomas, page 10.

[299] Coon, page 238.

[300] Bonvillain, pages 20-21.

[301] Coon, page 210.

[302] Thomas, e.g., pages 146-47, 199.

[303] Coon, page 253.

[304] *Ibid.*, page 251.

[305] Schebesta, I. Band, page 106.

[306] Turnbull, *Wayward Servants*, page 161.

[307] Turnbull, *Change and Adaptation*, page 18.

[308] Turnbull, *Forest People*, page 250.

[309] Coon, page 104.

[310] Holmberg, pages 63-64, 268.

[311] E.g., *Encycl. Brit.*, Vol. 14, article "Biosphere," pages 1191, 1197; Mercader, pages 2, 235, 238, 241, 282, 306, 309. On other reckless use of fire, see Coon, page 6.

[312] Mercader, page 233. *Encycl. Brit.*, Vol. 14, article "Biosphere," pages 1159, 1196; Vol. 23, article "Mammals," pages 435, 448.

[313] See Bill Joy, "Why the Future Doesn't Need Us," *Wired* magazine, April 2000; and *Our Final Century*, by the British Astronomer Royal, Sir Martin Rees.

LIST OF WORKS CITED

Due to the fact that I am a prisoner and have no direct access to library facilities, the bibliographical information given in this list is in some instances incomplete. In most cases, however, I do not think this will lead to any serious difficulty in locating the works cited.

WORKS LISTED ALPHABETICALLY BY AUTHOR'S LAST NAME

Barclay, Harold B., letter to editor, in *Anarchy: A Journal of Desire Armed*, spring/summer 2002, pages 70-71.

Black, Bob, "Primitive Affluence," in *The Abolition of Work/Primitive Affluence: Essays Against Work by Bob Black*, Green Anarchist Books, BCM 1715, London WC 1N 3XX, 1998.

Bonvillain, Nancy, *Women and Men: Cultural Constructs of Gender*, Second Edition, Prentice Hall, Upper Saddle River, New Jersey, 1998.

Cashdan, Elizabeth, "Hunters and Gatherers: Economic Behavior in Bands," in Stuart Plattner (editor), *Economic Anthropology*, Stanford University Press, 1989, pages 21-48.

Coon, Carleton S., *The Hunting Peoples*, Little, Brown and Company, Boston, Toronto, 1971.

Davidson, H. R. Ellis, *Gods and Myths of Northern Europe*, Penguin Books, 1990.

Debo, Angie, *Geronimo: The Man, His Time, His Place*, University of Oklahoma Press, 1976.

Elkin, A. P., *The Australian Aborigines*, Fourth Edition, Anchor Books, Doubleday, Garden City, New York, 1964.

Evans-Pritchard, E. E., *The Nuer*, Oxford University Press, 1972.

Fernald, Merritt Lyndon, and Alfred Charles Kinsey, *Edible Wild Plants of Eastern North America*, Revised Edition, Dover, New York, 1996.

Gibbons, Euell, *Stalking the Wild Asparagus*, Field Guide Edition, David McKay Company, New York, 1972.

Haviland, William A., *Cultural Anthropology*, Ninth Edition, Harcourt Brace College Publishers, 1999.

Holmberg, Allan R., *Nomads of the Long Bow: The Siriono of Eastern Bolivia*, The Natural History Press, Garden City, New York, 1969.

Joy, Bill, "Why the Future Doesn't Need Us," *Wired* magazine, April 2000, pages 238-262.

Leach, Douglas Edward, "Colonial Indian Wars," in *Handbook of North American Indians*, William C. Sturtevant, general editor; Vol. 4, *History of Indian-White Relations*, Wilcomb E. Washburn, volume editor.

Leakey, Richard E., *The Making of Mankind*, E. P. Dutton, New York, 1981.

Marquis, Thomas B. (interpreter), *Wooden Leg: A Warrior Who Fought Custer*, Bison Books, University of Nebraska Press, 1967.

Massola, Aldo, *The Aborigines of South-Eastern Australia: As They Were*, Heinemann, Melbourne, 1971.

Mercader, Julio (editor), *Under the Canopy: The Archaeology of Tropical Rain Forests*, Rutgers University Press, 2003.

Nietzsche, Friedrich, "The Antichrist," §55; in *Twilight of the Idols/The Antichrist*, translated by R. J. Hollingdale, Penguin Classics, 1990.

Nitzberg, Julien, "Back to the Future Primitive" (interview with John Zerzan), *Mean* magazine, April 2001, pages 68, 69, 78.

Pfeiffer, John E., *The Emergence of Man*, Harper & Row, New York, Evanston, and London, 1969.

Pfeiffer, John E., *The Emergence of Society*, New York, 1977.

Poncins, Gontran de, *Kabloona*, Time-Life Books Inc., Alexandria, Virginia, 1980.

Rees, Martin, *Our Final Century*, Heinemann, 2003.

Reichard, Gladys A., *Navaho Religion: A Study of Symbolism*, Princeton University Press, 1990.

Sahlins, Marshall, *Stone Age Economics*, Aldine Atherton, 1972.

Schebesta, Paul, *Die Bambuti-Pygmäen vom Ituri*, Institut Royal Colonial Belge, Brussels; I. Band, 1938; II. Band, I. Teil, 1941.

Thomas, Elizabeth Marshall, *The Harmless People*, Second Vintage Books Edition, Random House, New York, 1989.

Turnbull, Colin M., *The Forest People*, Simon and Schuster, text copyright 1961, Foreword copyright 1962.

Turnbull, Colin M., *Wayward Servants: The Two Worlds of the African Pygmies*, The Natural History Press, Garden City, New York, 1965.

Turnbull, Colin M., *The Mbuti Pygmies: Change and Adaptation*, Harcourt Brace College Publishers, 1983.

Vestal, Stanley, *Sitting Bull, Champion of the Sioux: A Biography*, University of Oklahoma Press, 1989.

Von Laue, Theodore H., *Why Lenin? Why Stalin?*, J. B. lippencott, Co., New York, 1971.

Wissler, Clark, *Indians of the United States*, Revised Edition, Anchor Books, Random House, New York, 1989.

Zerzan, John, "Future Primitive," in *Future Primitive and Other Essays*, by the same author, 1994 edition.

Zerzan, John, "Whose Future?" in *Species Traitor* No. 1.

WORKS WITHOUT NAMED AUTHOR

Encyclopedia Americana, International Edition, 1998.

The New Encyclopædia Britannica, Fifteenth Edition, 2003 (abbreviated as *Encycl. Brit.*). Note: copies of the *Encyclopædia Britannica* labeled "Fifteenth Edition" but bearing a copyright date other than 2003 are not necessarily identical to the *Britannica* of 2003.

The Unabomber Manifesto, *Industrial Society and Its Future*.

PERIODICALS

Anarchy: A Journal of Desire Armed, P. O. Box 3448, Berkeley CA 94703.

El Mundo.

Green Anarchy, P. O. Box 11331, Eugene, OR 97440.

Mean magazine.

Science News.

Species Traitor, P. O. Box 835, Greensburg, PA 15601.

Time magazine.

Wired magazine.

The System's Neatest Trick

The supreme luxury of the society of technical necessity will be to grant the bonus of useless revolt and of an acquiescent smile.

—Jacques Ellul[1]

The System has played a trick on today's would-be revolutionaries and rebels. The trick is so cute that if it had been consciously planned one would have to admire it for its almost mathematical elegance.

1. WHAT THE SYSTEM IS NOT

Let's begin by making clear what the System is not. The System is not George W. Bush and his advisors and appointees, it is not the cops who maltreat protesters, it is not the CEOs of the multinational corporations, and it is not the Frankensteins in their laboratories who criminally tinker with the genes of living things. All of these people are servants of the System, but in themselves they do not constitute the System. In particular, the personal and individual values, attitudes, beliefs, and behavior of any of these people may be significantly in conflict with the needs of the System.

To illustrate with an example, the System requires respect for property rights, yet CEOs, cops, scientists, and politicians sometimes steal. (In speaking of stealing we don't have to confine ourselves to actual lifting of physical objects. We can include all illegal means of acquiring property, such as cheating on income tax, accepting bribes, and any other form of graft or corruption.) But the fact that CEOs, cops, scientists, and politicians sometimes steal does not mean that stealing is part of the System. On the contrary, when a cop or a politician steals something he is rebelling against the System's requirement of respect for law and property. Yet, even when they are stealing, these people remain servants of the System as long as they publicly maintain their support for law and property.

Whatever illegal acts may be committed by politicians, cops, or CEOs as individuals, theft, bribery, and graft are not part of the System but diseases of the System. The less stealing there is, the better the System functions, and that is why the servants and boosters of the System always advocate obedience to the law in public, even if they may sometimes find it convenient to break the law in private.

Take another example. Although the police are the System's enforcers, police brutality is not part of the System. When the cops beat the crap out of a suspect they are not doing the System's work, they are only letting out their own anger and hostility. The System's goal is not brutality or the expression of anger. As far as police work is concerned, the System's goal is to compel obedience to its rules and to do so with the least possible amount of disruption, violence, and bad publicity. Thus, from the System's point of view, the ideal cop is one who never gets angry, never uses any more violence than necessary, and as far as possible relies on manipulation rather than force to keep people under control. Police brutality is only another disease of the System, not part of the System.

For proof, look at the attitude of the media. The mainstream media almost universally condemn police brutality. Of course, the attitude of the mainstream media represents, as a rule, the consensus of opinion among the powerful classes in our society as to what is good for the System.

What has just been said about theft, graft, and police brutality applies also to issues of discrimination and victimization such as racism, sexism, homophobia, poverty, and sweatshops. All of these are bad for the System. For example, the more that black people feel themselves scorned or excluded, the more likely they are to turn to crime and the less likely they are to educate themselves for careers that will make them useful to the System.

Modern technology, with its rapid long-distance transportation and its disruption of traditional ways of life, has led to the mixing of populations, so that nowadays people of different races, nationalities, cultures, and religions have to live and work side by side. If people hate or reject one another on the basis of race, ethnicity, religion, sexual preference, etc., the resulting conflicts interfere with the functioning of the System. Apart from a few old fossilized relics of the past like Jesse Helms, the leaders of the System know this very well, and that is why we are taught in school and through the media to believe that racism, sexism, homophobia, and so forth are social evils to be eliminated.

No doubt some of the leaders of the System, some of the politicians, scientists, and CEOs, privately feel that a woman's place is in the home, or that homosexuality and interracial marriage are repugnant. But even if the majority of them felt that way it would not mean that racism, sexism, and homophobia were part of the System—any more than the existence of stealing among the leaders means that stealing is part of the System. Just as the System must promote respect for law and property for the sake of its own security, the System must also discourage racism and other forms of victimization, for the same reason. That is why the System, notwithstanding any private deviations by individual members of the elite, is basically committed to suppressing discrimination and victimization.

For proof, look again at the attitude of the mainstream media. In spite of occasional timid dissent by a few of the more daring and reactionary commentators, media propaganda overwhelmingly favors racial and gender equality and acceptance of homosexuality and interracial marriage.[2]

The System needs a population that is meek, nonviolent, domesticated, docile, and obedient. It needs to avoid any conflict or disruption that could interfere with the orderly functioning of the social machine. In addition to suppressing racial, ethnic, religious, and other group hostilities, it also has to suppress or harness for its own advantage all other tendencies that could lead to disruption or disorder, such as machismo, aggressive impulses, and any inclination to violence.

Naturally, traditional racial and ethnic antagonisms die slowly, machismo, aggressiveness, and violent impulses are not easily suppressed, and attitudes toward sex and gender identity are not transformed overnight. Therefore there are many individuals who resist these changes, and the System is faced with the problem of overcoming their resistance.[3]

2. HOW THE SYSTEM EXPLOITS THE IMPULSE TO REBEL

All of us in modern society are hemmed in by a dense network of rules and regulations. We are at the mercy of large organizations such as corporations, governments, labor unions, universities, churches, and political parties, and consequently we are powerless. As a result of the servitude, the powerlessness, and the other indignities that the System inflicts on us, there is widespread frustration, which leads to an impulse to rebel. And this is where the System plays its neatest trick: Through a brilliant sleight of hand, it turns rebellion to its own advantage.

Many people do not understand the roots of their own frustration, hence their rebellion is directionless. They know that they want to rebel, but they don't know what they want to rebel against. Luckily, the System is able to fill their need by providing them with a list of standard and stereotyped grievances in the name of which to rebel: racism, homophobia, women's issues, poverty, sweatshops… the whole laundry-bag of "activist" issues.

Huge numbers of would-be rebels take the bait. In fighting racism, sexism, etc., etc., they are only doing the System's work for it. In spite of this, they imagine that they are rebelling against the System. How is this possible?

First, 50 years ago the System was not yet committed to equality for black people, women and homosexuals, so that action in favor of these causes really was a form of rebellion. Consequently these causes came to be conventionally regarded as rebel causes. They have retained that status today simply as a matter of tradition; that is, because each rebel generation imitates the preceding generations.

Second, there are still significant numbers of people, as I pointed out earlier, who resist the social changes that the System requires, and some of these people even are authority figures such as cops, judges, or politicians. These resisters provide a target for the would-be rebels, someone for them to rebel against. Commentators like Rush Limbaugh help the process by ranting against the activists: Seeing that they have made someone angry fosters the activists' illusion that they are rebelling.

Third, in order to bring themselves into conflict even with that majority of the System's leaders who fully accept the social changes that the System demands, the would-be rebels insist on solutions that go farther than what the System's leaders consider prudent, and they show exaggerated anger over trivial matters. For example, they demand payment of reparations to black people, and they often become enraged at any criticism of a minority group, no matter how cautious and reasonable.

In this way the activists are able to maintain the illusion that they are rebelling against the System. But the illusion is absurd. Agitation against racism, sexism, homophobia and the like no more constitutes rebellion against the System than does agitation against political graft and corruption. Those who work against graft and corruption are not rebelling but acting as the System's enforcers: They are helping to keep the politicians obedient to the rules of the System. Those who work against racism, sexism, and homophobia similarly are acting as the Systems' enforcers: They help the System to suppress the deviant racist, sexist, and homophobic attitudes that cause problems for the System.

But the activists don't act only as the System's enforcers. They also serve as a kind of lightning rod that protects the System by drawing public resentment away from the System and its institutions. For example, there were several reasons why it was to the System's advantage to get women out of the home and into the workplace. Fifty years ago, if the System, as represented by the government or the media, had begun out of the blue a propaganda campaign designed to make it socially acceptable for women to center their lives on careers rather than on the home, the natural human resistance to change would have caused widespread public resentment. What actually happened was that the changes were spearheaded by radical feminists, behind whom the System's institutions trailed at a safe distance. The resentment of the more conservative members of society was directed primarily against the radical feminists rather than against the System and its institutions, because the changes sponsored by the System seemed slow and moderate in comparison with the more radical solutions advocated by feminists, and even these relatively slow changes were seen as having been forced on the System by pressure from the radicals.

3. THE SYSTEM'S NEATEST TRICK

So, in a nutshell, the System's neatest trick is this:

(a) For the sake of its own efficiency and security, the System needs to bring about deep and radical social changes to match the changed conditions resulting from technological progress.

(b) The frustration of life under the circumstances imposed by the System leads to rebellious impulses.

(c) Rebellious impulses are co-opted by the System in the service of the social changes it requires; activists "rebel" against the old and outmoded values that are no longer of use to the System and in favor of the new values that the System needs us to accept.

(d) In this way rebellious impulses, which otherwise might have been dangerous to the System, are given an outlet that is not only harmless to the System, but useful to it.

(e) Much of the public resentment resulting from the imposition of social changes is drawn away from the System and its institutions and is directed instead at the radicals who spearhead the social changes.

Of course, this trick was not planned in advance by the System's leaders, who are not conscious of having played a trick at all. The way it works is something like this:

In deciding what position to take on any issue, the editors, publishers, and owners of the media must consciously or unconsciously balance several factors. They must consider how their readers or viewers will react to what they print or broadcast about the issue, they must consider how their advertisers, their peers in the media, and other powerful persons will react, and they must consider the effect on the security of the System of what they print or broadcast.

These practical considerations will usually outweigh whatever personal feelings they may have about the issue. The personal feelings of the media leaders, their advertisers, and other powerful persons are varied. They may be liberal or conservative, religious or atheistic. The only universal common ground among the leaders is their commitment to the System, its security, and its power. Therefore, within the limits imposed by what the public is willing to accept, the principal factor determining the attitudes propagated by the media is a rough consensus of opinion among the media leaders and other powerful people as to what is good for the System.

Thus, when an editor or other media leader sets out to decide what attitude to take toward a movement or a cause, his first thought is whether the movement includes anything that is good or bad for the System. Maybe he tells himself that his decision is based on moral, philosophical, or religious grounds, but it is an observable fact that in practice the security of the System takes precedence over all other factors in determining the attitude of the media.

For example, if a news-magazine editor looks at the militia movement, he may or may not sympathize personally with some of its grievances and goals, but he also sees that there will be a strong consensus among his advertisers and his peers in the media that the militia movement is potentially dangerous to the System and therefore should be discouraged. Under these circumstances he knows that his magazine had *better* take a negative attitude toward the militia movement. The negative attitude of the media presumably is part of the reason why the militia movement has died down.

When the same editor looks at radical feminism he sees that some of its more extreme solutions would be dangerous to the System, but he also sees that feminism holds much that is useful to the System. Women's participation in the business and technical world integrates them and their families better into the System. Their talents are of service to the System in business and technical matters. Feminist emphasis on ending domestic abuse and rape also serves the System's needs, since rape and abuse, like other forms of violence, are dangerous to the System. Perhaps most important, the editor recognizes that the pettiness and meaninglessness of modern housework and the social isolation of the modern housewife can lead to serious frustration for many women; frustration that will cause problems for the System unless women are allowed an outlet through careers in the business and technical world.

Even if this editor is a macho type who personally feels more comfortable with women in a subordinate position, he knows that feminism, at least in a relatively moderate form, is good for the System. He knows that his editorial posture must be favorable toward moderate feminism, otherwise he will face the disapproval of his advertisers and other powerful people. This is why the mainstream media's attitude has been generally supportive of moderate feminism, mixed toward radical feminism, and consistently hostile only toward the most extreme feminist positions.

Through this type of process, rebel movements that are dangerous to the System are subjected to negative propaganda, while rebel movements that are believed to be useful to the System are given cautious encouragement in the media. Unconscious absorption of media propaganda influences would-be rebels to "rebel" in ways that serve the interests of the System.

The university intellectuals also play an important role in carrying out the System's trick. Though they like to fancy themselves independent thinkers, the intellectuals are (allowing for individual exceptions) the most oversocialized, the most conformist, the tamest and most domesticated, the most pampered, dependent, and spineless group in America today. As a result, their impulse to rebel is particularly strong. But, because they are incapable of independent thought, real rebellion is impossible for them. Consequently they are suckers for the System's trick, which allows them to irritate people and enjoy the illusion of rebelling without ever having to challenge the System's basic values.

Because they are the teachers of young people, the university intellectuals are in a position to help the System play its trick on the young, which they do by steering young people's rebellious impulses toward the standard, stereotyped targets: racism, colonialism, women's issues, etc. Young people who are not college students learn through the media, or through personal contact, of the "social justice" issues for which students rebel, and they imitate the students. Thus a youth culture develops in which

there is a stereotyped mode of rebellion that spreads through imitation of peers—just as hairstyles, clothing styles, and other fads spread through imitation.

4. THE TRICK IS NOT PERFECT

Naturally, the System's trick does not work perfectly. Not all of the positions adopted by the "activist" community are consistent with the needs of the System. In this connection, some of the most important difficulties that confront the System are related to the conflict between the two different types of propaganda that the System has to use, integration propaganda and agitation propaganda.[4]

Integration propaganda is the principal mechanism of socialization in modern society. It is propaganda that is designed to instill in people the attitudes, beliefs, values, and habits that they need to have in order to be safe and useful tools of the System. It teaches people to permanently repress or sublimate those emotional impulses that are dangerous to the System. Its focus is on long-term attitudes and deep-seated values of broad applicability, rather than on attitudes toward specific, current issues.

Agitation propaganda plays on people's emotions so as to bring out certain attitudes or behaviors in specific, current situations. Instead of teaching people to suppress dangerous emotional impulses, it seeks to stimulate certain emotions for well-defined purposes localized in time.

The System needs an orderly, docile, cooperative, passive, dependent population. Above all it requires a nonviolent population, since it needs the government to have a monopoly on the use of physical force. For this reason, integration propaganda has to teach us to be horrified, frightened, and appalled by violence, so that we will not be tempted to use it even when we are very angry. (By "violence" I mean physical attacks on human beings.)

More generally, integration propaganda has to teach us soft, cuddly values that emphasize nonaggressiveness, interdependence, and cooperation.

On the other hand, in certain contexts the System itself finds it useful or necessary to resort to brutal, aggressive methods to achieve its own objectives. The most obvious example of such methods is warfare. In wartime the System relies on agitation propaganda: In order to win public approval of military action, it plays on people's emotions to make them feel frightened and angry at their real or supposed enemy.

In this situation there is a conflict between integration propaganda and agitation propaganda. Those people in whom the cuddly values and the aversion to violence have been most deeply planted can't easily be persuaded to approve a bloody military operation.

Here the System's trick backfires to some extent. The activists, who have been "rebelling" all along in favor of the values of integration propaganda, continue to do so during wartime. They oppose the war effort not only because it is violent but because it is "racist," "colonialist," "imperialist," etc., all of which are contrary to the soft, cuddly values taught by integration propaganda.

The System's trick also backfires where the treatment of animals is concerned. Inevitably, many people extend to animals the soft values and the aversion to violence that they are taught with respect to humans. They are horrified by the slaughter of animals for meat and by other practices harmful to animals, such as the reduction of chickens to egg-laying machines kept in tiny cages or the use of animals in scientific experiments. Up to a point, the resulting opposition to mistreatment of animals may be useful to the System: Because a vegan diet is more efficient in terms of resource-utilization than a carnivorous one is, veganism, if widely adopted, will help to ease the burden placed on the Earth's limited resources by the growth of the human population. But activists' insistence on ending the use of animals in scientific experiments is squarely in conflict with the System's needs, since for the foreseeable future there is not likely to be any workable substitute for living animals as research subjects.

All the same, the fact that the System's trick does backfire here and there does not prevent it from being on the whole a remarkably effective device for turning rebellious impulses to the System's advantage.

It has to be conceded that the trick described here is not the only factor determining the direction that rebellious impulses take in our society. Many people today feel weak and powerless (for the very good reason that the System really does make us weak and powerless), and therefore identify obsessively with victims, with the weak and the oppressed. That's part of the reason why victimization issues, such as racism, sexism, homophobia, and neocolonialism have become standard activist issues.

5. AN EXAMPLE

I have with me an anthropology textbook[5] in which I've noticed several nice examples of the way in which university intellectuals help the System with its trick by disguising conformity as criticism of modern society. The cutest of these examples is found on pages 132-36, where the author quotes, in "adapted" form, an article by one Rhonda Kay Williamson, an intersexed person (that is, a person born with both male and female physical characteristics).

Williamson states that the American Indians not only accepted intersexed persons but especially valued them.[6] She contrasts this attitude with the Euro-American attitude, which she equates with the attitude that her own parents adopted toward her.

Williamson's parents mistreated her cruelly. They held her in contempt for her intersexed condition. They told her she was "cursed and given over to the devil," and they took her to charismatic churches to have the "demon" cast out of her. She was even given napkins into which she was supposed to "cough out the demon."

But it is obviously ridiculous to equate this with the modern Euro-American attitude. It may approximate the Euro-American attitude of 150 years ago, but nowadays almost any American educator, psychologist, or mainstream clergyman would be horrified at that kind of treatment of an intersexed person. The media would never dream of portraying such treatment in a favorable light. Average middle-class Americans today may not be as accepting of the intersexed condition as the Indians were, but few would fail to recognize the cruelty of the way in which Williamson was treated.

Williamson's parents obviously were deviants, religious kooks whose attitudes and beliefs were way out of line with the values of the System. Thus, while putting on a show of criticizing modern Euro-American society,

Williamson really is attacking only deviant minorities and cultural laggards who have not yet adapted to the dominant values of present-day America.

Haviland, the author of the book, on page 12 portrays cultural anthropology as iconoclastic, as challenging the assumptions of modern Western society. This is so far contrary to the truth that it would be funny if it weren't so pathetic. The mainstream of modern American anthropology is abjectly subservient to the values and assumptions of the System. When today's anthropologists pretend to challenge the values of their society, typically they challenge only the values of the past—obsolete and outmoded values now held by no one but deviants and laggards who have not kept up with the cultural changes that the System requires of us.

Haviland's use of Williamson's article illustrates this very well, and it represents the general slant of Haviland's book. Haviland plays up ethnographic facts that teach his readers politically correct lessons, but he understates or omits altogether ethnographic facts that are politically incorrect. Thus, while he quotes Williamson's account to emphasize the Indians' acceptance of intersexed persons, he does not mention, for example, that among many of the Indian tribes women who committed adultery had their noses cut off,[7] whereas no such punishment was inflicted on male adulterers; or that among the Crow Indians a warrior who was struck by a stranger had to kill the offender immediately, else he was irretrievably disgraced in the eyes of his tribe;[8] nor does Haviland discuss the habitual use of torture by the Indians of the eastern United States.[9] Of course, facts of that kind represent violence, machismo, and gender-discrimination, hence they are inconsistent with the present-day values of the System and tend to get censored out as politically incorrect.

Yet I don't doubt that Haviland is perfectly sincere in his belief that anthropologists challenge the assumptions of Western society. The capacity for self-deception of our university intellectuals will easily stretch that far.

To conclude, I want to make clear that I'm not suggesting that it is good to cut off noses for adultery, or that any other abuse of women should be tolerated, nor would I want to see anybody scorned or rejected because they are intersexed or because of their race, religion, sexual orientation, etc., etc., etc. But in our society today these matters are, at most, issues of reform. The System's neatest trick consists in having turned powerful rebellious impulses, which otherwise might have taken a revolutionary direction, to the service of these modest reforms. •

ENDNOTES

[1] Jacques Ellul, *The Technological Society*, translated by John Wilkinson, published by Alfred A. Knopf, New York, 1964, page 427.

[2] Even the most superficial review of the mass media in modern industrialized countries, or even in countries that merely aspire to modernity, will confirm that the System is committed to eliminating discrimination in regard to race, religion, gender, sexual orientation, etc., etc., etc. It would be easy to find thousands of examples that illustrate this, but here we cite only three, from three disparate countries.

United States: "Public Displays of Affection," *U.S. News & World Report*, September 9, 2002, pages 42-43. This article provides a nice example of the way propaganda functions. It takes an ostensibly objective or neutral position on homosexual partnerships, giving some space to the views of those who oppose public acceptance of homosexuality. But anyone reading the article, with its distinctly sympathetic treatment of a homosexual couple, will be left with the impression that acceptance of homosexuality is desirable and, in the long run, inevitable. Particularly important is the photograph of the homosexual couple in question: A physically attractive pair has been selected and has been photographed attractively. No one with the slightest understanding of propaganda can fail to see that the article constitutes propaganda in favor of acceptance of homosexuality. And bear in mind that *U.S. News & World Report* is a right-of-center magazine.

Russia: "Putin Denounces Intolerance," *The Denver Post*, July 26, 2002, page 16A. "MOSCOW—President Vladimir Putin strongly denounced racial and religious prejudice on Thursday... 'If we let this chauvinistic bacteria of either national or religious intolerance develop, we will ruin the country', Putin said in remarks prominently replayed on Russian television on Thursday night." Etc., etc.

Mexico: "Persiste racismo contra indígenas" ("Racism against indigenous people persists"), *El Sol de México*, January 11, 2002, page 1/B. Photo caption: "In spite of efforts to give dignity to the indigenous people of our country, they continue to suffer discrimination...." The article reports on the efforts of the bishops of Mexico to combat discrimination, but says that the bishops want to "purify" indigenous customs in order to liberate the women from their traditionally inferior status. *El Sol de México* is reputed to be a right-of-center newspaper.

Anyone who wanted to take the trouble could multiply these examples a thousand times over. The evidence that the System itself is set on eliminating discrimination and victimization is so obvious and so massive that one boggles at the radicals' belief that fighting these evils is a form of rebellion. One can only attribute it to a phenomenon well known to professional propagandists: People tend to block out, to fail to perceive or to remember, information that conflicts with their ideology. See the interesting article, "Propaganda," in *The New Encyclopædia Britannica*, Volume 26, Macropædia, 15th Edition, 1997, pages 171-79, specifically page 176.

[3] In this section I've said something about what the System is *not*, but I haven't said what the System *is*. A friend of mine has pointed out that this may leave the reader nonplussed, so I'd better explain that for the purposes of this article it isn't necessary to have a precise definition of what the System is. I couldn't think of any way of defining the System in a single, well-rounded sentence and I didn't want to break the continuity of the article with a long, awkward, and unnecessary digression addressing the question of what the System is, so I left that question unanswered. I don't think my failure to answer it will seriously impair the reader's understanding of the point that I want to make in this article.

[4] The concepts of "integration propaganda" and "agitation propaganda" are discussed by Jacques Ellul in his book *Propaganda*, published by Alfred A. Knopf, 1965.

[5] William A. Haviland, *Cultural Anthropology*, Ninth Edition, Harcourt Brace & Company, 1999.

[6] I assume that this statement is accurate. It certainly reflects the Navaho attitude. See Gladys A. Reichard, *Navaho Religion: A Study of Symbolism*, Princeton University Press, 1990, page 141. This book was originally copyrighted in 1950, well before American anthropology became heavily politicized, so I see no reason to suppose that its information is slanted.

[7] This is well known. See, e.g., Angie Debo, *Geronimo: The Man, His Time, His Place*, University of Oklahoma Press, 1976, page 225; Thomas B. Marquis (interpreter), *Wooden Leg: A Warrior Who Fought Custer*, Bison Books, University of Nebraska Press, 1967, page 97; Stanley Vestal, *Sitting Bull, Champion of the Sioux: A Biography*, University of Oklahoma Press, 1989, page 6; *The New Encyclopædia Britannica*, Vol. 13, Macropædia, 15th Edition, 1997, article "American Peoples, Native," page 380.

[8] Osborne Russell, *Journal of a Trapper*, Bison Books edition, page 147.

[9] Use of torture by the Indians of the eastern U.S. is well known. See, e.g., Clark Wissler, *Indians of the United States*, Revised Edition, Anchor Books, Random House, New York, 1989, pages 131, 140, 145, 165, 282; Joseph Campbell, *The Power of Myth*, Anchor Books, Random House, New York, 1988, page 135; *The New Encyclopædia Britannica*, Vol. 13, Macropædia, 15th Edition, 1997, article "American Peoples, Native," page 385; James Axtell, *The Invasion Within: The Contest of Cultures in Colonial North America*, Oxford University Press, 1985, page citation not available.

The Coming Revolution

Our entire much-praised
technological progress, and
civilization generally,
could be compared to an ax
in the hand of a pathological
criminal.

-Albert Einstein[1]

1. A great revolution is brewing; a world revolution. Consider the origin of the two most important revolutions of modern times: the French and the Russian. During the 18th century France was ruled by a monarchical government and a hereditary aristocracy. This regime had originated in the Middle Ages and had been founded on feudal concepts and values—concepts and values suitable for a warlike agrarian society in which power was based principally on heavy cavalry that fought with lance and sword. The regime had been modified over the centuries as political power became increasingly concentrated in the hands of the king. But it retained certain traits that did not vary: It was a conservative regime in which a traditional and hereditary class enjoyed a monopoly on power and prestige.

Meanwhile, the rate of social evolution was accelerating, and by the 18th century it had become unusually rapid. New techniques, new economic structures, and new ideas were appearing with which the old regime in France did not know how to deal. The growing importance of commerce, industry, and technology demanded a regime that would be flexible and capable of adapting itself to rapid changes; therefore, a social and political structure in which power and prestige would belong not to those who had inherited them but to those who deserved them because of their talents and achievements. At the same time new knowledge, together with new ideas that reached Europe as a result of contact with other cultures, was undermining the old values and beliefs. The philosophers of the so-called Enlightenment were expressing and giving definite form to the new yearnings and anxieties, so that a new system of values incompatible with the old values was being developed. By 1789, France found itself in the grip of an obsolete regime that could not have yielded to the new values without destroying itself; for it was impossible to put these values into practice without throwing off the domination of a hereditary class. Human nature being what it is, it is not surprising that those who constituted the old regime refused to give up their privileges to make way for what was called "progress." Thus the tension between the old values and the new continued to rise until the breaking-point was reached and a revolution followed.

The prerevolutionary situation of Russia was similar to that of France, except that the Russian regime was even more outof-date, backward, and rigid than that of France; and in Russia, moreover, there was a revolutionary movement that worked persistently to undermine the regime and the old values. As in France, the old regime in Russia could not have yielded to the new values without ceasing to exist. Because the Tsars and others who constituted the regime naturally refused to give up their privileges, the conflict between the two systems of values was irreconcilable, and the resulting tension rose until a revolution broke out.

The world today is approaching a situation analogous to that of France and Russia prior to their respective revolutions.

The values linked with so-called "progress"—that is, with immoderate economic and technological growth—were those that in challenging the values of the old regimes created the tensions that led to the French and Russian Revolutions. The values linked with "progress" have now become the values of another dominating regime: the technoindustrial system that rules the world today. And other new values are emerging that are beginning to challenge in their turn the values of the technoindustrial system. The new values are totally incompatible with technoindustrial values, so that the tension between the two systems of values cannot be relieved through compromise. It is certain that the partisans of technology will not voluntarily give in to the new values. Doing so would entail the sacrifice of everything they live for; they would rather die than yield. If the new values spread and grow strong enough, the tension will rise to a point at which revolution will be the only possible outcome. And there is reason to believe that the new values will indeed spread and grow stronger.

2 • The naive optimism of the 18th century led some people to believe that technological progress would lead to a kind of utopia in which human beings, freed from the need to work in order to support themselves, would devote themselves to philosophy, to science, and to music, literature, and the other fine arts. Needless to say, that is not the way things have turned out.

In discussing the way things *have* turned out, I will refer especially to the United States, which is the country I know best. The United States is technologically the most advanced country in the world. As the other industrialized countries progress, they tend to follow trajectories parallel to that of the United States. So, speaking broadly and with some reservations, we can say that where the United States is today the other industrialized countries will be in the future.[2]

Instead of using their technological means of production to provide themselves with free time in which to undertake intellectual and artistic work, people today devote themselves to the struggle for status, prestige, and power, and to the accumulation of material goods that serve only as toys. The kind of art and literature in which the average modern American immerses himself is the kind provided by television, movies, and popular novels and magazines; and it is not exactly what the 18th-century optimists had in mind. In effect, American popular culture has been reduced to mere hedonism, and hedonism of a particularly contemptible kind. "Serious" art does exist, but it tends to neurosis, pessimism, and defeatism.

As was to be expected, hedonism has not brought happiness. The spiritual emptiness of the culture of hedonism has left many people deeply dissatisfied. Depression, nervous tension, and anxiety disorders are widespread,[3] and for that reason many Americans resort to drugs (legal or illegal) to alleviate these symptoms, or to modify their mental state in some other way. Other indications of American social sickness are, for example, child abuse and the frequent inability to sleep or to eat normally. And, even among those Americans who seem to have adapted best to modern life, a cynical attitude toward the institutions of their own society is prevalent.

This chronic dissatisfaction and the sickly psychological condition of modern man are not normal and inevitable parts of human existence. We need not idealize the life of primitive peoples or conceal facts that are unpleasant from a modern point of view, such as the high rate of infant mortality or, in some cultures, a violent and warlike spirit. There is nevertheless reason to believe that primitive man was better satisfied with his way of life than modern man is and suffered much less from psychological problems than modern man does. For example, among hunting-and-gathering cultures, *before they were disrupted by the intrusion of industrial society*, child abuse was almost nonexistent.[4] And there is evidence that in most of these cultures there was very little anxiety or nervous tension.[5]

But what is at stake is not only the harm that modern society, does to human beings. The harm done to nature must also be taken into account. Even today, and even though modern man only occasionally comes into contact with her, Nature, our mother, attracts and entrances him and offers him a picture of the greatest and most fascinating beauty. The destruction of the wild natural world is a sin that worries, disturbs, and even horrifies many people. But we don't need to dwell here on the devastation of nature, for the facts are well known: more and more ground covered with pavement instead of herbage, the abnormally accelerated rate of extinction of species, the poisoning of the water and of the atmosphere, and as a result of the latter the alteration even of the Earth's climate, the ultimate consequences of which cannot be foreseen and may turn out to be disastrous.[6]

Which reminds us that the unrestrained growth of technology threatens the very survival of the human race. Human society, together with its worldwide environment, constitutes a system of the greatest complexity, and in a system as complex as this the consequences of a given change cannot in general be predicted.[7] And modern technology is in the process of bringing about the most profound changes in human society as well as in its physical and biological environment. That the consequences of such changes are unpredictable has been demonstrated not only theoretically, but also through experience. For example, no one could have predicted in advance that modern changes, through mechanisms that still have not been definitely determined, would lead to an epidemic of allergies.[8]

When a complex and more-or-less stable system is disturbed through some important change, the results commonly are destabilizing and therefore harmful. For example, it is known that genetic mutations of living organisms (unless merely insignificant) are almost always harmful; only rarely are they beneficial to the organism. Thus, as technology introduces greater and greater "mutations" into the "organism" that is biosphere (the totality of all living things on Earth), the harm done by these "mutations" becomes correspondingly greater and greater. No one but a fool can deny that the continual introduction, through technological progress, of ever-greater changes in the system of Man-plus-Earth is in the highest degree dangerous, foolhardy, and rash.

Still, I am not one of those who predict a worldwide physical and biological disaster that will bring down the entire technoindustrial system within the next few decades. The risk of such a disaster is real and serious, but at present we do not know whether it will actually occur. Nevertheless, if a disaster of this kind does not come upon us, it is practically certain that there will be a disaster of another kind: the loss of our humanity.

Technological progress not only is changing man's environment, his culture, and his way of life; it is changing man himself. For a human being is in large part a product of the conditions in which he lives. In the future, assuming that the technological system continues its development, the conditions in which man lives will be so profoundly different from the conditions in which he has lived previously that they will have to transform man himself.

The yearning for freedom, attachment to nature, courage, honor, honesty, morality, friendship, love and all of the other social instincts...even free will itself: all of these human qualities, valued in the highest degree from the dawn of the human race, evolved through the millennia because they were appropriate and useful in the primitive circumstances in which people lived. But today, so-called "progress" is changing the circumstances of human life to such an extent that these formerly advantageous qualities are becoming obsolete and useless. Consequently, they will disappear or will be transformed into something totally different and to us alien. This phenomenon can already be observed: Among the American middle class, the concept of honor has practically vanished, courage is little valued, friendship almost always lacks depth, honesty is decaying,[9] and freedom seems to be identified, in the opinion of some people, with obedience to the rules. And bear in mind that this is only the beginning of the beginning.

It can be assumed that the human being will continue to change at an accelerating rate, because the evolution of an organism is very swift when its environment is suddenly transformed. Beyond that, man is transforming himself, as well as other living organisms, through the agency of biotechnology. Today, so-called "designer babies" are in fashion in the United States. A woman who wants a baby having certain characteristics, for example, intelligence, athletic ability, blond hair, or tall stature, comes to an agreement with another woman who has the desired characteristics. The latter donates an egg (usually in exchange for a sum of money—there are women who make a business of this) which is implanted in the uterus of the first woman so that nine months later she will give birth to a child having—it is hoped—the desired traits.[10] There is no room for doubt that, as biotechnology advances, babies will be designed more and more effectively through genetic modification of eggs and sperm cells,[11] so that human beings will come more and more to resemble planned and manufactured products instead of free creations of Nature. Apart from the fact that this is extremely offensive to our sense of what a person should be, its social and biological consequences will be profound and unforeseeable; therefore in all probability disastrous.

But maybe this won't matter in the long run, because it is quite possible that human beings will some day become obsolete. There are distinguished scientists who believe that within a few decades computer experts will have succeeded in producing machines more intelligent than human beings. If this actually happens, then human beings will be superfluous and obsolete, and it is likely that the system will dispense with them.[12]

Although it is not certain that this will happen, it *is* certain that immoderate economic growth and the mad, headlong advance of technology are overturning everything, and it is hardly possible to conceive how the final result can be anything other than disastrous.

3. In the countries that have been industrialized longest, such as England, Germany, and above all the United States, there is a growing understanding that the technological system is taking us down the road to disaster.

When I was a boy in the 1950s, practically everyone gladly or even enthusiastically welcomed progress, economic growth, and above all technology, and believed without reservation that they were purely beneficial. A German I know has told me that the same attitude toward technology was prevalent in Germany at that time, and we may assume that the same was true throughout the industrialized world.

But with the passage of time this attitude has been changing. Needless to say, most people don't even have an attitude toward technology because they don't take the trouble to apply their minds to it; they just accept it unthinkingly. But in the United States and among thoughtful people—those who do take the trouble to reflect seriously on the problems of the society in which they live—attitudes toward technology have changed profoundly and continue to change. Those who are enthusiastic about technology are in general those who expect to profit from it personally in some way, such as scientists, engineers, military men, and corporation executives. The attitude of many other people is apathetic or cynical: they know of the dangers and the social decay that so-called progress brings with it, but they think that progress is inevitable and that any attempt to resist it is useless.

All the same, there are growing numbers of people, especially young people, who are not so pessimistic or so passive. They refuse to accept the destruction of their world, and they are looking for new values that will free them from the yoke of the present technoindustrial system.[13] This movement is still formless and has hardly begun to jell; the new values are still vague and poorly defined. But as technology advances along its mad and destructive path, and as the damage it does becomes ever more obvious and disturbing, it is to be expected that the movement will grow and acquire firmness, and will reinforce its values , making them more precise. These values, to judge by present appearances and also by what such values logically ought to be, will probably take a form somewhat like the following:

(i) Rejection of all modern technology. This is logically necessary, because modern technology is a whole in which all parts are interconnected; you can't get rid of the bad parts without also giving up those parts that seem good. Like a complex living organism, the technological system either lives or dies; it can't remain half alive and half dead for any length of time.

(ii) Rejection of civilization itself. This too is logical, because the present technological civilization is only the most recent stage of the ongoing process of civilization, and earlier civilizations already contained the seed of the evils that today are becoming so great and so dangerous.

(iii) Rejection of materialism,[14] and its replacement with a conception of life that values moderation and self-sufficiency while deprecating the acquisition of property or of status. The rejection of materialism is a necessary part of the rejection of technological civilization, because only technological civilization can provide the material goods to which modern man is addicted.

(iv) Love and reverence toward nature, or even worship of nature. Nature is the opposite of technological civilization, which threatens death to nature. It is therefore logical to set up nature as a positive value in opposition to the negative value of technology. Moreover, reverence toward or adoration of nature may fill the spiritual vacuum of modern society.

(v) Exaltation of freedom. Of all the things of which modern civilization deprives us, freedom and intimacy with nature are the most precious. In fact, ever since the human race submitted to the servitude of civilization, freedom has been the most frequent and most insistent demand of rebels and revolutionaries throughout the ages.

(vi) Punishment of those responsible for the present situation. The scientists, engineers, corporation executives, politicians, and so forth who consciously and intentionally promote technological progress and economic growth are criminals of the worst kind. They are worse than Stalin or Hitler, who never even dreamed of anything approaching what today's technophiles are doing. Therefore justice and punishment will be demanded.

The movement in opposition to the technoindustrial system should develop something more or less similar to the foregoing set of values; and in fact there is much evidence of the emergence of such values. Clearly these values are totally incompatible with the survival of technological civilization, just as the values that emerged prior to the French and Russian Revolutions were totally incompatible with the survival of the old regimes of those countries. As the damage done by the technoindustrial system grows worse, it is to be expected that the new values that oppose it will spread and become stronger. If the tension between technological values and the new values rises high enough, and if a suitable occasion presents itself, what happened in France and Russia will happen again: A revolution will break out.

4. But I don't predict a revolution; it remains to be seen whether one will occur. There are several factors that may stand in the way of revolution , among them the following:

(a) Lack of belief in the possibility of revolution. Most people take it for granted that the existing system is invulnerable and that nothing can divert it from its appointed path. It never occurs to them that revolution might be a real possibility. History shows that human beings commonly will submit to any injustice, however outrageous, if the people around them submit and everyone believes there is no way out. On the other hand, once the hope of a way out has arisen, in many cases a revolution follows.

Thus, paradoxically, the greatest obstacle to a revolution against the technoindustrial system is the very belief that such a revolution cannot happen. If enough people come to believe that a revolution is possible, then it will be possible in reality.

(b) Propaganda. The technological society possesses a system of propaganda, made possible by modern media of communications, that is more powerful and effective than that of any earlier society.[15] This system of propaganda makes more difficult the revolutionary task of undermining technoindustrial values.

(c) The pseudorevolutionaries. At present there are too many people who pride themselves on being rebels without really being committed to the overthrow of the existing system. They only play at rebellion or revolution in order to satisfy their own psychological needs. These pseudorevolutionaries may form an obstacle to the emergence of an effective revolutionary movement.

(d) Cowardice. Modern society has taught us to be passive and obedient, and to be horrified at physical violence. Moreover, the conditions of modern life are conducive to laziness, softness, and cowardice. Those who want to be revolutionaries will have to overcome these weaknesses. •

NOTES

I wrote "The Coming Revolution" several years ago at the suggestion of a young Spanish man, and I wrote it in Spanish. Here, obviously, I've translated it into English.

As I originally wrote the notes to "The Coming Revolution" many of them contained direct quotations, translated into Spanish, from English language sources. If I translated these quotations back into English, the results certainly would not be identical with the original English-language versions. Therefore, where possible, I have returned to the original English-language sources in order to quote them accurately. However, in several cases I no longer have access to the English-language materials in question, and in such cases I've had to use paraphrases in these notes rather than direct quotations. But material enclosed in quotation marks always is quoted verbatim.

[1] Quoted by Gordon A. Craig, *The New York Review of Books*, November 4, 1999, page 14.

[2] My correspondent who writes under the pseudonym "Último Reducto" disagrees. he says that the United States, with its "hard capitalism," is in a certain sense backward: The path of the future is that of Western Europe, which, with its more advanced social-welfare programs, seduces and weakens the average citizen by

making his life too soft and easy. This is a plausible opinion, and Último Reducto may well be right. But it is also possible that he is wrong. As technology increasingly frees the system from the need for human work, growing numbers of people will become superfluous and will then constitute no more than a useless burden. The system will have no reason to waste its resources in taking care of the superfluous people, and therefore may find it more efficient to treat them ruthlessly. Thus, possibly, it is the "hard" capitalism of the United States rather than the softer capitalism of Western Europe that points to the future. Only time will tell.

[3] In regard to the sickly psychological state of modern man, see, e.g.: "The Science of Anxiety," *Time*, June 10, 2002, pages 46-54 (anxiety is spreading and afflicts 19 million Americans, page 48; drugs have proven very useful in the treatment of anxiety, page 54); "The Perils of Pills," *U.S. News & World Report*, March 6, 2000, pages 45-50 (almost 21 percent of children 9 years old or older have a mental disorder, page 45); "On the Edge on Campus," *U.S. News & World Report*, February 18, 2002, pages 56-57 (the mental health of college students continues to worsen); *Funk & Wagnalls New Encyclopedia*, 1996, Volume 24, page 423 (in the United States the suicide rate of persons between 15 and 24 years old tripled between 1950 and 1990; some psychologists think that growing feelings of isolation and rootlessness, and that life is meaningless, have contributed to the rising suicide rate); "Americanization a Health Risk, Study Says," *Los Angeles Times*, September 15, 1998, pages A1, A19 (a new study reports that Mexican immigrants in the United States have only half as many psychiatric disorders as persons of Mexican descent born in the United States, page A1).

[4] E.g.: Gontran de Poncins, *Kabloona*, Time-Life Books, Alexandria, Virginia, 1980, pages 32-33, 36, 157 ("no Eskimo has ever punished a child," page 157); Allan R. Holmberg, *Nomads of the Long Bow: The Siriono of Eastern Bolivia*, The Natural History Press, New York, 1969, pages 204-05 (an unruly child is never beaten; children generally are allowed great latitude for physical expression of aggressive impulses against their parents, who are patient and long-suffering with them); John E. Pfeiffer, *The Emergence of Man*, Harper & Row, New York, 1969, page 317 (The Australian Aborigines practiced infanticide, but: "Nothing is denied to the children who are reared. Whenever they want food...they get it. Aborigine mothers rarely spank or otherwise punish their offspring, even under the most provoking circumstances.")

On the other hand, the Mbuti of Africa did not hesitate to give their children hard slaps. Colin Turnbull, *The Forest People*, Simon And Schuster, 1962, pages 65, 129, 157. But this is the only example that I know of among hunting-and-gathering cultures of what by present standards could be considered child abuse. And I don't think that it was abuse in the context of Mbuti culture, because the Mbuti had little hesitation about hitting one another and they often did hit one another, so that among them a blow did not have the same psychological significance that it has among us: a blow did not humiliate. Or so it seems to me on the basis of what I've read about the Mbuti.

[5] E.g., Gontran de Poncins, *op. cit.*, pages 212, 273, 292 ("their minds were at rest, and they slept the sleep of the unworried," page 273; "Of course he would not worry. He was an Eskimo," page 292). Still, there have existed hunting-and-gathering cultures in which anxiety was indeed a serious problem; for example, the Ainu of Japan. Carleton S. Coon, *The Hunting Peoples*, Little, Brown and Company, Boston, 1971, pages 372-73.

[6] See, e.g., Elizabeth Kolbert, "Ice Memory," *The New Yorker*, January 7, 2002, pages 30-37.

[7] Roberto Vacca, *The Coming Dark Age*, translated by J. S. Whale, Doubleday, 1973, page 13 ("Jay W. Forrester of the Massachusetts Institute of Technology has shown that in the field of complex systems, cause-to-effect relationships are very difficult to analyse: hardly ever does one given parameter depend on just one other factor. What happens is that all factors and parameters are interrelated by multiple feedback loops, the structure of which is far from obvious....")

[8] "Allergy Epidemic," *U.S. News & World Report*, May 8, 2000, pages 47-53. "Allergies: A Modern Epidemic," *National Geographic*, May 2006, pages 116-135.

[9] In regard to the decay of honesty in the United States, see an interesting article by Mary McNamara, *Los Angeles Times*, August 27, 1998, pages E1, E4.

[10] Rebecca Mead, "Eggs for Sale," *The New Yorker*, August 9, 1999, pages 56-65.

[11] "Redesigning Dad," *U.S. News & World Report*, November 5, 2001, pages 62-63 (sperm cells may be the best place in which to repair defective genes; the technology is nearly ready).

[12] See Bill Joy, "Why the Future Doesn't Need Us," *Wired*, April 2000, pages 238-262. One should not have too much confidence in predictions of miraculous advances such as the development of intelligent machines. For example, in 1970 scientists predicted that within 15 years there would be machines more intelligent than human beings. *Chicago Daily News*, November 16, 1970 (page citation not available). Obviously this prediction did not come true. Nonetheless, it would be foolish to discount the possibility of machines more intelligent than human beings. In fact, there is reason to believe that such machines will indeed exist some day if the technological system continues to develop.

[13] See Bruce Barcott, "From Tree-hugger to Terrorist," *New York Times Sunday Magazine*, April 7, 2002, pages 56-59, 81. This article describes the development of

what may become within a few years a real and effective revolutionary movement committed to the overthrow of the technoindustrial system. (Since writing the foregoing several years ago, I've had to conclude that no effective movement of this kind is emerging in the United States. Capable leadership is lacking, and the real revolutionaries have failed to separate themselves from the pseudo-revolutionaries. But Bruce Barcott's article, along with information from other sources, shows that the raw material for a real revolutionary movement does exist: There are people with sufficient passion and commitment who are willing to take risks and make great sacrifices. Only a few able leaders would be needed to form this raw material into an effective movement.)

[14] Último Reducto has pointed out a possible ambiguity in this phrase. To eliminate it, I need to explain that the word "materialism" here refers not to philosophical materialism but to values that exalt the acquisition of material possessions.

[15] See the interesting article "Propaganda"; *The New Encyclopædia Britannica*, Volume 26, 15th edition, 1997, pages 171-79. This article reveals the impressive sophistication of modern propaganda.

The Road to Revolution

A revolution is not
a dinner party...

-Mao Zedong[1]

A great revolution is brewing. What this means is that the necessary preconditions for revolution are being created. Whether the revolution will become a reality will depend on the courage, determination, persistence, and effectiveness of revolutionaries.

The necessary preconditions for revolution[2] are these: There must be a strong development of values that are inconsistent with the values of the dominant classes in society, and the realization of the new values must be impossible without a collapse
of the existing structure of society.

When these conditions are present, there arises an irreconcilable conflict between the new values and the values that are necessary for the maintenance of the existing structure. The tension between the two systems of values grows and can be resolved only through the eventual defeat of one of the two. If the new system of values is vigorous enough, it will prove victorious and the existing structure of society will be destroyed.

This is the way in which the two greatest revolutions of modern times—the French and Russian Revolutions—came about. Just such a conflict of values is building up in our society today. If the conflict becomes sufficiently intense, it will lead to the greatest revolution that the world has ever seen.

The central structure of modern society, the key element on which everything else depends, is technology. Technology is the principal factor determining the way in which modern people live and is the decisive force in modern history. This is the expressed opinion of various learned thinkers,[3] and I doubt that many serious historians could be found who would venture to disagree with it. However, you don't have to rely on learned opinions to realize that technology is the decisive factor in the modern world. Just look around you and you can see it yourself. Despite the vast differences that formerly existed between the cultures of the various industrialized countries, all of these countries are now converging rapidly toward a common culture and a common way of life, and they are doing so because of their common technology.

Because technology is the central structure of modern society—the structure on which everything else depends—the strong development of values totally inconsistent with the needs of the technological system would fulfill the preconditions for revolution. This kind of development is taking place right now.

Fifty years ago, when I was a kid, warm approval or even enthusiasm for technology were almost universal. By 1962 I had become hostile toward technology myself, but I wouldn't have dared to express that opinion openly, for in those days nearly everyone assumed that only a kook, or maybe a Bible-thumper from the backwoods of Mississippi, could oppose technology. I now know that even at that time there were a few thinkers who wrote critically about technology. But they were so rare and so little heard from that until I was almost 30 years old I never knew that anyone but myself opposed technological progress.

Since then there has been a profound change in attitudes toward technology. Of course, most people in our society don't *have* an attitude toward technology, because they never bother to think about technology as such. If the advertising industry teaches them to buy some new techno-gizmo, then they will buy it and play with it, but they won't think about it. The change in attitudes toward technology has occurred among the minority of people who think seriously about the society in which they live.

As far as I know, almost the only thinking people who remain enthusiastic about technology are those who stand to profit from it in some way, such as scientists, engineers, corporate executives and military men. A much larger number of people are cynical about modern society and have lost faith in its institutions. They no longer respect a political system in which the most despicable candidates can be successfully sold to the public through sophisticated propaganda techniques. They are contemptuous of an electronic entertainment industry that feeds us garbage. They know that schoolchildren are being drugged (with Ritalin, etc.) to keep them docile in the classroom, they know that species are becoming extinct at an abnormal rate, that environmental catastrophe is a very real possibility, and that technology is driving us all into the unknown at reckless speed, with consequences that may be utterly disastrous. But, because they have no hope that the technological juggernaut can be stopped, they have grown apathetic. They simply accept technological progress and its consequences as unavoidable evils, and they try not to think about the future.

But at the same time there are growing numbers of people, especially young people, who are willing to face squarely the appalling character of what the technoindustrial system is doing to the world. They are prepared to reject the values of the technoindustrial system and replace them with opposing values. They are willing to dispense with the physical security and comfort, the Disney-like toys, and the easy solutions to all problems that technology provides. They don't need the kind of status that comes from owning more and better material goods than one's neighbor does. In place of these spiritually empty values they are ready to embrace a lifestyle of moderation that rejects the obscene level of consumption that characterizes the technoindustrial way of life; they are capable of opting for courage and independence in place of modern man's cowardly servitude; and above all they are prepared to discard the technological ideal of human control over nature and replace it with reverence for the totality of all life on Earth—free and wild as it was created through hundreds of millions of years of evolution.

How can we use this change of attitude to lay the foundation for a revolution?

One of our tasks, obviously, is to help promote the growth of the new values and spread revolutionary ideas that will encourage active opposition to the technoindustrial system. But spreading ideas, by itself, is not very effective. Consider the response of a person who is exposed to revolutionary ideas. Let's assume that she or he is a thoughtful person who is sickened on hearing or reading of the horrors that technology has in store for the world, but feels stimulated and hopeful on learning that better, richer, more fulfilling ways of life are possible. What happens next?

Maybe nothing. In order to maintain an interest in revolutionary ideas, people have to have hope that those ideas will actually be put into effect, and they need to have an opportunity to participate personally in carrying out the ideas. If a person who has been exposed to revolutionary ideas is not offered anything practical that she can do against the techosystem, and if nothing significant is going on to keep her hope alive, she will probably lose interest. Additional exposures to the revolutionary message will have less and less effect on her the more times they are repeated, until eventually she becomes completely apathetic and refuses to think any further about the technology problem.

In order to hold people's interest, revolutionaries have to show them that things are *happening*—significant things—and they have to give people an opportunity to participate actively in working toward revolution. For this reason an effective revolutionary movement is necessary, a movement that is capable of making things happen, and that interested people can join or cooperate with so as to take an active part in preparing the way for revolution. Unless such a movement grows hand-in-hand with the spread of ideas, the ideas will prove relatively useless.

For the present, therefore, the most important task of revolutionaries is to build an effective movement.

The effectiveness of a revolutionary movement is not measured only by the number of people who belong to it. Far more important than the numerical strength of a movement are its cohesiveness, its determination, its commitment to a well-defined goal, its courage, and its stubborn persistence. Possessing these qualities, a surprisingly small number of people can outweigh the vacillating and uncommitted majority. For example, the Bolsheviks were never a numerically large party, yet it was they who determined the course that the Russian Revolution took. (I hasten to add that I am NOT an admirer of the Bolsheviks. To them, human beings were of value only as gears in the technological system. But that doesn't mean we can't learn lessons from the history of Bolshevism.)

An effective revolutionary movement will not worry too much about public opinion. Of course, a revolutionary movement should not offend public opinion when it has no good reason to do so. But the movement should never sacrifice its integrity by compromising its basic principles in the face of public hostility. Catering to public opinion may bring short-term advantage, but in the long run the movement will have its best chance of success if it sticks to its principles through thick and thin, no matter how unpopular those principles may become, and if it is willing to go head-to-head against the system on the fundamental issues even when the odds are all against the movement. A movement that backs off or compromises when the going gets tough is likely to lose its cohesiveness or turn into a wishy-washy reform movement. Maintaining the cohesion and integrity of the movement, and proving its courage, are far more important than keeping the goodwill of the general public. The public is fickle, and its goodwill can turn to hostility and back again overnight.

A revolutionary movement needs patience and persistence. It may have to wait several decades before the occasion for revolution arrives, and during those decades it has to occupy itself with preparing the way for revolution.

This was what the revolutionary movement in Russia did. Patience and persistence often pay off in the long run, even contrary to all expectation. History provides many examples of seemingly lost causes that won out in the end because of the stubborn persistence of their adherents, their refusal to accept defeat.

On the other hand, the occasion for revolution may arrive unexpectedly, and a revolutionary movement has to be well prepared in advance to take advantage of the occasion when it does arrive. It is said that the Bolsheviks never expected to see a revolution in their own lifetimes, yet, because their movement was well constituted for decisive action at any time, they were able to make effective use of the unforeseen breakdown of the Tsarist regime and the ensuing chaos.

Above all, a revolutionary movement must have courage. A revolution in the modern world will be no dinner party. It will be deadly and brutal. You can be sure that when the technoindustrial system begins to break down, the result will not be the sudden conversion of the entire human race into flower children. Instead, various groups will compete for power. If the opponents of technology prove toughest, they will be able to assure that the breakdown of the technosystem becomes complete and final. If other groups prove tougher, they may be able to salvage the technosystem and get it running again. Thus, an effective revolutionary movement must consist of people who are willing to pay the price that a real revolution demands: They must be ready to face disaster, suffering, and death.

There already is a revolutionary movement of sorts, but it is of low effectiveness.

First, the existing movement is of low effectiveness because it is not focused on a clear, definite goal. Instead, it has a hodgepodge of vaguely-defined goals such as an end to "domination," protection of the environment, and "justice" (whatever that means) for women, gays, and animals.

Most of these goals are not even revolutionary ones. As was pointed out at the beginning of this article, a precondition for revolution is the development of values that can be realized only through the destruction of the existing structure of society. But, to take an example, feminist goals such as equal status for women and an end to rape and domestic abuse are perfectly compatible with the existing structure of society. In fact, realization of these goals would even make the technoindustrial system function more efficiently. The same applies to most other "activist" goals. Consequently, these goals are reformist.

Among so many other goals, the one truly revolutionary goal— namely, the destruction of the technoindustrial system itself— tends to get lost in the shuffle. For revolution to become a reality, it is necessary that there should emerge a movement that has a distinct identify of its own, and is dedicated solely to eliminating the technosystem. It must not be distracted by reformist goals such as justice for this or that group.

Second, the existing movement is of low effectiveness because too many of the people in the movement are there for the wrong reasons. For some of them, revolution is just a vague and indefinite hope rather than a real and practical goal. Some are concerned more with their own special grievances than with the overall problem of technological civilization. For others, revolution is only a kind of game that they play as an outlet for rebellious impulses. For still others, participation in the movement is an ego-trip. They compete for status, or they write "analyses" and "critiques" that serve more to feed their own vanity than to advance the revolutionary cause.

To create an effective revolutionary movement it will be necessary to gather together people for whom revolution is not an abstract theory, a vague fantasy, a mere hope for the indefinite future, or a game played as an outlet for rebellious impulses, but a real, definite, and practical goal to be worked for in a practical way. •

ENDNOTES

[1] "Report on an Investigation of the Peasant Movement in Hunan," in *Selected Readings from the Works of Mao Tsetung* [=Zedong], Foreign Languages Press, Peking, 1971, page 30.

[2] As used in this article, the term "revolution" means a radical and rapid collapse of the existing structure of a society, intentionally brought about from within the society rather than by some external factor, and contrary to the will of the dominant classes of the society. An armed rebellion, even one that overthrows a government, is not a revolution in this sense of the word unless it sweeps away the existing structure of the society in which the rebellion occurs.

[3] Karl Marx maintained that the means of production constituted the decisive factor in determining the character of a society, but Marx lived in a time when the principal problem to which technology was applied was that of production. Because technology has so brilliantly solved the problem of production, production is no longer the decisive factor. More critical today are other problems to which technology is applied, such as processing of information and the regulation of human behavior (e.g., through propaganda). Thus Marx's conception of the force determining the character of a society must be broadened to include all of technology and not just the technology of production. If Marx were alive today he would undoubtedly agree.

Morality and Revolution

"Morality, guilt and fear of condemnation act as cops in our heads, destroying our spontaneity, our wildness, our ability to live our lives to the full.... I try to act on my whims, my spontaneous urges without caring what others think of me.... I want no constraints on my life; I want the opening of all possibilities.... This means...destroying all morality." —Feral Faun, "The Cops in Our Heads: Some Thoughts on Anarchy and Morality."[1]

It is true that the concept of morality as conventionally understood is one of the most important tools that the system uses to control us, and we must liberate ourselves from it.

But suppose you're in a bad mood one day. You see an inoffensive but ugly old lady; her appearance irritates you, and your "spontaneous urges" impel you to knock her down and kick her. Or suppose you have a "thing" for little girls, so your "spontaneous urges" lead you to pick out a cute 4-year-old, rip off her clothes, and rape her as she screams in terror.

I would be willing to bet that there is not one anarchist reading this who would not be disgusted by such actions, or who would not try to prevent them if he saw them being carried out. Is this only a consequence of the moral conditioning that our society imposes on us?

I argue that it is not. I propose that there is a kind of natural "morality" (note the quotation marks), or a conception of fairness, that runs as a common thread through all cultures and tends to appear in them in some form or other, though it may often be submerged or modified by forces specific to a particular culture. Perhaps this conception of fairness is biologically predisposed. At any rate it can be summarized in the following Six Principles:

1. Do not harm anyone who has not previously harmed you, or threatened to do so.
2. (Principle of self-defense and retaliation) You can harm others in order to forestall harm with which they threaten you, or in retaliation for harm that they have already inflicted on you.

3. One good turn deserves another: If someone has done you a favor, you should be willing to do her or him a comparable favor if and when he or she should need one.
4. The strong should have consideration for the weak.
5. Do not lie.
6. Abide faithfully by any promises or agreements that you make.

To take a couple of examples of the ways in which the Six Principles often are submerged by cultural forces, among the Navajo, traditionally, it was considered "morally acceptable" to use deception when trading with anyone who was not a member of the tribe (W. A. Haviland, *Cultural Anthropology*, 9th ed., p. 207), though this contravenes principles 1, 5, and 6. And in our society many people will reject the principle of retaliation: Because of industrial society's imperative need for social order and because of the disruptive potential of personal retaliatory action, we are trained to suppress our retaliatory impulses and leave any serious retaliation (called "justice") to the legal system.

In spite of such examples, I maintain that the Six Principles *tend* toward universality. But whether or not one accepts that the Six Principles are to any extent universal, I feel safe in assuming that almost all readers of this article will agree with the principles (with the possible exception of the principle of retaliation) in some shape or other. Hence the Six Principles can serve as a basis for the present discussion.

I argue that the Six principles should not be regarded as a moral code, for several reasons.

First. The principles are vague and can be interpreted in such widely varying ways that there will be no consistent agreement as to their application in concrete cases. For instance, if Smith insists on playing his radio so loud that it prevents Jones from sleeping, and if Jones smashes Smith's radio for him, is Jones's action unprovoked harm inflicted on Smith, or is it legitimate self-defense against harm that Smith is inflicting on Jones? On this question Smith and Jones are not likely to agree! (All the same, there are limits to the interpretation of the Six Principles. I imagine it would be difficult to find anyone in any culture who would interpret the principles in such a way as to justify brutal physical abuse of unoffending old ladies or the rape of 4-year-old girls.)

Second. Most people will agree that it is sometimes "morally" justifiable to make exceptions to the Six Principles. If your friend has destroyed logging equipment belonging to a large timber corporation, and if the police come around to ask you who did it, any green anarchist will agree that it is justifiable to lie and say, "I don't know."

Third. The Six Principles have not generally been treated as if they possessed the force and rigidity of true moral laws. People often violate the Six Principles even when there is no "moral" justification for doing so. Moreover, as already noted, the moral codes of particular societies frequently conflict with and override the Six Principles. Rather than laws, the principles are only a kind of guide, an expression of our more generous impulses that reminds us not to do certain things that we may later look back on with disgust.

Fourth. I suggest that the term "morality" should be used only to designate socially imposed codes of behaviour that are specific to certain societies, cultures, or subcultures. Since the Six Principles, in some form or other, tend to be universal and may well be biologically predisposed, they should not be described as morality.

Assuming that most anarchists will accept the Six Principles, what the anarchist (or, at least, the anarchist of individualistic type) does is claim the right to interpret the principles for himself in any concrete situation in which he is involved and decide for himself when to make exceptions to the principles, rather than letting any authority make such decisions for him.

However, when people interpret the Six Principles for themselves, conflicts arise because different individuals interpret the principles differently. For this reason among others, practically all societies have evolved rules that restrict behavior in more precise ways than the Six Principles do. In other words, whenever a number of people are together for an extended period of time, it is almost inevitable that some degree of morality will develop. Only the hermit is completely free. This is not an attempt to debunk the idea of anarchy. Even if there is no such thing as a society perfectly free of morality, still there is a big difference between a society in which the burden of morality is light and one in which it is heavy. The pygmies of the African rain forest, as described by Colin Turnbull in his books *The Forest People* and

Wayward Servants: The Two Worlds of the African Pygmies, provide an example of a society that is not far from the anarchist ideal. Their rules are few and flexible and allow a very generous measure of personal liberty. (Yet, even though they have no cops, courts or prisons, Turnbull mentions no case of homicide among them.)

In contrast, in technologically advanced societies the social mechanism is complex and rigid, and can function only when human behavior is closely regulated. Consequently such societies require a far more restrictive system of law and morality. (For present purposes we don't need to distinguish between law and morality. We will simply consider law as a particular kind of morality, which is not unreasonable, since in our society it is widely regarded as immoral to break the law.) Old-fashioned people complain of moral looseness in modern society, and it is true that in *some* respects our society is relatively free of morality. But I would argue that our society's relaxation of morality in sex, art, literature, dress, religion, etc., is in large part a reaction to the severe tightening of controls on human behavior in the practical domain. Art, literature and the like provide a harmless outlet for rebellious impulses that would be dangerous to the system if they took a more practical direction, and hedonistic satisfactions such as overindulgence in sex or food, or intensely stimulating modern forms of entertainment, help people to forget the loss of their freedom.

At any rate, it is clear that in any society some morality serves practical functions. One of these functions is that of forestalling conflicts or making it possible to resolve them without recourse to violence. (According to Elizabeth Marshall Thomas's book *The Harmless People*, Vintage Books, Random House, New York, 1989, pages 10, 82, 83, the Bushmen of Southern Africa own as private property the right to gather food in specified areas of the veldt, and they respect these property rights strictly. It is easy to see how such rules can prevent conflicts over the use of food resources.)

Since anarchists place a high value on personal liberty, they presumably will want to keep morality to a minimum, even if this costs them something in personal safety or other practical advantages. It's not my purpose here to try to determine where to strike the balance between freedom and the practical advantages of morality, but I do want to call attention to a point that is often overlooked: the practical or materialistic benefits of morality are counterbalanced by the psychological cost of repressing our "immoral" impulses. Common among moralists is a concept of "progress" according to which the human race is supposed to become ever more moral. More and more "immoral" impulses are to be suppressed and replaced by "civilized" behavior. To these people morality apparently is an end in itself. They never seem to ask *why* human beings should become more moral. What *end* is to be served by morality? If the end is anything resembling human well-being then an ever more sweeping and intensive morality can only be counterproductive, since it is certain that the psychological cost of suppressing "immoral" impulses will eventually outweigh any advantages conferred by morality (if it does not do so already). In fact, it is clear that, whatever excuses they may invent, the real motive of the moralists is to satisfy some psychological need of their own by imposing their morality on other people. Their drive toward morality is not an outcome of any rational program for improving the lot of the human race.

This aggressive morality has nothing to do with the Six Principles of fairness. It is actually inconsistent with them. By trying to impose their morality on other people, whether by force or through propaganda and training, the moralists are doing them unprovoked harm in contravention of the first of the Six Principles. One thinks of 19th-century missionaries who made primitive people feel guilty about their sexual practices, or modern leftists who try to suppress politically incorrect speech.

Morality often is antagonistic toward the Six Principles in other ways as well. To take just a few examples:

In our society private property is not what it is among the Bushmen—a simple device for avoiding conflict over the use of resources. Instead, it is a system whereby certain persons or organizations arrogate control over vast quantities of resources that they use to exert power over other people. In this they certainly violate the first and fourth principles of fairness. By requiring us to respect property, the morality of our society helps to perpetuate a system that is clearly in conflict with the Six Principles.

Among many primitive peoples, deformed babies are killed at birth (see, e.g., Paul Schebesta, *Die Bambuti-Pygmäen vom Ituri*, I. Band, Institut Royal Colonial Belge, Brussels, 1938, page 138), and a similar practice apparently was widespread in the United States up to about the middle of the 20th century. "Babies who were born malformed or too small or just blue and not breathing well were listed [by doctors] as stillborn, placed out of sight and left to die." Atul Gawande, "The Score," *The New Yorker*, October 9, 2006, page 64. Nowadays any such practice would be regarded as shockingly immoral. But mental-health professionals who study the psychological problems of the disabled can tell us how severe these problems often are. True, even among the grossly deformed—for example, those born without arms or legs—there may be occasional individuals who achieve satisfying lives. But most persons with such a degree of disability are condemned to lives of inferiority and helplessness, and to rear a baby with extreme deformities until it is old enough to be conscious of its own helplessness is usually an act of cruelty. In any given case, of course, it may be difficult to balance the likelihood that a deformed baby will lead a miserable existence, if reared, against the chance that it will achieve a worthwhile life. The point is, however, that the moral code of modern society does not permit such balancing. It *automatically* requires every baby to be reared, no matter how extreme its physical or mental disabilities, and no matter how remote the chances that its life can be anything but wretched. This is one of the most ruthless aspects of modern morality.

The military is expected to kill or refrain from killing in blind obedience to orders from the government; policemen and judges are expected to imprison or release persons in mechanical obedience to the law. It would be regarded as "unethical" and "irresponsible" for soldiers, judges, or policemen to act according to their own sense of fairness rather than in conformity with the rules of the system. A moral and "responsible" judge will send a man to prison if the law tells him to do so, even if the man is blameless according to the Six Principles.

A claim of morality often serves as a cloak for what would otherwise be seen as the naked imposition of one's own will on other people. Thus, if a person said, "I am going to prevent you from having an abortion (or from having sex or eating meat or something else) just because I personally find it offensive," his attempt to impose his will would be considered arrogant and unreasonable. But if he claims to have a moral basis for what he is doing, if he says, "I'm going to prevent you from having an abortion because it's immoral," then his attempt to impose his will acquires a certain legitimacy, or at least tends to be treated with more respect than it would be if he made no moral claim.

People who are strongly attached to the morality of their own society often are oblivious to the principles of fairness. The highly moral and Christian businessman John D. Rockefeller used underhand methods to achieve success, as is admitted by Allan Nevin in his admiring biography of Rockefeller. Today, screwing people in one way or another is almost an inevitable part of any large-scale business enterprise. Willful distortion of the truth, serious enough so that it amounts to lying, is in practice treated as acceptable behavior among politicians and journalists, though most of them undoubtedly regard themselves as moral people.

I have before me a flyer sent out by a magazine called *The National Interest*. In it I find the following:

"Your task at hand is to defend our nation's interests abroad, and rally support at home for your efforts.

"You are not, of course, naive. You believe that, for better or worse, international politics remains essentially power politics— that as Thomas Hobbes observed, when there is no agreement among states, clubs are always trumps."

This is a nearly naked advocacy of Machiavellianism in international affairs, though it is safe to assume that the people responsible for the flyer I've just quoted are firm adherents of conventional morality within the United States. For such people, I suggest, conventional morality serves as a *substitute* for the Six Principles. As long as these people comply with conventional morality, they have a sense of righteousness that enables them to disregard the principles of fairness without discomfort.

Another way in which morality is antagonistic toward the Six Principles is that it often serves as an excuse for mistreatment or exploitation of persons who have violated the moral code or the laws of a given society. In the United States, politicians promote their careers by "getting tough on crime" and advocating harsh penalties for people who have broken the law. Prosecutors often seek personal advancement by being as hard on defendants as the law allows them to be. This satisfies certain sadistic and authoritarian impulses of the public and allays the privileged classes' fear of social disorder. It all has little to do with the Six Principles of fairness. Many of the "criminals" who are subjected to harsh penal-ties—for example, people convicted of possessing marijuana—have in no sense violated the Six Principles. But even where culprits have violated the Six Principles their harsh treatment is motivated not by a concern for fairness, or even for morality, but politicians' and prosecutors' personal ambitions or by the public's sadistic and punitive appetites. Morality merely provides the *excuse*.

In sum, anyone who takes a detached look at modern society will see that, for all its emphasis on morality, it observes the principles of fairness very poorly indeed. Certainly less well than many primitive societies do.

Allowing for various exceptions, the main purpose that morality serves in modern society is to facilitate the functioning of the technoindustrial system. Here's how it works:

Our conception both of fairness and of morality is heavily influenced by self-interest. For example, I feel strongly and sincerely that it is perfectly fair for me to smash up the equipment of someone who is cutting down the forest. Yet part of the reason why I feel this way is that the continued existence of the forest serves my personal needs. If I had no personal attachment to the forest I might feel differently. Similarly, most rich people probably feel sincerely that the laws that protect their property are both fair and moral, and that laws that restrict the ways in which they use their property are unfair. There can be no doubt that, however sincere these feelings may be, they are motivated largely by self-interest.

People who occupy positions of power within the system have an interest in promoting the security and the expansion of the system. When these people perceive that certain moral ideas strengthen the system or make it more secure, then, either from conscious self-interest or because their moral feelings are influenced by self-interest, they apply pressure to the media and to educators to promote these moral ideas. Thus the requirements of respect for property, and of orderly, docile, rule-following, cooperative behavior, have become moral values in our society (even though these requirements can conflict with the principles of fairness) because they are necessary to the functioning of the system. Similarly, harmony and equality between different races and ethnic groups is a moral value of our society because interracial and interethnic conflict impede the functioning of the system. Equal treatment of all races and ethnic groups may be required by the principles of fairness, but this is not why it is a moral value of our society. It is a moral value of our society because it is good for the technoindustrial system. Traditional moral restraints on sexual behavior have been relaxed because the people who have power see that these restraints are not necessary to the functioning of the system and that maintaining them produces tensions and conflicts that are harmful to the system.

Particularly instructive is the moral prohibition of violence in our society. (By "violence" I mean physical attacks on human beings or the application of physical force to human beings.) Several hundred years ago, violence per se was not considered immoral in European society. In fact, under suitable conditions, it was admired. The most prestigious social class was the nobility, which was then a warrior caste. Even on the eve of the Industrial Revolution violence was not regarded as the greatest of all evils, and certain other values—personal liberty for example—were felt to be more important than the avoidance of violence. In America, well into the 19th century, public attitudes toward the police were negative, and police forces were kept weak and inefficient because it was felt that they were a threat to freedom. People preferred to see to their own defense and accept a fairly high level of violence in society rather than risk any of their personal liberty.[2]

Since then, attitudes toward violence have changed dramatically. Today the media, the schools, and all who are committed to the system brainwash us to believe that violence is the one thing above all others that we must never commit. (Of course, when the system finds it convenient to to use violence—via the police or the military—for its own purposes, it can always find an excuse for doing so.)

It is sometimes claimed that the modern attitude toward violence is a result of the gentling influence of Christianity, but this makes no sense. The period during which Christianity was most powerful in Europe, the Middle Ages, was a particularly violent epoch. It has been during the course of the Industrial Revolution and the ensuing technological changes that attitudes toward violence have been altered, and over the same span of time the influence of Christianity has been markedly weakened. Clearly it has not been Christianity that has changed attitudes toward violence.

It is necessary for the functioning of modern industrial society that people should cooperate in a rigid, machine-like way, obeying rules, following orders and schedules, carrying out prescribed procedures. Consequently the system requires, above all, human docility and social order. Of all human behaviors, violence is the one most disruptive of social order, hence the one most dangerous to the system. As the Industrial Revolution progressed, the powerful classes, perceiving that violence was increasingly contrary to their interest, changed their attitude toward it. Because their influence was predominant in determining what was printed by the press and taught in the schools, they gradually transformed the attitude of the entire society, so that today most middle-class people, and even the majority of those who think themselves rebels against the system, believe that violence is the ultimate sin. They imagine that their opposition to violence is the expression of a moral decision on their part, and in a sense it is, but it is based on a morality that is designed to serve the interest of the system and is instilled through propaganda. In fact, these people have simply been brainwashed.

It goes without saying that in order to bring about a revolution against the technoindustrial system it will be necessary to discard conventional morality. One of the two main points that I've tried to make in this article is that even the most radical rejection of conventional morality does not necessarily entail the abandonment of human decency: There is a "natural" (and in some sense perhaps universal) morality—or, as I have preferred to call it, a concept of fairness—that tends to keep our conduct toward other people "decent" even when we have discarded all formal morality.

The other main point I've tried to make is that the concept of morality is used for many purposes that have nothing to do with human decency or with what I've called "fairness." Modern society in particular uses morality as a tool in manipulating human behavior for purposes that often are completely inconsistent with human decency.

Thus, once revolutionaries have decided that the present form of society must be eliminated, there is no reason why they should hesitate to reject existing morality; and their rejection of morality will by no means be equivalent to a rejection of human decency.

There's no denying, however, that revolution against the technoindustrial system *will* violate human decency and the principles of fairness. With the collapse of the system, whether it is spontaneous or a result of revolution, countless innocent people will suffer and die. Our current situation is one of those in which we have to decide whether to commit injustice and cruelty in order to prevent a greater evil.

For comparison, consider World War II. At that time the ambitions of ruthless dictators could be thwarted only by making war on a large scale, and, given the conditions of modern warfare, millions of innocent civilians inevitably were killed or mutilated. Few people will deny that this constituted an extreme and inexcusable injustice to the victims, yet fewer still will argue that Hitler, Mussolini, and the Japanese militarists should have been allowed to dominate the world.

If it was acceptable to fight World War II in spite of the severe cruelty to millions of innocent people that that entailed, then a revolution against the technoindustrial system should be acceptable too. Had the fascists come to dominate the world, they doubtless would have treated their subject populations with brutality, would have reduced millions to slavery under harsh conditions, and would have exterminated many people outright. But, however horrible that might have been, it seems almost trivial in comparison with the disasters with which the technoindustrial system threatens us. Hitler and his allies merely tried to repeat on a larger scale the kinds of atrocities that have occurred again and again throughout the history of civilization. What modern technology threatens is absolutely without precedent. Today we have to ask ourselves whether nuclear war, biological disaster, or ecological collapse will produce casualties many times greater than those of World War II; whether the human race will continue to exist or whether it will be replaced by intelligent machines or genetically engineered freaks; whether the last vestiges of human dignity will disappear, not merely for the duration of a particular totalitarian regime but for all time; whether our world will even be inhabitable a couple of hundred years from now. Under these circumstances, who will claim that World War II was acceptable but that a revolution against the technoindustrial system is not?

Though revolution will necessarily involve violation of the principles of fairness, revolutionaries should make every effort to avoid violating those principles any more than is really necessary—not only from respect for human decency, but also for practical reasons. By complying with the principles of fairness to the extent that doing so is not incompatible with revolutionary action, revolutionaries will win the respect of nonrevolutionaries, will be able to recruit better people to be revolutionaries, and will increase the self-respect of the revolutionary movement, thereby strengthening its esprit de corps. •

ENDNOTES

[1] *The Quest for the Spiritual: A Basis for a Radical Analysis of Religion, and Other Essays by Feral Faun*, published by Green Anarchist, BCM 1715, London WC 1N 3XX, United Kingdom.

[2] See Hugh Davis Graham and Ted Robert Gurr (editors), *Violence in America: Historical and Comparative Perpectives*, Bantam Books, New York, 1970, Chapter 12, by Roger Lane; also, *The New Encyclopædia Britannica*, 15th Edition, 2003, Volume

25, article "Police," pages 959-960. On medieval attitudes toward violence and the reasons why those attitudes changed, see Norbert Elias, *The Civilizing Process*, Revised Edition, Blackwell Publishing, 2000, pages 161-172.

AFTERWORD

"Morality and Revolution" was originally written in 1999, was published in *Green Anarchist*, and was addressed specifically to anarchists, but I think it may be of interest to a much wider readership. The essay is presented here in heavily revised form.

Because it was written for anarchists, who are not generally religious, this essay discusses morality in purely secular terms; the whole question of a religious basis for morality is left out. That question of course is a formidable one in itself, and I'm not going to undertake a discussion of it here. I will only point out that no one has yet succeeded in demonstrating that the particular moral code prescribed by his own religion is in fact the one ordained by the Deity, assuming that there is a Deity. All we have are the conflicting and unproven claims of the various religions. •

Hit Where It Hurts

1. The Purpose of This Article

The purpose of this article is to point out a very simple principle of human conflict, a principle that opponents of the technoindustrial system seem to be overlooking. The principle is that in any form of conflict, if you want to win, you must hit your opponent where it hurts.

I have to explain that when I talk about "hitting where it hurts" I am not necessarily referring to physical blows or to any other form of physical violence. For example, in a debate, "hitting where it hurts" would mean making the arguments to which your opponent is most vulnerable. In a presidential election, "hitting where it hurts" would mean winning from your opponent the states that have the most electoral votes. Still, in discussing this principle I will use the analogy of physical combat, because it is vivid and clear.

If a man punches you, you can't defend yourself by hitting back at his fist, because you can't hurt the man that way. In order to win the fight, you have to hit him where it hurts. That means you have to go behind the fist and hit the sensitive and vulnerable parts of the man's body.

Suppose a bulldozer belonging to a logging company has been tearing up the woods near your home and you want to stop it. It is the blade of the bulldozer that rips the earth and knocks trees over, but it would be a waste of time to take a sledgehammer to the blade. If you spent a long, hard day working on the blade with the sledge, you *might* succeed in damaging it enough so that it became useless. But in comparison with the rest of the bulldozer the blade is relatively inexpensive and easy to replace. The blade is only the "fist" with which the bulldozer hits the earth. To defeat the machine you must go behind the "fist" and attack the bulldozer's vital parts. The engine, for example, can be ruined with very little expenditure of time and effort by means well known to many radicals.

At this point I must make clear that I am not recommending that anyone should damage a bulldozer (unless it is his own property). Nor should anything in this article be interpreted as recommending illegal activity of any kind. I am a prisoner, and if I were to encourage illegal activity this article would not even be allowed to leave the prison. I use the bulldozer analogy only because it is clear and vivid and will be appreciated by radicals.

2. Technology Is the Target

It is widely recognized that "the basic variable which determines the contemporary historic process is provided by technological development" (Celso Furtado).[1] Technology, above all else, is responsible for the current condition of the world and will control its future development. Thus, the "bulldozer" that we have to destroy is modern technology itself. Many radicals are aware of this, and therefore realize that their task is to eliminate the entire technoindustrial system. But unfortunately they have paid little attention to the need to hit the system where it hurts.

Smashing up McDonald's or Starbuck's is pointless. Not that I give a damn about McDonald's or Starbuck's. I don't care whether anyone smashes them up or not. But that is not a revolutionary activity. Even if every fast-food chain in the world were wiped out the technoindustrial system would suffer only minimal harm as a result, since it could easily survive without fast-food chains. When you attack McDonald's or Starbuck's, you are not hitting where it hurts.

Some months ago I received a letter from a young man in Denmark who believed that the technoindustrial system had to be eliminated because, as he put it, "What will happen if we go on this way?" Apparently, however, his form of "revolutionary" activity was raiding fur farms. As a means of weakening the technoindustrial system this activity is utterly useless. Even if animal liberationists succeeded in eliminating the fur industry completely they would do no harm at all to the system, because the system can get along perfectly well without furs.

I agree that keeping wild animals in cages is intolerable, and that putting an end to such practices is a noble cause. But there are many other noble causes, such as preventing traffic accidents, providing shelter for the homeless, recycling, or helping old people cross the street. Yet no one is foolish enough to mistake these for revolutionary activities, or to imagine that they do anything to weaken the system.

3. The Timber Industry Is a Side Issue

To take another example, no one in his right mind believes that anything like real wilderness can survive very long if the technoindustrial system continues to exist. Many environmental radicals agree that this is the case and hope for the collapse of the system. But in practice all they do is attack the timber industry.

I certainly have no objection to their attack on the timber industry. In fact, it's an issue that is close to my heart and I'm delighted by any successes that radicals may have against the timber industry. In addition, for reasons that I need not explain here, I think that opposition to the timber industry should be one component of the effort to overthrow the system.

But, by itself, attacking the timber industry is not an effective way of working against the system, for even in the unlikely event that radicals succeeded in stopping all logging everywhere in the world, that would not bring down the system. And it would not permanently save wilderness. Sooner or later the political climate would change and logging would resume. Even if logging never resumed, there would be other avenues through which wilderness would be destroyed, or if not destroyed then tamed and domesticated. Mining and mineral exploration, acid rain, climate change, and species extinction destroy wilderness; wilderness is tamed and domesticated through recreation, scientific study, and resource management, including among other things electronic tracking of animals, stocking of streams with hatchery-bred fish, and planting of genetically engineered trees.

Wilderness can be saved permanently only by eliminating the technoindustrial system, and you cannot eliminate the system by attacking the timber industry. The system would easily survive the death of the timber industry because wood products, though very useful to the system, can if necessary be replaced with other materials.

Consequently, when you attack the timber industry you are not hitting the system where it hurts. The timber industry is only the "fist" (or one of the fists) with which the system destroys wilderness and, just as in a fistfight, you can't win by hitting at the fist. You have to go behind the fist and strike at the most sensitive and vital organs of the system. By legal means, of course, such as peaceful protests.

4. Why the System Is Tough

The technoindustrial system is exceptionally tough due to its so-called "democratic" structure and its resulting flexibility. Because dictatorial systems tend to be rigid, social tensions and resistance can build up in them to the point where they damage and weaken the system and may lead to revolution. But in a "democratic" system, when social tension and resistance build up dangerously the system backs off enough, it compromises enough, to bring the tensions down to a safe level.

During the 1960s people first became aware that environmental pollution was a serious problem, the more so because the visible and smellable filth in the air over our major cities was beginning to make people physically uncomfortable. Enough protest arose so that an Environmental Protection Agency was established and other measures were taken to alleviate the problem. Of course, we all know that our pollution problems are a long, long way from being solved. But enough was done so that public complaints subsided and the pressure on the system was reduced for a number of years.

Thus, attacking the system is like hitting a piece of rubber. A blow with a hammer can shatter cast iron, because cast iron is rigid and brittle. But you can pound a piece of rubber without hurting it because it is flexible. It gives way before the hammer and bounces back as soon as the force of the hammer is expended. That's how it is with the "democratic" industrial system: It gives way before protest, just enough so that the protest loses its force and momentum. Then the system bounces back.

So, in order to hit the system where it hurts, you need to select issues on which the system will *not* back off, on which it will fight to the finish. For what you need is not compromise with the system but a life-and-death struggle.

5. It Is Useless to Attack the System in Terms of Its Own Values

It is absolutely essential to attack the system not in terms of its own technologically-oriented values, but in terms of values that are inconsistent with the values of the system. As long as you attack the system in terms of its own values, you do not hit the system where it hurts, and you allow the system to deflate protest by giving way, by backing off.

For example, if you attack the timber industry primarily on the basis that forests are needed to preserve water resources and recreational opportunities, then the system can give ground to defuse protest without compromising its own values. Water resources and recreation are fully consistent with the values of the system, and if the system backs off, if it restricts logging in the name of water resources and recreation, then it only makes a tactical retreat and does not suffer a strategic defeat for its code of values.

If you push victimization issues (such as racism, sexism, homophobia, or poverty) you are not challenging the system's values and you are not even forcing the system to back off or compromise. You are directly helping the system. All of the wisest *proponents* of the system recognize that racism, sexism, homophobia, and poverty are harmful to the system, and this is why the system itself works to combat these and similar forms of victimization.

"Sweatshops," with their low pay and wretched working conditions, may bring profit to certain corporations, but wise *proponents* of the system know very well that the system as a whole functions better when workers are treated decently. In making an issue of sweatshops, you are helping the system, not weakening it.

Many radicals fall into the temptation of focusing on nonessential issues, like racism, sexism, and sweatshops, because it is easy. They pick an issue on which the system can afford to compromise and on which they will get support from people like Ralph Nader, Winona LaDuke, the labor unions, and all the other pink reformers. Perhaps the system, under pressure, will back off a bit, the activists will see some visible result from their efforts, and they will have the satisfying illusion that they have accomplished something. But in reality they have accomplished nothing at all toward eliminating the technoindustrial system.

The globalization issue is not completely irrelevant to the technology problem. The package of economic and political measures termed "globalization" does promote economic growth and, consequently, technological progress. Still, globalization is an issue of marginal importance and not a well-chosen target for revolutionaries. The system can afford to give ground on the globalization issue. Without giving up globalization as such, the system can take steps to mitigate the negative environmental and economic consequences of globalization so as to defuse protest. At a pinch, the system could even afford to give up globalization altogether. Growth and progress would still continue, only at a slightly slower rate. And when you fight globalization you are not attacking the system's fundamental values. Opposition to globalization is motivated in terms of securing decent wages for workers and protecting the environment, both of which are completely consistent with the values of the system. (The system, for its own survival, can't afford to let environmental degradation go too far.) Consequently, in fighting globalization you do not hit the system where it really hurts. Your efforts may promote reform, but they are useless for the purpose of overthrowing the technoindustrial system.

6. Radicals Must Attack the System at the Decisive Points

To work effectively toward the elimination of the technoindustrial system, revolutionaries must attack the system at points at which it cannot afford to give ground. They must attack the vital organs of the system. Of course, when I use the word "attack," I am not referring to physical attack but only to legal forms of protest and resistance.

* * *

The rest of Hit Where it Hurts *is omitted, because it is considered unsuitable for inclusion in this book.* •

ENDNOTE

[1] In *Latin American Radicalism,* edited by Irving Louis Horowitz, Josué de Castro, and Jon Gerassi, Vintage Books, 1969, page 64.

Extracts from

9

Letters to David Skrbina

Letter to David Skrbina, January 2, 2004

I've been able to identify only three ways (apart from modest reforms) in which human beings' intentions concerning the future of their own society can be realized successfully: (i) Intelligent administration can prolong the life of an existing social order. (E.g., if 19th-century Russian Tsars had been a great deal less competent than they were, tsarism might have broken down earlier than it did. If Nicholas II had been a great deal more competent than he was, tsarism might have lasted a few decades longer.) (ii) Revolutionary action can bring about, or at least hasten, the breakdown of an existing social order. (E.g., if there had been no revolutionary movement in Russia, a new Tsar would doubtless have been appointed on the abdication of Nicholas II and tsarism would have survived for a while.) (iii) An existing social order can sometimes be extended to encompass additional territory. (E.g., the social order of the West was successfully extended to Japan following World War II.)

If I'm right, and if we want to exert any rational influence (beyond modest reforms) on the future of our own society, then we have to choose one of the foregoing alternatives.

Letter to David Skrbina, August 29, 2004

You sent me a copy of Bill Joy's article "Why the Future Doesn't Need Us," and you said you would be interested in my assessment of it. I read the article soon after it came out. I had already read elsewhere of most of the technological hazards described by Joy, but I considered his article useful because it gave further information about such hazards. Also, the fact that even a distinguished technophile like Bill Joy is scared about where technology is taking us should help to persuade people that the dangers of technology are real. Apart from that I was unimpressed by Joy's article. I assume that his technical expertise is solid, but it seems to me that his understanding of human nature and of how human societies work is at a naive level. A couple of people who wrote to me about the article expressed similarly unenthusiastic opinions of it.

To give an example of what I consider to be Joy's naiveté, he writes:

"Verifying compliance will also require that scientists and engineers adopt a strong code of ethical conduct...and that they have the courage to whistleblow as necessary, even at high personal cost.... [T]he Dalai Lama argues that the most important thing is for us to conduct our lives with love and compassion for others, and that our societies need to develop a stronger notion of universal responsibility and of our interdependency...."

If Bill Joy thinks that anything will be accomplished by this kind of preaching, then he is out of touch with reality. This part of his article would be funny if what is at stake weren't so desperately serious.

I've reread Joy's article to see if I had been missing anything, but I found that my impression of it was the same as before. Of course, it's possible that the article has merits that I've overlooked.

# # #

I don't particularly consider small-scale technology to be acceptable; it's simply inevitable. See ISAIF, paragraphs 207-212. I see no way of getting rid of it. People can't use organization-dependent technology if the social organization breaks down. E.g., you can't drive a car if the refineries aren't producing gasoline. But how could people be prevented from using small-scale technology? E.g., working steel, building a water-wheel, or ploughing and planting fields?

You ask whether I would consider a primitive steam-engine to be small-scale technology. To give a confident answer I would have to know more than I do about primitive steam-engines and their possible applications, but I think that steam-engines probably cannot be small-scale technology. "[Newcomen steam-engines'] heavy fuel consumption made them uneconomical when used where coal was expensive, but in the British coalfields they performed an essential service by keeping deep mines clear of water...."[1] An autonomous local community, without outside assistance, would find it very difficult to build an adequate steam-engine, and the engine probably would be of little use to such a community. Considering the effort required to build and maintain the engine, to produce oil to lubricate it, and to collect firewood to fuel it, any work the engine might do for a small community could probably be done more efficiently with human or animal muscle-power. Steam engines very likely could have been invented much earlier than they were, but—I would guess—they would have been of little use until certain 17th- and 18th-century economic and technological developments offered work for which steam engines were appropriate.

# # #

I'm quite sure that it will be impossible to control post-revolution conditions, but I think you're quite right in saying that a "positive social vision" is necessary. However, the social ideal I would put forward is that of the nomadic hunting-and-gathering society.

First, I would argue that in order to be successful a revolutionary movement *has* to be extremist. Jacques Ellul says somewhere that a revolution must take as its ideal the opposite of what it intends to overthrow.[2] Trotsky wrote: "The different stages of a revolutionary process [are] certified by a change of parties in which the more extreme always supersedes the less...."[3] The nomadic hunting-and-gathering society recommends itself as a social ideal because it is at the opposite extreme of human culture from the technological society.

Second, if one takes the position that certain appurtenances of civilization must be saved, e.g., cultural achievements up to the 17th century, then one will be tempted to make compromises when it comes to eliminating the technoindustrial system, with the possible or probable result that one will not succeed in eliminating the system at all. If the system breaks down, what will happen to the art museums with their priceless paintings and statues? Or to the great libraries with their vast stores of books? Who will take care of the artworks and books when there are no organizations large enough and rich enough to hire curators and librarians, as well as policemen to prevent looting and vandalism? And what about the educational system? Without an organized system of education, children will grow up uncultured and perhaps illiterate. Clearly, anyone who feels it is important to preserve human cultural achievements up to the 17th century will be very reluctant to see a complete breakdown of the system, hence will look for a compromise solution and will not take the frankly reckless measures that are necessary to knock our society off its present technological-determined course of development. Hence, only those can be effective revolutionaries who are prepared to dispense with the achievements of civilization.

Third, to most people, a hunting-and-gathering existence will appear much more attractive than that offered by preindustrial civilization. Even many modern people enjoy hunting, fishing, and gathering wild fruits and nuts. I think few would enjoy such tasks as ploughing, hoeing, or threshing. And in civilized societies the majority of the population commonly have been exploited in one way or another by the upper classes: If they were not slaves or serfs, then they often were hired laborers or tenant-farmers subject to the domination of landowners. Preindustrial civilized societies often suffered from disastrous epidemics or famines, and the common people in many cases had poor nutrition. In contrast, hunter-gatherers, except in the far north, generally had good nutrition.[4] Famines among them were probably rare.[5] They were relatively little troubled by infectious diseases until such diseases were introduced among them by more "advanced" peoples.[6] Slavery and well-developed social hierarchies could exist among *sedentary* hunter-gatherers, but (apart from the tendency of women to be in some degree subordinate to men), *nomadic* hunter-gatherer societies typically (not always) were characterized by social equality, and normally did not practice slavery. (Though I know of one exception: Apparently some Cree Indians who were probably hunter-gatherers did take slaves.)[7]

Just in case you've read anarcho-primitivist writings that portray the hunter-gatherer lifestyle as a kind of politically correct Garden of Eden where no one ever had to work more than 3 hours a day, men and women were equal, and all was love, cooperation and sharing, that's just a lot of nonsense, and at your request I'll prove it with numerous citations to the literature. But even when one discounts the anarcho-primitivists' idealized version and takes a hard-headed look at the facts, nomadic hunter-gatherer societies seem a great deal more attractive than preindustrial civilized ones.

I imagine that your chief objection to hunter-gatherer societies as opposed to (for example) late medieval or Renaissance European civilization would be their relatively very modest level of cultural achievement (in terms of art, music, literature, scholarship, etc.). But I seriously doubt that more than a small fraction of the population of modern industrial society cares very much about that kind of cultural achievement.

Hunter-gatherer society moreover has proved its appeal as a social ideal: Anarcho-primitivism seems to have gained wide popularity. One can hardly imagine equal success for a movement taking as its ideal—for example—late medieval society. Of course, one has to ask to what extent the success of anarcho-primitivism is dependent on its idealized portrayal of hunter-gatherer societies. My guess, or at least my hope, is that certain inconvenient aspects of hunter-gatherer societies (e.g., male dominance, hard work) would turn off the leftists, the neurotics, and the lazies but that such societies, depicted realistically, would remain attractive to the kind of people who could be effective revolutionaries.

I don't think that a worldwide return to a hunting-and-gathering economy would actually be a plausible outcome of a collapse of industrial society. No ideology will persuade people to starve when they can feed themselves by planting crops, so presumably agriculture will be practiced wherever the soil and climate are suitable for it. Reversion to hunting and gathering as the sole means of subsistence could occur only in regions unsuitable for agriculture, e.g., the subarctic, arid plains, or rugged mountains.

# # #

I'm not terribly interested in questions of values of the kind you discuss here, such as "herd values" versus the "will to power." As I see it, the overwhelmingly dominant problem of our time is that technology threatens either to destroy the world or to transform it so radically that all past questions of human values will simply become irrelevant, because the human race, as we have known it, will no longer exist. I don't mean that the human race necessarily will become physically extinct (though that is a possibility), but that the way human beings function socially and psychologically will be transformed so radically as to make traditional questions of values practically meaningless. The old-fashioned conformist will become as obsolete as the old-fashioned individualist.

Since this is the most critical juncture in the history of the human race, all other issues must be subordinated to the problem of stopping the technological juggernaut before it is too late. If I advocate a break with conventional morality, I do so not because I disapprove of the herd mentality, but because conventional morality acts as a brake on the development of an effective revolutionary movement. Furthermore, any effective revolutionary movement probably has to make use of the herd mentality. Imitativeness is part of human nature, and one has to work with it rather than preach against it.

Possibly you misinterpret my motives for emphasizing the "power process." The purpose of doing so is not to exalt the "will to power." There are two main reasons for discussing the power process. First, discussion of the power process is necessary for the analysis of the psychology of the people whom I call "leftists." Second, it is difficult to get people excited about working to avoid a future evil. It is less difficult to get people excited about throwing off a *present* evil. Discussion of the power process helps to show people how a great deal of *present* dissatisfaction and frustration results from the fact that we live in a technological society.

I should admit, though, that I personally am strongly inclined to individualism. Ideally, I shouldn't allow my individualistic predilections to influence my thinking on revolutionary strategy but should arrive at my conclusions objectively. The fact that you have spotted my individualistic leanings may mean that I have not been as objective as I should have been.

But even leaving aside all questions of "political" utility and considering only my personal predilections, I have little interest in philosophical questions such as the desirability or undesirability of the "herd mentality." The mountains of Western Montana offered me nearly everything I needed or wanted. If those mountains could have remained just as they were when I first moved to Montana in 1971, I would have been satisfied. The rest of the world could have had a herd mentality, or an individualistic mentality or whatever, and it would have been all the same to me. But, of course, under modern conditions there was no way the mountains could have remained isolated from the rest of the world. Civilization moved in and squeezed me, so.....

# # #

Yes, growth in the population of nations and increasing racial/ethnic diversity no doubt affected social values. But increasing racial/ethnic diversity was unquestionably a consequence of technological events, namely, the development of relatively safe and efficient sailing ships, along with economic (therefore also technological) factors that provided incentives to trade, travel, and migrate widely. Presumably, population growth too was dependent on technological factors, such as improvements in agriculture that made it possible to feed more people.

# # #

I'll draw a distinction between a revolutionary movement and a reform movement. The distinction is not valid in all situations, but I think it is valid in the present situation.

The objective of a revolutionary movement, as opposed to a reform movement, is not to make piecemeal corrections of various evils of the social order. The objectives of a revolutionary movement are (i) to build its own strength, and (ii) to increase the tension within the social order until those tensions reach the breaking point.

Correcting this or that social evil is likely to *decrease* the tensions within the social order. This is the reason for the classic antagonism between revolutionary movements and reform movements.

Generally speaking, correction of a given social evil serves the purposes of a revolutionary movement only if it (a) constitutes a victory for the revolutionary movement that enhances the movement's prestige, (b) represents humiliating defeat for the existing social order, (c) is achieved by methods that, if not illegal, are at least offensive to the existing order, and (d) is widely perceived as a step toward dissolution of the existing order.

In the particular situation that the world faces today, there may be also another case in which partial or piecemeal correction of a social evil may be useful: It may buy us time. For example, if progress in biotechnology is slowed, a biological catastrophe will be less likely to occur before we have time to overturn the system.

# # #

To address specifically your argument that a focus on population reduction is appropriate, at least as an "ancillary approach," I disagree for two reasons: (I) An effort to reduce population would be futile. (II) Even if it could be achieved, population reduction would accomplish nothing against the system. For these reasons, a focus on population reduction would waste time and energy that should be devoted to efforts that are more useful.

(I) If you were as old as I am and had watched the development of our society for 50 years, I don't think you would suggest a campaign against population growth. It has been tried and it has failed. Back in the 1960s and early 1970s, concern about "the population problem" was "in." There was even a national organization called "Zero Population Growth" whose goal was its name. Of course, it never accomplished anything. In those days, the fact that population was a problem was a new discovery, but nowadays it's "old hat," people are blasé, and it's much harder to get people aroused about population than it was back in the 1960s. Especially since the latest predictions are that world population will level off at about 9 billion some time around the middle of this century. Such predictions are unreliable, but they nevertheless reduce anxiety about runaway population growth.

In any case, you could never get large numbers of people to have fewer children simply by pointing out to them the problems caused by overpopulation. As professional propagandists are well aware, reason by itself is of little use for influencing people on a mass basis.[8] To have any substantial effect, you would have to resort to the system's own techniques of propaganda. By dirtying its hands in this way, an anti-system movement would perhaps discredit itself. Anyhow, it's wildly improbable that such a movement could be rich enough to mount an effective worldwide or even nationwide campaign to persuade people to have fewer children. "Propaganda that aims to induce major changes is certain to take great amounts of time, resources, patience, and indirection, except in times of revolutionary crisis when old beliefs have been shattered....[9] The *Encyclopædia Britannica* Macropædia article "Propaganda" provides a good glimpse of the technical basis of modern propaganda, hence an idea of the vast amount of money you would need in order to make any substantial impression on the birthrate through persuasion. "Many of the bigger and wealthier propaganda agencies...conduct 'symbol campaigns' and 'image-building' operations with mathematical calculation, using quantities of data that can be processed only by computers...,"[10] etc.,etc.. (This should lay to rest your suggestion that "Propaganda can be opposed by counter-propaganda." Unless you have billions of dollars at your disposal, there's no way you can defeat the system in a head-on propaganda contest. A revolutionary movement has to find other means of making an impact.)

How difficult it would be to reduce the birthrate can be seen from the fact that the Chinese government has been trying to do that for years. According to the latest reports I've heard (several years ago), they've had only very limited success, even though they have vastly greater resources than any revolutionary movement could hope to have.

Furthermore, a campaign against having children could be a kind of suicide for a movement. The people who were with you wouldn't have children, your opponents *would* have children. Since the political orientation of children tends statistically to resemble that of their parents, your movement would get weaker with each generation.

And, to put it bluntly, a revolutionary movement needs an enemy, it needs someone or something to hate. If you are working against overpopulation, then who is your enemy? Pregnant women? I don't think that would work very well.

(II) Even assuming you could reduce the birthrate, a population decline would be of little use and might well be counterproductive. I fail to understand your statement (page 7 of your letter) that population growth "seems to drive the whole technoindustrial process forward at an accelerating rate." Population increase no doubt is an important stimulus for economic growth, but it's hardly a decisive factor. In developed countries, economic growth probably occurs more through increasing demand for goods and services on the part of each individual than through an increase in the number of individuals. In any case, do you seriously believe that scientists would stop developing supercomputers and biological technology if the population started to decline? Of course, scientists need financial support from large organizations such as corporations and governments. But the large organizations' support for research is driven not by population growth but by competition for power among the large organizations.

So I think we can say that population is a dependent variable, technology is the independent variable. It's not primarily population growth that drives technology, but technology that makes population growth possible. Furthermore, because overcrowding makes people uncomfortable and increases stress and aggression, a reduction of population would tend to decrease the tensions in our society, hence would be contrary to the interests of a revolutionary movement, which, as already noted, needs to *increase* social tension. Even in the unlikely event that a victory on the population issue could be achieved, I don't think it would satisfy any of the conditions (b), (c), (d) that I listed earlier in this letter. Arguably, population decline could "buy us time" in the sense I've mentioned, but when this is weighed against the other factors I've just described I think the balance comes down decisively against an effort to reduce population. But a revolutionary movement can make use of the population issue by pointing to overpopulation as one of the negative consequences of technological progress.

# # #

I don't think the U.S. situation is as unique as you do. In any case, I wouldn't emphasize the U.S. situation, because there are too many people who are too ready to focus on the U.S. as the world's villain. I'm not a patriot and not particularly interested in defending the U.S. But obsessive anti-Americanism distracts attention from the technology problem just as the issues of sexism, racism, etc., do. Given the present global technological and economic situation, if the U.S. weren't playing the role of the world's bully then probably some other country or group of countries would be doing so. And if the Russians, for example, were playing that role, I suspect they would play rougher than the U.S. does.

I'm not sure exactly what you mean by your final remark that there are "many roads to revolution." But I would argue that a revolutionary movement can't afford to be diverse and eclectic. It must be flexible, and up to a point must allow for dissent within the movement. But a revolutionary movement needs to be unified, with a clear doctrine and goals. I believe that a catchall movement that tries to embrace simultaneously all roads to revolution will fail. A couple of cases in point:

A. Under the Roman Empire there were several salvational religious movements analogous to Christianity. You'll find a discussion of this in Jerome Carcopino's *Daily Life in Ancient Rome*. It seems that, with the exception of Christianity, all of these religious movements were syncretistic and mutually tolerant; one could belong to more than one of them.[11] Only

Christianity required exclusive devotion. And I don't have to tell you which religion became in the end the dominant religion of Europe.

B. In the early stage of the Russian Revolution of 1917, the Social Revolutionary Party was dominant; the Bolshevik Party was small and isolated. But the Social Revolutionary Party was a catch-all party that took in everyone who was vaguely in favor of the revolution. "To vote for the Social Revolutionaries meant to vote for the revolution in general, and involved no further obligation."[12] The Bolsheviks, in contrast, were reasonably unified and developed a program of action with clear goals. "The Bolsheviks acted, or strove to act...like uncompromising revolutionists."[13] And in the end it was the Bolsheviks, not the Social Revolutionaries, who determined the outcome of the revolution.

Letter to David Skrbina, September 18, 2004

I think that as a preliminary to answering your letter of July 27, it would be a good idea for me to give a more detailed outline of the "road to revolution" that I envision. The "road" is of course speculative. It's impossible to foretell the course of events, so any movement aspiring to get rid of the technoindustrial system will have to be flexible and proceed by trial and error. It's nevertheless necessary to give a *tentative* indication of the route to be followed, because without some idea of where it is going the movement will flounder around aimlessly. Also, an outline of at least a *possible* route to revolution helps to make the idea of revolution seem plausible. Probably the biggest current obstacle to the creation of an effective revolutionary movement is the mere fact that most people (at least in the U.S.) don't see revolution as a plausible possibility.

In the first place, I believe that illegal action will be indispensable. I wouldn't be allowed to mail this letter if I appeared to be trying to incite illegal action, so I will say only this much about it: A revolutionary movement should consist of two separate and independent sectors, an illegal, underground sector, and a legal sector. I'll say nothing about what the illegal sector should do. The legal sector (if only for its own protection) should carefully avoid any connection with the illegal sector.

With the possible exceptions listed in my letter of 08/29/04, the function of the legal sector would not be to correct any evils of technology. Instead, its function would be to prepare the way for a future revolution, to be carried out when the right moment arrives.

Advance preparation is especially important in view of the fact that the occasion for revolution may arrive at any time and quite unexpectedly. The spontaneous insurrection in St. Petersburg in February 1917 took all of Russia by surprise. It is safe to say that this insurrection (if it had occurred at all) would have been no more than a massive but purposeless outburst of frustration if the way to revolution had not been prepared in advance. As it happened, there was already in existence a strong revolutionary movement that was in a position to provide leadership, and the revolutionaries moreover had for a long time been educating (or indoctrinating) the workers of St. Petersburg so that when the latter revolted they were not merely expressing senseless anger, but were acting purposefully and more or less intelligently.[14]

In order to prepare the way for revolution, the legal sector of the movement should:

(I) Build its own strength and cohesiveness. Increasing its *numbers* will be far less important than collecting members who are loyal, capable, deeply committed, and prepared for practical action. (The example of the Bolsheviks is instructive here.)[15]

(II) Develop and disseminate an ideology that will (a) show people how many dangers the advance of technology presents for the future; (b) show people how many of their present problems and frustrations derive from the fact that they live in a technological society; (c) show people that there have existed past societies that have been more or less free of these problems and frustrations; (d) offer as a positive ideal a life close to nature; and (e) present revolution as a realistic alternative.[16]

The utility of (II) is as follows:

As matters stand at the moment, revolution in the stable parts of the industrialized world is impossible. A revolution could occur only if something happened to shake the stability of industrial society. It is easy to imagine events or developments that could shake the system in this way. To take just one example, suppose a virus created in an experimental laboratory escaped and wiped out, say, a third of the population of the industrialized world. But if this happened *now*, it hardly seems possible that it could lead to revolution. Instead of blaming the technoindustrial system as a whole for the disaster, people would blame only the carelessness of a particular laboratory. Their reaction would be not to dump technology, but to try to pick up the pieces and get the system running again—though doubtless they would enact laws requiring much stricter supervision of biotechnological research in the future.

The difficulty is that people see problems, frustrations, and disasters in isolation rather than seeing them as manifestations of the one central problem of technology. If Al Qaeda should set off a nuclear bomb in Washington, D.C., people's reaction will be, "Get those terrorists!" They will forget that the bomb could not have existed without the previous development of nuclear technology. When people find their culture or their economic welfare disrupted by the influx of large numbers of immigrants, their reaction is to hate the immigrants rather than take account of the fact that massive population movements are an inevitable consequence of economic developments that result from technological progress. If there is a worldwide depression, people will blame it merely on someone's economic mismanagement, forgetting that in earlier times when small communities were largely self-sufficient, their welfare did not depend on the decisions of government economists. When people are upset about the decay of traditional values or the loss of local autonomy, they preach against "immorality" or get angry at "big government," without any apparent awareness that the loss of traditional values and of local autonomy is an unavoidable result of technological progress.

But, if a revolutionary movement can show a sufficient number of people how the foregoing problems and many others all are outgrowths of one central problem, namely, that of technology, and if the movement can successfully carry out the other tasks listed under (II), then, in case of a shattering event such as the epidemic mentioned above,[17] or a worldwide depression, or an accumulation of diverse factors that make life difficult or insecure, a revolution against the technoindustrial system may be possible.

Furthermore, the movement does not have to wait passively for a crisis that may weaken the system. Quite apart from any activities of the illegal sector, the dissention sown by the legal sector of the movement may help to bring on a crisis. For example, the Russian Revolution was precipitated by the tsarist regime's military disasters in World War I, and the revolutionary movement may have helped to create those disasters, since "[i]n no other belligerent country were political conflicts waged as intensively during the war as in Russia, preventing the effective mobilization of the rear."[18]

In carrying out the task (II) described above, the movement will of course use rational argument. But as I pointed out in my letter of 8/29/04, reason by itself is a very weak tool for influencing human behavior on a mass basis. You have to work also with the nonrational aspects of human behavior. But in doing so you can't rely on the system's own techniques of propaganda. As I argued in my letter of 8/29/04, you can't defeat the system in a head-on propaganda contest. Instead, you have to circumvent the system's superiority in psychological weaponry by making use of certain advantages that a revolutionary movement will have over the system. These advantages would include the following:

(i) It seems to be felt by many people that there is a kind of spiritual emptiness in modern life. I'm not sure exactly what this means, but "spiritual emptiness" would include at least the system's apparent inability to provide any positive values of wide appeal other than hedonistic ones or the simple worship of technological progress for its own sake. Evidence that many people find these values unsatisfactory is provided by the existence within modern society of groups that offer alternative systems of values—values that sometimes are in conflict with those of the system. Such groups would include fundamentalist churches and other, smaller cults that are still farther from the mainstream, as well as deviant political movements on the left and on the right. A successful revolutionary movement would have to do much better than these groups and fill the system's spiritual vacuum with values that can appeal to rational, self-disciplined people.

(ii) Wild nature still fascinates people. This shown by the popularity of magazines like National Geographic, tourism to such (semi-)wild places as remain, and so forth. But, notwithstanding all the nature magazines, the guided wilderness tours, the parks and preserves, etc., the system's propaganda is unable to disguise the fact that "progress" is destroying wild nature. I think that many people continue to find this seriously disturbing, even apart from the practical consequences of environmental destruction, and their feelings on this subject provide a lever that a revolutionary movement can utilize.

(iii) Most people feel a need for a sense of community, or for belonging to what sociologists call a "reference group." The system tries to satisfy this need to the extent that it is able: Some people find their reference group in a mainstream church, a Boy Scout troop, a "support group," or the like. That these system-provided reference groups are for many people unsatisfactory is indicated by the proliferation of independent groups that lie outside the mainstream or even are antagonistic toward it. These include, inter alia, cults, gangs, and politically dissident groups. Possibly the reason why many people find the system-provided reference groups unsatisfactory is the very fact that these groups are appendages of the system. It may be that people need groups that are "their own thing," i.e., that are autonomous and independent of the system.

A revolutionary movement should be able to form reference groups that would offer values more satisfying than the system's hedonism. Wild nature perhaps would be the central value, or one of the central values.

In any case, where people belong to a close-knit reference group, they become largely immune to the system's propaganda to the extent that that propaganda conflicts with the values and beliefs of the reference group.[19] The reference group thus is one of the most important tools by means of which a revolutionary movement can overcome the system's propaganda.

(iv) Because the system needs an orderly and docile population, it must keep aggressive, hostile, and angry impulses under firm restraint. There is a good deal of anger toward the system itself, and the system needs to keep this kind of anger under especially tight control. Suppressed anger therefore is a powerful psychological force that a revolutionary movement should be able to use against the system.

(v) Because the system relies on cheap propaganda and requires willful blindness to the grim prospect that continued technological progress offers, a revolutionary movement that develops its ideas carefully and rationally may gain a decisive advantage by having reason on its side. I've pointed out previously that reason *by itself* is a very weak tool for influencing people in the mass. But I think nevertheless that if a movement gives ample attention to the non-rational factors that affect human behavior, it may profit enormously in the long run by having its key ideas established on a solidly rational foundation. In this way the movement will attract rational, intelligent people who are repelled by the system's propaganda and its distortion of reality. Such a movement may draw a smaller number of people than one that relies on a crude appeal to the irrational, but I maintain that a modest number of high-quality people will accomplish more in the long haul than a large number of fools. Bear in mind that rationality does not preclude a deep commitment or a powerful emotional investment.

Compare Marxism with the irrational religious movements that have appeared in the U.S. The religious movements achieved little or nothing of lasting importance, whereas Marxism shook the world. Marxism to be sure had its irrational elements: To many people belief in Marxism served as an equivalent of religious faith. But Marxism was far from being wholly irrational, and even today historians recognize Marx's contribution to the understanding of the effect of economic factors on history. From the perspective of the 19th and early 20th centuries, Marxism was plausible and highly relevant to the problems of the time, hence it attracted people of an entirely different stamp from those who were drawn to religious revivals.

It's possible however that faith in Marxism as dogma may have played an essential role in the success of the Russian Revolutionary movement. I read somewhere years ago that Lenin himself did not believe dogmatically in Marxist doctrine, but considered it inexpedient to challenge the faith of the true believers,[20] and I suspect that the same must have been true of others among the more rational and intelligent Marxists of Lenin's time. It may be that a movement should not try to impose too rigid a rationality on its adherents, but should leave room for faith. If the movement's ideology has an underlying rational basis, I would guess that it should be able to attract rational and intelligent people notwithstanding a certain amount of nonrational or irrational ideological superstructure. This is a delicate question, and the answer to it can be worked out only through trial and error. But I still maintain that a largely rational basis for its position should give a revolutionary movement a powerful advantage vis-à-vis the system.

In any case, the kind of people who constitute the movement will be of decisive importance. The biggest mistake that such a movement could make would be to assume that the more people it has, the better, and to encourage everyone who might be interested to join it. This is exactly the mistake that was made by the original Earth First! As it was originally constituted in the early 1980s, Earth First! may have had the makings of a genuine revolutionary movement. But it indiscriminately invited all comers, and—of course!—the majority of comers were leftish types. These swamped the movement numerically and then took it over, changing its character. The process is documented by Martha F. Lee, *Earth First!: Environmental*

Apocalypse, Syracuse University Press, 1995. I do not believe that Earth First! as *now* constituted is any longer a potentially revolutionary movement.

The green anarchist/anarcho-primitivist movement, in addition to attracting leftish types, manifests another kind of personnel problem: It has attracted too many people who are mentally disorganized and seriously deficient in self-control, so that the movement as a whole has an irrational and sometimes childish character, as a result of which I think it is doomed to failure.

Actually there are some very good ideas in the green anarchist/anarcho-primitivist movement, and I believe that in certain ways that movement takes the right approach. But the movement has been ruined by an excessive influx of the wrong kinds of people.

So a critically important problem facing a nascent revolutionary movement will be to keep out the leftists, the disorganized, irrational types, and other unsuitable persons who come flocking to any rebel movement in America today.

Probably the hardest part of building a movement is the very first step: One has to collect a handful of strongly committed people of the right sort. Once that small nucleus has been formed, it should be easier to attract additional adherents.

A point to bear in mind, however, is that a group will not attract and hold adherents if it remains a mere debating society. One has to get people involved in practical projects if one wants to hold their interest. This is true whether one intends to build a revolutionary movement or one directed merely toward reform. The first project for the initial handful of people would be library research and the collection of information from other sources. Information to be collected would include, for example, historical data about the ways in which social changes have occurred in past societies, and about the evolution of political, ideological, and religious movements in those societies, information about the development of such movements in our own society during recent decades; results of scholarly studies of collective behavior; and data concerning the kinds of people involved in Earth First!, green anarchism, anarcho-primitivism, and related movements today. Once the group had gathered sufficient information it could design a provisional program of action, perhaps modifying or discarding many of the ideas I've outlined on the preceding pages.

But for anyone who seriously wants to do something about the technology problem, the initial task is quite clear: It is to build a nucleus for a new movement that will keep itself strictly separate from the leftists and the irrational types who infest the existing anti-technological movement.

Letter to David Skrbina, October 12, 2004

1. I'll begin by summarizing some information from Martin E. P. Seligman, *Helplessness: On Depression, Development, and Death*. Here I have to rely on memory, because I do not have a copy of Seligman's book, nor do I have extensive notes on it. Seligman arrived at the following conclusions through experiments with animals:

Take an animal, subject it repeatedly to a painful stimulus, and each time block its efforts to escape from the stimulus. The animal becomes frustrated. Repeat the process enough times, and the state of frustration gives way to one of depression. The animal just gives up. The animal has now acquired "learned helplessness." If, at a later time, you subject the animal to the same painful stimulus, it will not try to escape from the stimulus even if it could easily do so.

Learned helplessness can be unlearned. I don't recall the details, but the general idea is that the animal gets over learned helplessness by making *successful* efforts.

Both learning and unlearning of helplessness occur within the specific area of behavior in which the animal is trained. For example, if an animal acquires learned helplessness through repeated frustration of its efforts to escape from electrical shocks, it will not necessarily show learned helplessness in relation to efforts to get food. But learned helplessness does to some extent carry over from one area to another: If an animal acquires learned helplessness in relation to electrical shocks, subsequently it will more easily become discouraged when its efforts to get food are frustrated. The same principles apply to *un*learning of helplessness.

An animal can be partly "immunized" to learned helplessness: If an animal is given prior experience in overcoming obstacles through effort, it will be much more resistant to learned helplessness (hence also to depression) than an animal that has not had such experience. For example, if caged pigeons are able to get food only by pushing a lever on an apparatus that gives them one grain of wheat or the like for each push of the lever, then they will later acquire learned helplessness much less easily than pigeons that have not had to work for their food.

My memory of the following is not very clear, but I think Seligman indicates that laboratory rats and wild rats differ in that wild rats are far more energetic and persistent than laboratory ones in trying to save themselves in a desperate situation. Presumably the wild rats have been immunized to learned helplessness through successful efforts made in the course of their earlier lives.

At any rate, it does appear that purposeful effort plays an essential role in the psychological economy of animals.

I first read Seligman's book in the late 1980s. The book originally came out in the early 1970s, and I haven't had much opportunity to read later work on learned helplessness. But the theory is believed to be valid also for human beings, and I believe it is the subject of continuing work.

I don't necessarily accept a psychological theory just because some psychologists say it's true. There's a lot of nonsense in the field, and even experimental psychologists sometimes draw silly conclusions from their data. But the theory of learned helplessness squares very neatly with my own personal experience and with my impressions of human nature gained from observation of others.

The need for purposeful, successful effort implies a need for competence, or a need to be able to exercise control, because one's goals can't be attained if one does not have the competence, or the power to exercise control, that is necessary to reach the goals. Seligman writes:

"Many theorists have talked about the need or drive to master events in the environment. In a classic exposition, R. W. White (1959) proposed the concept of *competence*. He argued that the basic drive for control had been overlooked by learning theorists and psychoanalytic thinkers alike. The need to master could be more pervasive than sex, hunger, and thirst in the lives of animals and men.... J. L. Kavanau (1967) has postulated that the drive to resist compulsion is more important to wild animals than sex, food, or water. He found that captive white-footed mice spent inordinate time and energy just resisting experimental manipulation. If the experimenters turned the lights up, the mouse spent his time setting them down. If the experimenters turned the lights down, the mouse turned them up."[21]

This suggests a need not only for power but for autonomy. In fact, such a need would seem to be implied by the need to attain goals through effort; for if one's efforts are undertaken in subordination to another person, then those efforts will be directed toward the other person's goals rather than toward one's own goals.

Yet the inconvenient fact is that human individuals seem to differ greatly in the degree of autonomy that they need. For some people the drive for autonomy is very powerful, while at the other extreme there are people who seem to need no autonomy at all, but prefer to have someone else do their thinking for them. It may be that these people, automatically and without even willing it, accept as their own goals whatever goals are set up for them by those whose authority they recognize. Another view might be that for some reason certain people need purposeful effort that exercises their powers of thinking and decision-making, while other people need only to exercise their physical and their strictly routine mental capacities. Yet another hypothesis would be that those who prefer to have others set their goals for them are persons who have acquired learned helplessness in the area of thinking and decision-making.

So the question of autonomy remains somewhat problematic. In any case, it's clear how ISAIF's concept of the power process is related to the foregoing discussion. As ISAIF explains in §33, the need for the power process consists in a need to have goals, to make efforts toward those goals, and to succeed in attaining at least some of the goals; and most people need a greater or lesser degree of autonomy in pursuing their goals.

If one has had insufficient experience of the power process, then one has not been "immunized" to learned helplessness, hence one is more susceptible to helplessness and consequently to depression. Even if one has been immunized, long-continued inability to attain goals will cause frustration and will lead eventually to depression. As any psychologist will tell you, frustration causes anger, and depression tends to produce guilt feelings, self-hatred, anxiety, sleep disorders, eating disorders, and other symptoms. (See ISAIF, §44 and Note 6.) Thus, if the theory of learned helplessness is correct, then ISAIF's definition of "freedom" in terms of the power process is not arbitrary but is based on biological needs of humans and of animals.

This picture has support in other quarters. The zoologist Desmond Morris, in his book *The Human Zoo*, describes some of the abnormal behavior shown by wild animals when they are confined in cages, and he explains the prevalence of abnormal behavior (e.g., child abuse and sexual perversion) among modern people by comparing present-day humans to zoo animals: Modern society is our "cage." Morris shows no awareness of the theory of learned helplessness, but much of what he says dovetails very nicely with that theory. He even mentions "substitute activities" that are equivalent to ISAIF's "surrogate activities."

The need for power, autonomy, and purposeful activity is perhaps implicit in some of Ellul's work. Shortly after my trial, a Dr. Michael Aleksiuk sent me a copy of his book *Power Therapy*, which contains ideas closely related to that of the power process. A major theme of Kenneth Keniston's study *The Uncommitted* is the sense of purposelessness that afflicts many people in the modern world. I think he mentions an "instinct of workmanship," meaning a need to do purposeful work. In the first part of his book *Growing Up Absurd*, Paul Goodman discusses as a source of social problems the fact that men no longer need to do hard, demanding work that is essential for survival. Reviewing a book by Gerard Piel, Nathan Keyfitz wrote:

"Among other signs of the lack of adaptation [in modern society] is... purposelessness. Our ancestors, whose work was hard and often dangerous, always necessary simply to keep alive, seemed to know what they were here for. Now 'anomie and preoccupation with the isolated self recur as a central theme of U.S. popular culture. That they find resonance in every other industrial country suggests that the solving of the economic problem brings on these quandaries everywhere.'"[22]

Thus, I argue that the power process is not a luxury but a fundamental need in human psychological development, and that disruption of the power process is a critically important problem in modern society.

Because of my lack of access to good library facilities I haven't been able to explore the relevant psychological literature to any significant extent, but for anyone interested in modern social problems such an exploration should be well worth the time it would cost.

In answering your letters I'm not going to stick rigidly with the definition of freedom given in ISAIF, §94, but I will assume throughout that the kind of freedom that really matters is the freedom to do things that have important practical consequences, and that the freedom to do things merely for pleasure, or for "fulfillment," or in pursuit of surrogate activities, is relatively insignificant. See ISAIF, §72.

"Human dignity" is a very vague term and a broadly inclusive one. But I will assume that one essential element of human dignity is the capacity to exert oneself in pursuit of important, practical goals that one has selected either by oneself or as a member of a small, autonomous group. Thus, both freedom and dignity, as I will use those terms, are closely involved with the power process and with the associated biological need.

II. You ask for a "core reason" why things are getting worse. There are two core reasons.

A. Until roughly ten thousand years ago, all people lived as hunter-gatherers, and that is the way of life to which we are adapted physically and mentally. Many of us, including some Europeans,[23] lived as hunter-gatherers much more recently than ten thousand years ago. We may have undergone some genetic changes since becoming agriculturalists, but those changes are not likely to have been massive.[24] Hunter-gatherers who survived into modern times were people very much like ourselves.

As technology has advanced over the millennia, it has increasingly altered our way of life, so that we've had to live under conditions that have diverged more and more from the conditions to which we are adapted. This growing maladaptation subjects us to an ever-increasing strain. The problem has become particularly acute since the Industrial Revolution, which has been changing our lives more profoundly than any earlier development in human history. Consequently, we are suffering more acutely than ever from maladaptation to the circumstances in which we live. (Robert Wright has developed this thesis in an article that you might be interested to read.)[25]

I argue that the most important single maladaptation involved derives from the fact that our present circumstances deprive us of the opportunity to experience the power process properly. In other words, we lack freedom as the term is defined in ISAIF, §94.

The argument that "people now have more freedom than ever" is based on the fact that we are allowed to do almost anything we please *as long as it has no practical consequences*. See ISAIF, §72. Where our actions have practical consequences that may be of concern to the system (and few important practical consequences are not of concern to the system), our behavior, generally speaking, is closely regulated. Examples: We can believe in any religion we like, have sex with any consenting adult partner, take a plane to China or Timbuktu, have the shape of our nose changed, choose any from a huge variety of books, movies, musical recordings, etc., etc., etc. But these choices normally have no important practical consequences. Moreover, they do not require any serious effort on our part. We don't change the shape of our own nose, we pay a surgeon to do it for us. We don't go to China or Timbuktu under our own power, we pay someone to fly us there.

On the other hand, within our own home city we can't go from point A to point B without our movement being controlled by traffic regulations, we can't buy a firearm without undergoing a background check, we can't change jobs without having our background scrutinized by prospective employers, most people's jobs require them to work according to rules, procedures, and schedules prescribed by their employers, we can't start a business without getting licenses and permits, observing numerous regulations, and so forth.

Moreover, we live at the mercy of large organizations whose actions determine the circumstances of our existence, such as the state of the economy and the environment, whether there will be a war or a nuclear accident, what kind of education our children will receive and what media influences they will be exposed to. Etc., etc., etc.

In short, we have more freedom than ever before to *have fun*, but we can't intervene significantly in the life-and-death issues that hang over us. Such issues are kept firmly under the control of large organizations. Hence our deprivation with respect to the power process, which requires that we have *serious* goals and the power to reach those goals through our own effort.

B. The second "core reason" why things are getting worse is that there is no way to prevent technology from being used in harmful ways, especially because the ultimate consequences of any given application of technology commonly cannot be predicted. Therefore, harm cannot be foreseen until it is too late.

Of course, the consequences of primitive man's actions may often have been unpredictable, but because his powers were limited, the negative consequences of his actions also were limited. As technology becomes more and more powerful, even the unforeseeable consequences of its well-intentioned use,—let alone the consequences of its irresponsible or malicious use— become more and more serious, and introduce into the world a growing instability that is likely to lead eventually to disaster. See Bill Joy's article, "Why the Future Doesn't Need Us," *Wired* magazine, April 2000, and Martin Rees, *Our Final Century*.

III. A. *"Objective" factors in history.* I assert that the course of history, in the large, is normally determined primarily by "objective" factors rather than by human intentions or by the decisions of individuals. Human intentions or the decisions of individuals may occasionally make a major, long-term difference in the course of history, but when this happens the results do not fulfill the intentions of the individuals or groups that have made the decisions. Some exceptions, however, can be identified. Human intentions can sometimes be realized in the following three ways (see my letter of 1/2/04): (i) Intelligent administration may prolong the life of an existing social order. (ii) It may be possible to cause, or at least to hasten, the breakdown of an existing social order. (iii) An existing social order can sometimes be extended so as to encompass additional territory.[26]

I need to explain what the foregoing means. Human intentions often are realized, even for a long period, with respect to some particular factor in society. But, in such cases, human intentions for the society as a whole are not realized.

For example, in the Soviet Union the Communists achieved some of their goals, such as rapid industrialization, full employment, and a significant reduction in social inequality, but the society they created was very different from what the Bolsheviks had originally intended. (And in the *long* run the socialist system failed altogether.) Since the onset of the Industrial Revolution in the 18th century, people have succeeded in achieving material abundance, but the result is certainly not the kind of society that was envisioned by 18th-century proponents of progress. (And today people like Bill Joy and Martin Rees fear that industrial society may not survive much longer.) The Prophet Mohammed succeeded in establishing his new religion as the faith of millions of people; that religion has flourished for nearly fourteen centuries and may well do so for many centuries more. But: "At the end of the rule of the 'rightly guided' caliphs, the Prophet's dream of ushering in a new era of equality and social justice remained unfulfilled…,"[27] nor has that dream been fulfilled today.

To explain further what I mean when I say that history is generally guided by "objective" factors and not by human intentions or human will, I'll use an example that presents the issue in simplified form.

Given three factors:

(i) the presence of hunting-and-gathering bands at the eastern extremity of Siberia;

(ii) the presence of good habitat for humans at the western extremity of Alaska; and

(iii) the existence of a land-bridge across what is now the Bering strait, the occupation of the Americas by human beings was a historical inevitability and was in a certain sense independent of human intention and of human will.

Of course, human intentions were involved. In order for the Americas to be occupied, some hunting-and-gathering band at some point had to choose intentionally to move eastward across the land-bridge. But the occupation of the Americas did not depend on the intentions of any one hunting-and-gathering band—or any dozen bands—because, given the three conditions listed above, it was inevitable that *some* band sooner or later would move across the land-bridge. It is in this sense that major, long-term historical developments normally result from the operation of "objective" factors and are independent of human intentions.

The foregoing does not mean that history is rigidly deterministic in the sense that the actions of individuals and small groups can *never* have an important, long-term effect on the course of events. For example, if the period during which the Bering Strait could be crossed had been short, say 50 or 100 years, then the decision of a single hunting-and-gathering band to cross or not to cross to Alaska might have determined whether Columbus would find the Americas populated or uninhabited. But even in this case the occupation of the Americas would not have been a realization of the intentions of the single band that made the crossing. The intention of that band would have been only to move into one particular patch of desirable habitat, and it could have had no idea that its action would lead to the occupation of two great continents.

B. *Natural selection*. A principle to bear in mind in considering the "objective" factors in history is the law of what I call "natural selection": Social groups (of any size, from two or three people to entire nations) having the traits that best suit them to survive and propagate themselves, are the social groups that best survive and propagate themselves. This of course is an obvious tautology, so it tells us nothing new. But it does serve to call our attention to factors that we might otherwise overlook. I have not seen the term "natural selection" used elsewhere in connection with this principle, but the principle itself has not gone unnoticed. In the *Encyclopædia Britannica* we find:

"These processes were not inevitable in the sense that they corresponded to any 'law' of social change. They had the tendency, however, to spread whenever they occurred. For example, once the set of transformations known as the agrarian revolution had taken place anywhere in the world, their extension over the rest of the world was predictable. Societies that adopted these innovations grew in size and became more powerful. As a consequence, other societies had only three options: to be conquered and incorporated by a more powerful agrarian society; to adopt the innovations; or to be driven away to the marginal places of the globe. Something similar might be said of the Industrial Revolution and other power-enhancing innovations, such as bureaucratization and the introduction of more destructive weapons."[28]

Notice that there is a difference between the "natural selection" that operates among human groups and the natural selection that we are familiar with in biology. In biology, more successful organisms simply replace less successful ones and are not imitated by them. But in human affairs less successful groups tend to try to imitate more successful ones. That is, they try to adopt the social forms or practices that appear to have made the latter groups successful. Thus, certain social forms and practices propagate themselves not only because groups having those forms and practices tend to replace other groups, but also because other groups adopt those forms and practices in order to avoid being replaced. So it is probably more correct to describe natural selection as operating on social forms and practices rather than as operating on groups of people.

The principle of natural selection is beyond dispute because it is a tautology. But the principle could produce misleading conclusions if applied carelessly. For example, the principle does not a priori exclude human will as a factor guiding history.

C. *Human will versus "objective" forces of history.* In Western Europe, until recently, bellicosity—a readiness and ability to make war—was an advantageous trait in terms of "natural selection": Militarily successful nations increased their power and their territory at the expense of other nations that were less successful in war. However, I think this is no longer true, because there is a strong consensus in Western Europe today that war between two Western European nations is absolutely unacceptable. Any nation that initiated such a war would be pounced upon by all the rest of Western Europe and soundly defeated. Thus, in Western Europe, bellicosity (at least as directed against other Western European nations), is now a *dis*advantageous trait in terms of natural selection, and it is so because of the human will to avoid war in Western Europe. This shows that human will can be a "selective force" involved in the process of "natural selection" as it operates in human affairs.

However (to the extent that it does not rely on the U.S. for protection) Western Europe as a whole still needs to be prepared for war, because outside Western Europe there exist other entities (nations or groups of nations) that might well make war on Western Europe if they thought they could get away with it. As it is, if any nation outside Western Europe made war on a Western European nation, and if the latter were unable to defend itself adequately, the rest of Western Europe would help it to defeat the aggressor. Thus, by eliminating *internal* warfare and acquiring a certain degree of unity, Western Europe has become more formidable in war against any outside entity.

What has happened in Western Europe is simply a continuation of a process that has been going on for thousands of years: Smaller political entities group together (whether voluntarily or through conquest) to form a larger political entity that eliminates internal warfare and thereby becomes a more successful competitor in war against other political entities. Size does not always guarantee survival (e.g., consider the breakup of the Roman Empire), but in the course of history smaller political entities generally have tended to coalesce to form larger and therefore militarily more powerful ones; and this process is not dependent on human intention but results from "natural selection."

Thus, when we take a relatively localized view of history and consider only Western Europe over the last several decades, human will appears to be an important factor in the process of natural selection, but when we take a broader view and look at the whole course of history, human will appears insignificant: "Objective" factors have determined the replacement of smaller political entities by larger ones.

Of course, it's conceivable that human will might some day eliminate war altogether. A world government might not even be necessary. It would be enough that there should exist a strong worldwide consensus, similar to the consensus now existing in Western Europe, that war was unacceptable and that any nation initiating a war should be promptly crushed by all the other nations. Bellicosity would then become a highly disadvantageous trait in terms of natural selection. And, since the whole world would be encompassed by the consensus, there would be no outside competitor left against whom it might be necessary to make war.

But you can see how difficult it is to reach the necessary consensus. Efforts to end war have been going on at least since the end of World War I with the League of Nations, and outside of Western Europe there has been little progress in that regard. Moreover, even if conventional warfare could be ended through an international consensus, organized violence might well continue, because there are forms of organized violence (e.g., guerrilla warfare, terrorism) that would be extremely difficult to suppress even if vigorously opposed by every nation on Earth.

The purpose of the foregoing discussion is not to prove that it is never possible for human will to change the course of history. If I didn't believe it were possible, then I wouldn't waste my time writing letters like this one. But we have to recognize how powerful the "objective" forces of history are and how limited is the scope for human choice. A realistic appraisal will help us to discard solutions that appear desirable but are impossible to put into practice, and concentrate our attention on solutions that may be less than ideal but perhaps have a chance of success.

D. *Democracy as a product of "objective" forces.* In your letter of 7/27/04, you and your colleague offer "democracy" as an example of an improvement in the human condition brought about by "human action." I assume that by "democracy" you mean representative democracy, i.e., a system of government in which people elect their own leaders. And I assume that in referring to "human action" you mean that representative democracy became the dominant form of government in the modern world through a process that more or less fits the following model: problem perceived—solution devised—solution implemented— problem solved. If this is what you mean, then I think you are wrong.

I think the problem of political oppression has been perceived for thousands of years. Presumably, people have resented political oppression ever since the beginning of civilization; this is indicated by numerous peasant revolts and the like that have been recorded in history. If representative democracy is the solution to the problem of political oppression, then the solution, too, has long been known and sometimes implemented. The idea and the practice of representative democracy go back at least to ancient Athens, and may well go back to prehistoric times, for some of the aborigines of southeastern Australia practiced representative democracy.[29]

Sixteenth-century Cossacks had "a military organization of a peculiarly democratic kind, with a general assembly (*rada*) as the supreme authority and elected officers, including the commander in chief...."[30] Seventeenth-century buccaneers elected their own captains, who could be deposed by the crew at any time when an enemy was not in sight.[31] Fifteenth-century Geneva had a democratic government, though perhaps not strictly speaking a representative democracy since the legislative body consisted of all citizens.[32] In addition to fully democratic systems, there have been some partially democratic ones. Under the Roman Republic, for example, public officials were elected by the assembled people, but the aristocratic Senate was the dominant political force.[33]

Thus, representative democracy has been tried with varying degrees of success at many times and places. Nevertheless, among preindustrial civilized societies the dominant forms of government remained the monarchical, oligarchic, aristocratic, and feudal ones, and representative democracy was only a sporadic phenomenon. Clearly, under the conditions of preindustrial civilization, democracy was not as well adapted for survival and propagation as other forms of government were. This could have been due to internal weakness (instability, or a tendency to transmute into other forms of government), or to external weakness (a democratic government may have been unsuccessful in competing economically or militarily with its more authoritarian rivals).

Whatever it was that made preindustrial democracy weak, the situation changed with the advent of the Industrial Revolution. Suddenly people began to admire the (semi-)democratic systems of Britain and the United States, and attempts were made to imitate those systems. If Britain had been economically poor and militarily weak, and if the United States had been a stagnant backwater, would their systems have been admired and imitated? Not likely! Britain was economically and militarily the most successful nation in Europe, and the United States was a young but dynamically growing country, hence these two countries excited the admiration and envy of the propertied classes in other countries. It was the propertied classes, not the laboring classes, who were primarily responsible for the spread of democracy. That's why Marxists always referred to the democratic revolutions as "bourgeois revolutions."

The democracies had to survive repeated contests with authoritarian systems, and they did survive, largely because of their economic and technological vigor. They won World Wars I and II, and they didn't do so because soldiers were more willing to fight for a democratic than for an authoritarian government. No one has ever questioned the bravery or the fighting spirit of the German and Japanese soldiers. The democracies won largely because of their industrial might.[34]

Notice that fascism was popular, even to some extent in the U.S.,[35] between the two World Wars. (Here I use the term "fascism" in its generic sense, not referring specifically to Mussolini's Fascists.) After World War II, fascism lost its popularity. Why? Because the fascists lost the war. If the fascists had won, fascism undoubtedly would have been admired and imitated.

During much of the Cold War, "socialism" was the watchword throughout the Third World. It represented the state of bliss to which most politically-conscious people there aspired. But that lasted only as long as the Soviet Union appeared to be more dynamic and vigorous than the U.S. When it became clear that the Soviet Union and other socialist countries could not keep up with the West economically or technologically, socialism lost its popularity, and the new watchwords were "democracy" and "free market."

Thus, democracy has become the dominant political form of the modern world not because someone decided that we needed a more humane form of government, but because of an "objective" fact, namely, that under the conditions created by industrialization, democratic systems are more vigorous technologically and economically than other systems.

Bear in mind that, as technology continues to progress, there is no guarantee that representative democracy will always be the political form best adapted to survive and propagate itself. Democracy may be replaced by some more successful political system. In fact, it could be argued that this has already happened. It could plausibly be maintained that, notwithstanding the continuation of democratic forms such as reasonably honest elections, our society is really governed by the elites that control the media and lead the political parties. Elections, it might be claimed, have been reduced to contests between rival groups of propagandists and image-makers.

Letter to David Skrbina, November 23, 2004

Are things bad and getting worse, and is technology primarily responsible?

A. Arguments that technology has made things bad and is making them worse are presented throughout ISAIF (the Manifesto), as well as in the writings of Jacques Ellul, Lewis Mumford, Kirkpatrick Sale, and others. Your colleague has not addressed these arguments in any specific way. The only substantive arguments that he offers are the four examples of ways in which things are allegedly getting better. I would be perfectly justified in dismissing these four examples by pointing out that neither I nor any responsible commentator has claimed that technology makes *everything* worse—everyone knows that technology does some good things. I could then simply refer your colleague to ISAIF, Ellul, etc., for arguments that the evil done by technology outweighs the good, and challenge him to answer those arguments, which so far he has not attempted to do.

Nevertheless, I will consider the four examples in detail (below) because they offer scope for interesting discussion, and I will make your colleague's question about whether things are bad and getting worse into an opportunity to supplement some of the arguments offered in ISAIF and elsewhere.

B. Obviously, any determination as to whether things are bad and getting worse, and, if so, how bad, involves value judgments, so the question will have no answer that will be provably correct independently of the system of values that is applied.

I should mention by the way that in order to justify revolution it is not necessary, in my opinion, to prove that things will get worse: With respect to concerns that could be grouped under the very broad rubric of "freedom and dignity," things are *already* bad enough to justify revolution. This is another value-judgment, and I feel safe in assuming that it would be a waste of time to try to persuade your colleague to agree with it. Even so, I do not think it will be an idle exercise to call attention here to some facts that are relevant to the questions of whether things are bad and whether they are getting worse.

C. First let me point out that the answers to your questions as to whether there is a core reason why things are getting worse, and when the downhill trend began, are found in my letter of 10/12/04.

D. Your colleague suggests that "things have *always* been bad for human society, and that we have no rational reason to expect anything better than simply staying one step ahead of death." This is a highly pessimistic attitude, even a defeatist one, and on the basis of my readings about primitive societies I would be rather surprised if such an attitude had been current in any primitive society prior to the time when the society was damaged by the intrusion of civilization. But I actually agree that we have no rational reason to expect anything better than simply staying one step ahead of death—because simply staying one step ahead of death is just fine. We've been adapted by a couple of million years of evolution to a life in which our survival has depended on the success of our daily efforts—efforts that typically were strenuous and demanded considerable skill. Such efforts represented the perfect fulfillment of the power process, and, though the evidence admittedly is anecdotal, such evidence as I've encountered strongly suggests that people thrive best under rugged conditions in which their survival demands serious efforts—provided that their efforts are reasonably successful, and that they make those efforts as free and independent men and women, not under the demeaning conditions of servitude. A few examples:

W. A. Ferris, who lived in the Rocky Mountains as a fur trapper during the 1840s, wrote that the "Free Men" (hunters and trappers not connected with an organized fur-company) "lead[ ] a venturous and dangerous life, governed by no laws save their own wild impulses, and bound[ ] their desires and wishes to what their own good rifles and traps may serve them to procure.... [T]he toil, the danger, the loneliness, the deprivation of this condition of being, fraught with all its disadvantages, and replete with peril, is, they think, more than compensated by the lawless freedom, and the stirring excitement, incident to their situation and pursuits.... Yet so attached to [this way of life] do they become, that few ever leave it, and they deem themselves, nay are, ...far happier than the indwellers of towns and cities...."[36]

Ferris reported that during his own rugged and dangerous life in the mountains he usually felt "resolute, cheerful, contented."[37]

Gontran de Poncins wrote of the Eskimos with whom he lived about 1939–1940:

"[T]he Eskimo is constantly on the march, driven by hunger..."[38]

"[T]hese Eskimos afforded me decisive proof that happiness is a disposition of the spirit. Here was a people living in the most rigorous climate in the world,...haunted by famine...; shivering in their tents in the autumn, fighting the recurrent blizzard in the winter, toiling and moiling fifteen hours a day merely in order to get food and stay alive. ...[T]hey ought to have been melancholy men, men despondent and suicidal; instead, they were a cheerful people, always laughing, never weary of laughter."[39]

The 19th-century Argentine thinker Sarmiento wrote of the gaucho of his time:

"His moral character shows the effects of his habit of overcoming obstacles and the power of nature; he is strong, haughty, energetic...he is happy in the midst of his poverty and his privations, which are not such for him, who has never known greater enjoyments or desired anything higher..."[40]

Sarmiento was not romanticizing the gaucho. On the contrary, he wanted to replace what he called the "barbarism" of the gaucho with "civilization."

These examples are by no means exceptional. There's plenty more in the literature that suggests that people thrive when they have to exert themselves in order to "stay one step ahead of death," and I've encountered very little that indicates the opposite.

E. It would be instructive to compare the psychological state of primitive man with that of modern man, but such a comparison is difficult because, to my knowledge, there were hardly any systematic studies of psychological conditions in primitive societies prior to the time when the latter were disrupted by the intrusion of civilization. The evidence known to me is almost exclusively anecdotal and/or subjective.

Osborne Russell, who lived in the Rocky Mountains in the 1830s and 1840s, wrote:

"Here we found a few Snake Indians comprising 6 men 7 women and 8 or 10 children who were the only Inhabitants of this lonely and secluded spot. They were all neatly clothed in dressed deer and sheep skins of the best quality and seemed to be perfectly contented and happy. ...I almost wished I could spend the remainder of my days in a place like this where happiness and contentment seemed to reign in wild romantic splendor...."[41]

Such impressions of very primitive peoples are not uncommon, and are worth noting. But they represent only superficial observations and almost certainly overlook interpersonal conflicts that would not be evident to a traveler merely passing through. Colin Turnbull, who studied the

Mbuti pygmies of Africa thoroughly, found plenty of quarreling and fighting among them.[42] Nevertheless, his impression of their social and psychological life was on the whole very favorable; he apparently believed that hunter-gatherers were "untroubled by the various neuroses that accompany progress."[43] He also wrote that the Mbuti "were a people who had found in the forest something that made their life more than just worth living, something that made it, with all its hardships and problems and tragedies, a wonderful thing full of joy and happiness and free of care."[44] Turnbull's book *The Forest People* has been called "romantic," but Schebesta, who studied the Mbuti a couple of decades earlier than Turnbull, and who as far as I know has never been accused of romanticism, expressed a similar opinion of the pygmies:

"How many and varied are the dangers, but also the joyous experiences, on their hunting excursions and their innumerable travels through the primeval forest!"[45]

"Thus the pygmies stand before us as one of the most natural of human races, as people who live exclusively in accord with nature and without any violation of their organism. In this they show an unusually sturdy naturalness and heartiness, an unparalleled cheerfulness and freedom from care."[46]

This "freedom from care," or as we would say nowadays, freedom from stress, seems to have been generally characteristic of peoples at the hunting-and-gathering stage or not far beyond it. Poncins's account makes evident the absence of psychological stress among the Eskimos with whom he lived:

"[The Eskimo] had proved himself stronger than the storm. Like the sailor at sea, he had met it tranquilly, it had left him unmoved. ...In mid-tempest this peasant of the Arctic, by his total impassivity, had lent me a little of his serenity of soul."[47]

"Of course he would not worry. He was an Eskimo."[48]

"[My Eskimos'] minds were at rest, and they slept the sleep of the unworried."[49]

In discussing the reasons why many whites during colonial times voluntarily chose to live with the Indians, the historian James Axtell quotes two white converts to Indian life who referred to "the absence [among the Indians] of those cares and corroding solicitudes which so often prevail [among the whites]."[50] As we would put it, the absence of anxiety and stress. Axtell notes that while many whites chose to live as Indians, very few

Indians made the transition in the opposite direction.[51] Information from other sources confirms the attractiveness of Indian life to many whites.[52]

What I've just said about anxiety and stress probably applies to depression as well, though here I'm on shaky ground since I've encountered very little explicit information about depression in primitive societies. Robert Wright, without citing his source, states that "when a Western anthropologist tried to study depression among the Kaluli of New Guinea, he couldn't find any."[53] Though Schebesta met thousands of Mbuti pygmies,[54] he heard of only one case of suicide among them, and he never found or heard of any case of mental illness (Geisteskrankheit), though he did find three persons who were either feeble-minded (schwachsinnig) or peculiar (Sonderling).[55]

Even in classical (Greek & Roman) civilization, depression may have been rare: "Harris illuminatingly comments on the virtual absence of reference to anything like depression in [classical] antiquity."[56]

Needless to say, stress and depression were not completely absent from every hunting-and-gathering society. Depression and suicide could occur among Poncins's Eskimos, at least among the old people.[57] The Ainu (hunter-gatherers who were nearly sedentary)[58] suffered from such anxiety about following correct ritual procedure that it often led to serious psychological disorders.[59] But look at the psychological condition of modern man:

"About 45 percent of Australian men said they 'often' or 'almost always' felt stress."[60]

"There is certainly a lot of anxiety going around. Anxiety disorder...is the most common mental illness in the U.S. In its various forms...it afflicts 19 million Americans...."[61]

"According to the surgeon general, almost 21 percent of children age 9 and up have a mental disorder, including depression, attention deficit hyperactivity disorder, and bipolar disorder."[62]

"The state of college students' mental health continues to decline. ...The number of freshmen reporting less than average emotional health has been steadily rising since 1985...76 percent of students felt 'overwhelmed' last year while 22 percent were sometimes so depressed they couldn't function. ...85 percent of [college counseling-center] directors surveyed noted an increase in severe psychological problems over the past five years...."[63]

"Rates of major depression in every age group have steadily increased in several of the developed countries since the 1940s. ...Rates of depression, mania and suicide continue to rise as each new birth cohort ages...."[64]

"In the U.S., ...the suicide rate in the age group between 15 and 24 tripled between 1950 and 1990; suicide is the third leading cause of death in this age group."[65]

"A new UC Berkeley study reports that Mexican immigrants to the United States have only about half as many psychiatric disorders as U.S.-born Mexican Americans."[66]

One could go on and on.

F. Psychological problems of course represent only one of the ways in which "things are bad and getting worse." I will discuss a few of the other ways later. I want to make clear, though, that statistics on mental disorders, environmental damage, or other such problems fail to touch certain central issues. Though improbable, it's conceivable that the system might some day succeed in eliminating most mental disorders, cleaning up the environment, and solving all its other problems. But the human individual, however well the system may take care of him, will be powerless and dependent. In fact, the better the system takes care of him, the more dependent he will be. He will have been reduced to the status of domestic animal. See ISAIF, §174 & Note 12. A conscientious owner may keep his house-dog in perfect physical and psychological health. But would you want to be a well cared-for domestic pet? Maybe your colleague would be willing to accept that status, but I would choose an independent and autonomous existence, no matter how hard, in preference to comfortable dependence and servitude.

G. Your colleague's argument that things are getting better because "Humanity is 'flourishing'...based on sheer numbers" makes no sense. One of the principal objections to the technological society is that its food-producing capacity has allowed the world to become grotesquely overcrowded. I don't think I need to explain to you the disadvantages of overcrowding.

H. As for you colleague's claim that the "overall material standard of living seems to be increasing," the way that works is that the technoindustrial system simply defines the term "high standard of living" to mean the kind of living that the system itself provides, and the system then "discovers" that the standard of living is high and increasing. But to me and to many, many other people a high material standard of living consists not in cars, television sets, computers, or fancy houses, but in open spaces, forests, wild plants and animals, and clear-flowing streams. As measured by that criterion our material standard of living is falling rapidly.

IV. Your colleague claims that reform offers a better chance of success than revolution. He claims that "we...would act... to restrict technology as it becomes necessary," and that such action represents "the general pattern." You and your colleague offer four examples to illustrate this general pattern: "slavery," "political oppression," "sanitation and waste disposal," and "air and water pollution."

A. Let's take "political oppression" first.

1. As I argued in my letter to you of 10/12/04, representative democracy replaced authoritarian systems not through human choice or human planning but as a result of "objective" factors that were not under rational human control. Thus the spread of democracy is not an instance of the "general pattern" that you propose.

2. Political oppression has existed virtually since the beginning of civilization, i.e., for several thousand years. An alternative to authoritarian political systems—representative democracy—has been known at least since the days of ancient Athens. Yet, even under the most generous view, the time at which democracy became the world's dominant political form could not possibly be placed earlier than the 19th century. Thus, even after a workable solution was known, it took well over 2,000 years for the problem of political oppression to be (arguably) solved. If it takes 2,000 years for our present technology-related problems to be solved, we may as well forget about it, because it will be far, far too late. So your example of political oppression gives us no reason whatever to be hopeful that our technology-related problems can be solved in a peaceful and orderly way, and in time.

3. You admit that the replacement of authoritarian systems by democratic ones often occurred through revolution, but you claim that "many times it did not (e.g. England, Spain, S. Africa, Eastern European communist bloc)." However, you're wrong about England and S. Africa; or, at best, you can claim you are right about them only by insisting on strict adherence to a technical definition of the term "revolution."

England developed into a full-fledged democracy through a process that took roughly 6 1/2 centuries. Since the process took so long, one can't say it was a revolution. But the process certainly did involve violence and armed insurrection. The first step toward democracy in England was Magna Carta, which became law ca 1225 only through a revolt of the barons and an ensuing civil war (arguably a revolution).[67] At least one other step toward democracy in England required a very violent insurrection, 1642–49 (again, arguably a revolution), and the "revolution" of 1688 was nonviolent only because of the accidental fact that James II declined to fight.[68]

As for South Africa, democracy there *for whites only* goes back to the 19th century and was peacefully established,[69] but whites never comprised more than a fifth of the population,[70] and I assume that what you have in mind is the recent extension of democracy to the entire population. This, however, occurred at least in part through violent revolutionary action.[71] "Resistance by black workers continued, and saboteurs caused an increasing number of deaths and injuries."[72] If the process was not a revolution, then it was saved from being one only by the fact that the government decided to grant democracy to all races through a negotiated settlement rather than let the situation get further out of hand.[73]

In most of the principal nations of Western Europe, democracy was established through revolution and/or war: In England, partly through violent insurrection, as noted above; in France, through revolution (1789, 1830, 1848) and war (1870); in Germany and Italy democracy was imposed from the outside through warfare (World War II). Among the larger Western European nations, only Spain achieved democracy peacefully, in 1976, after Franco's death in 1975. But Spanish democracy clearly was only a spin-off of the democracy that had been established by violence throughout the rest of Western Europe. Spain was an outlier of a thoroughly democratized, powerful, and economically highly successful Western Europe, so it was only to be expected that Spain would follow the rest of Western Europe and become democratic. Would Spain have become democratic if the rest of Western Europe had been fascist? Probably not. So you can't maintain that the democratization of Spain occurred independently of the violence that established democracy throughout the rest of Western Europe.

The same can be said of much of that part of the "Eastern European communist bloc" that actually has become democratic and done so peacefully. Countries like Poland and the Czech Republic lie on the fringes of Western Europe and are very heavily influenced by it. When one looks at Eastern European countries less closely linked with Western Europe, the status of democracy there seems considerably less secure. As far as I know, Serbia has become democratic, but it did not achieve democracy peacefully. I suppose you realize what is happening in Russia: "President Putin continues to move his country away from democracy...," etc.[74] As for Belarus: "Belarussian President Alexander Lukashenko said...that he won a mandate from voters to stay in power in a...referendum scrapping presidential term limits. But foreign observers said the vote process was marred by violations.... That allows the authoritarian president...who has led the nation since 1994, to run again in 2006."[75] "Lukashenko [is] often branded as Europe's last dictator...."[76] In Ukraine, the future of democracy is still uncertain.[77]

So your purported examples of democracy peacefully achieved look rather unimpressive. You would have done better to cite the Netherlands and the Scandinavian countries.[78] The Netherlands' evolution toward democracy was quite peaceful,[79] though seemingly influenced by the violence elsewhere in Europe in 1848.[80] Sweden's evolution toward democracy began early in the 18th century and apparently was entirely peaceful.[81] Norway's democratization seems to have been equally nonviolent;[82] though Norway much of the time was not an independent nation. In Denmark on the other hand I think the absolute monarchy was abolished only as a result of the 1848 revolutions; however, Denmark's progress toward democracy thereafter was reasonably orderly.[83] Note that all of the foregoing countries, as well as England, are Germanic countries. Predominantly Germanic Switzerland, too, adopted democracy readily,[84] though the 1848 revolutions apparently played an important role.[85] Compare this with the often violent and for a long time unsuccessful struggles toward democracy of the Latin and Slavic countries. Germanics seem to take to democracy relatively easily, a point that I will have occasion to mention later. (It's true that in Germany itself the first attempt at democracy—the Weimar Republic—failed, but this can be attributed to peculiarly difficult conditions, namely, the Versailles treaty and disastrous economic problems.)

But what happened in particular countries is somewhat beside the point. Consider the worldwide democratization process as a whole: Democracy was an indigenous and partly violent development in England. It was established in America through a violent insurrection. As I pointed out in my letter of 10/12/04, democracy became the world's dominant political form only because of the economic and technological success of the democracies, especially the English-speaking countries. And this economic and technological success was achieved not only through industrialization at home but also through worldwide expansion that involved violent displacement of native peoples in North America, Australia, and New Zealand, and economic exploitation elsewhere that was often enforced by violence. The democracies repeatedly had to defend themselves in war against authoritarian systems, notably in World Wars I and II, and they won those wars only because of the vast economic and industrial power that they had built, and built in part through violent conquest and exploitation all over the world.

Thus, democracy became the world's dominant political form through a process that involved violent insurrection and extensive warfare, including predatory warfare against weaker peoples who were to be displaced or exploited.

It should also be noted that democracy, as a political form, cannot be viewed in isolation; it is just one element of a whole cultural complex that is associated with industrialization and that we call "modernity." Usually democracy (in its present-day form) can be successfully and lastingly implanted in a country only when that country has become culturally modernized. (India and Costa Rica are probable exceptions.) In my letter of 10/12/04, I maintained that democracy had become the world's dominant political form because it was the political form most conducive to economic and technological success under conditions of industrialization. It might possibly be argued that it is not democracy itself, but other elements of the associated cultural complex that are mainly responsible for economic and technological success. Singapore achieved outstanding economic success without democracy; Spain achieved good and Taiwan achieved excellent economic success even before they were democratized. I still think that democracy as a political form is an important element of the cultural complex that confers success in an industrialized world. But whether it is or not, the fact remains that modern democracy is not a detached phenomenon but a part of a cultural complex that *tends* to be transmitted as a whole.

When a country becomes democratized peacefully, what typically happens is that either the country is so impressed by the success and dominance of the leading democracies that it willingly tries to absorb their culture, including democracy;[86] or else, due to the economic dominance of the democracies, economic forces compel the country to permit the infiltration of modern culture, and once the country has become sufficiently assimilated culturally and economically, it will be capable of democracy.

But in either case the peaceful advent of democracy in any country in modern times (say, since 1900) is usually a consequence of the fact that the cultural complex of which democracy is a part has already become economically and technologically dominant throughout the world. And, as noted above, democracy and modernity have achieved this dominance, in important part, through violence.

So your example of democracy—as an allegedly *nonviolent* reform designed to solve the problem of political oppression—is clearly invalid. I want to make clear that my intention in the foregoing discussion has not been to indict democracy morally, but simply to show that it does not serve your purpose as an example of nonviolent reform.

B. Much of what I've said about the spread of democracy applies also to the elimination of slavery. Since the arguments applicable to slavery are analogous to those I've given in the case of democracy, I'll only sketch them briefly. First note that rejection of slavery, like democracy and industrialization, is a feature of the cultural complex that we call "modernity."

1. I would argue that slavery was (partly)[87] eliminated only because, in the modern world, there are more efficient means of getting people to work. In other words slavery, due to its economic inefficiency, has been eliminated from the industrialized world by "natural selection" (see my letter of 10/12/04), not primarily by human will. True, much slavery was eliminated through conscious humanitarian efforts,[88] but those efforts could not have had much success if slave societies had been more efficient economically than the industrializing countries where the antislavery efforts originated. Hence, the basic cause of the elimination of slavery was economic, not humanitarian.

2. Slavery was widespread for thousands of years before it was (partly) eliminated in modern times. As I pointed out above, we can't afford to wait thousands of years for a solution to our technology-related problems, so your example of slavery gives us no reason to hope for a timely and peaceful solution to those problems.

3. The elimination of slavery was by no means a nonviolent process. Slavery was expunged from Haiti through bloody revolution.[89] Slave revolts occurred repeatedly in at least some slave societies,[90] and, while these revolts rarely achieved lasting success, it seems safe to assume that they contributed to the economic inefficiency of slavery that led to its eventually being superseded by more efficient systems. When slavery was eliminated in modern times, it was often eliminated through *violent* intervention from outside. For example, slavery in the American South was ended by the Civil War, the bloodiest conflict in U.S. history, and the Arab slave trade in Southeast Africa was closed down in 1889 only after war between the slave-dealers and the colonial powers.[91]

So your example of slavery gives us no reason to hope for a *peaceful* solution to anything.

C. Before I address your other two examples, I want to point out that in focusing on isolated, formal features of societies— on whether governments were representative democracies or whether human beings were technically owned as property—you distract attention from more important questions: How much personal freedom did people have in practice and how satisfactory were their lives?

If I had to live in a specified society, would I rather live as a slave or as a non-slave? Of course, I would rather live as a non-slave. Would I prefer that the society's government should be democratic or authoritarian? *All else being equal*, I would prefer that the government should be democratic. For example, if I were to live in Spain I would rather live in Spain as it was in 1976, after democratization, than in Spain as it was in 1974, when Franco was still alive. If I had to live in Rome in AD 100, I would rather live there as a freeman than as a slave.

When the questions are framed as above, democracy and the elimination of slavery appear to be unequivocally beneficial. But, as we've seen, democracy and the elimination of slavery have prevailed not as isolated and detached features but as part of the cultural complex that we call "modernity." So what we really need to ask is: How does the quality of life in modern society compare with that in earlier societies that may have had authoritarian governments or practiced slavery? Here the answer is not so obvious.

Slavery has taken a wide variety of forms, some of which were very brutal, as everyone knows. But: "Various Greek and Roman authors report on how Etruscan slaves dressed well and how they often owned their own homes. They easily became liberated and rapidly rose in status once they were freed."[92] In as much of Spanish America as came under Simón Bolívar's observation, the slave-owner "has made his slave the companion of his indolence"; he "does not oppress his domestic servant with excessive labor: he treats him as a comrade...."[93] "The slave...vegetates in a state of neglect...enjoying, so to speak, his idleness, the estate of his lord, and many of the advantages of liberty; ...he considers himself to be in his natural condition, as a member of his master's family...."[94] Such examples are not rare exceptions,[95] and it will immediately occur to you to ask whether under these conditions slaves might not have been better off than modern wage-workers. But I would go farther and argue that even under the harsher forms of servitude many slaves and serfs had more freedom—the kind of freedom that really counts (see my letter of 10/12/04)—than modern man does. This, however, is not the place to make that argument.

I could make a much stronger argument that nominally free (non-slave, non-serf, etc.) people living under authoritarian systems of past ages often had greater personal freedom—of the kind that counts—than the average citizen of a modern democracy does. Again, this is not the place to make such an argument.

But I do want to suggest here that democracy (in the modern sense of the word) could actually be regarded as a sign of servitude in the following sense: A modern democracy is able to maintain an adequate level of social order with a relatively decentralized power structure and relatively mild instruments of physical coercion only because sufficiently many people are willing to abide by the rules more or less voluntarily. In other words, democracy demands an orderly and obedient population. As the historian Von Laue put it, "Industrial society...requires an incredible docility at the base of its freedoms."[96] I suggest that this is why the Germanic countries adjusted to democracy so easily: Germanic cultures tended to produce more disciplined, obedient, authority-respecting people than the comparatively unruly Latin and Slavic cultures did. The Latins of Europe achieved stable democracies only after experience of industrialized living trained them to a sufficient level of social discipline, and over part of the Slavic world there still is insufficient social discipline for stable democracy. Social discipline is even more insufficient in Latin America, Africa, and the Arabic countries. Democracy succeeded so well in Japan precisely because the Japanese are an especially obedient, conforming, orderly people.

Thus, it could be argued that modern democracy represents not freedom but subjection to a higher level of social discipline,[97] a discipline that is more psychological and based less on physical coercion than old-fashioned authoritarian systems were.

I can't leave the subject of democracy without inviting you to comment on this passage of Nietzsche: "Liberal institutions immediately cease to be liberal as soon as they are attained: subsequently there is nothing more thoroughly harmful to freedom than liberal institutions. ...As long as they are still being fought for, these same institutions produce quite different effects; they then in fact promote freedom mightily. ...For what is freedom? That one has the will to self-responsibility. ...That one has become more indifferent to hardship, toil, privation, even to life. That one is ready to sacrifice men to one's cause, oneself not excepted," *Twilight of the Idols (Götzen-Dämmerung)*, §38 (translation of R. J. Hollingdale).[98]

D. Now let's look at your third example, "Sanitation and waste disposal." It's not clear to me why you chose this particular example. It's just another one of the innumerable technical improvements that have been devised during the last few centuries, and you could equally well have cited any of the others. Of course, none of the responsible opponents of technology has ever denied that technology does some good things, so your example tells us nothing new.

Poor sanitation and inefficient waste disposal were bad for the system *and* bad for people, so the interests of the system coincided with the interests of human beings and it was therefore only to be expected that an effective solution to the problem would be developed.

But the fact that solutions are found in cases where the interests of the system *coincide* with the interests of human beings gives us no reason to hope for solutions in cases where the interests of the system *conflict* with those of human beings.

For instance, consider what happens when skilled craftsmen are put out of work by technical improvements that make them superfluous. I recently received a letter from a professional gravestone sculptor who provided me with a concrete example of this. He had spent years developing skills that were rendered useless a few years ago by some sort of laser-guided device that carved gravestones automatically. He's in his forties, unable to find work, and obviously depressed. This sort of thing has been going on ever since the beginning of the Industrial Revolution, and it will continue to go on because in this situation the interests of the system conflict with those of human beings, so human beings have to give way. Where is the solution that, according to your theory, society is supposed to have developed? As far as I know, only two solutions have been implemented: (i) welfare; and (ii) retraining programs. My guess is that organized retraining programs cover only a fraction of all workers displaced by technology; at any rate, they apparently hadn't covered the gravestone sculptor who wrote to me. But what if they did cover him? "Okay, John, you're 45 years old and the craft you've practiced all your life has just been rendered obsolete by Consolidated Colossal Corporation's new laser-guided stonecutter. But smile and be optimistic, because we're going to put you through a training program to teach you how to operate a ball-bearing-polishing machine...." Your colleague may think this is consistent with human dignity, but I don't, and I'm pretty sure the above-mentioned gravestone sculptor wouldn't think it was consistent with human dignity either.

It's worth mentioning, by the way, that improved sanitation too seems to have had unanticipated negative consequences. Sanitation no doubt is one of the most important factors in the dramatic, worldwide reduction in infant mortality rates, which presumably has played a major role in the population explosion. In addition, improved sanitation may be responsible for allergies and inflammatory bowel disease. There has been a "sharp increase" in allergies over the past few decades, and it is hypothesized that modern sanitation is responsible for this.[99] The idea is that because we are too clean, children's immune systems don't get enough "exercise," so to speak, and therefore fail to develop properly. Though I can't cite the source, I've read something similar about Crohn's disease, a form of inflammatory bowel disease that was virtually unknown until modern times. It is hypothesized that the disease is caused by lack of exposure to intestinal parasites, and one experimental treatment has been based on intentionally infecting patients with certain intestinal worms. I don't know whether the latest research has confirmed these hypotheses and I'm not in a position to dig up the relevant literature.

E. Your fourth example is "air and water pollution." You claim that the (partial) solution to this problem has been acceptable "as defined by the majority."

1. Assuming for the sake of argument that the solution actually has been acceptable to the majority, that means nothing. The great majority of Germans supported Hitler "until the very end."[100]

The majority's opinions about society's problems are to a great extent irrational, for at least two reasons: (i) the majority's outlook is shaped, to a considerable degree, by propaganda. (ii) Most people put very little serious effort into thinking about society's problems. This is not an elitist sneer at the "unthinking masses." The average man's refusal to think seriously about large-scale problems is quite sensible: Such thought is useless to him personally because he himself can't do anything to solve such problems. In fact, some psychologists and physicians have advised people to avoid thinking about problems that they are powerless to solve, because such thinking only causes unnecessary stress and anxiety. It could be argued that people like us, who put substantial time and effort into studying social problems while having only a minimal chance of contributing measurably to the solutions, are freaks. And our thinking may be influenced by propaganda more than we realize or would like to admit.

The point is, however, that the majority's putative acceptance of existing levels of air and water pollution is largely irrelevant.

2. And how do you know that existing levels of air and water pollution are acceptable to the majority? Have you taken a survey? Maybe you simply assume that existing levels of pollution are acceptable to the majority because there currently is very little public agitation over pollution. Though the meaning of the term "acceptable" is not at all clear in this context, it can by no means be assumed that the level of active public resistance is an accurate index of what the public feels is "acceptable." I think most historians would agree that active, organized public resistance is most likely to occur not necessarily when conditions are worst, but when people find new hope that resistance will bring success, or when some other new circumstance or event prods them into action.[101] So the absence of public resistance by no means proves that the majority is satisfied.

3. What the system has done is to alleviate the most visible and obvious signs of pollution, such as murky, stinking rivers and air darkened by smog. Since these symptoms are directly experienced by the average man, they presumably are the ones most likely to arouse public discontent; and while their (partial) cure may inconvenience certain industries it does not significantly impede the progress of the system as a whole. The most successful industrialized countries, for the present, have easily enough economic surplus to cover the cost of controlling the aforementioned visible forms of pollution. But this may not be true of backward countries that are struggling to catch up with the more advanced ones. For example, the air pollution over Mexico City is notoriously horrible.

In fact, if you look beyond the comforting improvements in air-pollution indices over our cities as reported by the EPA and consider the worldwide pollution situation as a whole, it appears that what the system has done to alleviate the problem is almost negligible. The following by the way goes also to support the argument that things are bad and getting worse:

Acid rain (due to certain forms of air pollution) is still damaging our forests. At least up to a few years ago (and perhaps even today) the Russians were still dumping their nuclear waste in the Arctic Ocean. The public (in the U.S.) has been warned not to eat too much fish, because fish are contaminated with mercury and PCB's (from water pollution, obviously). For the foregoing I can't cite a source; I'm depending on memory. But:

"The indigenous populations of Greenland and Arctic Canada are being poisoned by toxic industrial chemicals that drift north by wind and water, polluting their food supplies. On January 13, 2004, *The Los Angeles Times* told its readers that the pollutants, which include PCBs and 200 other hazardous compounds, get into the native food chains through zooplankton. 'The bodies of Arctic people...contain the highest human concentrations of industrial chemicals and pesticides found anywhere on Earth—levels so extreme that the breast milk and tissues of some Greenlanders could be classified as hazardous wastes,' the *Times*' Marla Cone reports."[102]

"In the mid-1980s, some researchers in the northern Midwest, Canada, and Scandinavia began reporting alarming concentrations of mercury in freshwater fish. ...[T]he skies already hold so much mercury that even if industrial emissions of the metal ended tomorrow, significant fallout of the pollutant might persist for decades...."[103]

"Measurable levels of cancer-causing pesticides have been found in the drinking water of 347 towns and cities. Creation and use of toxic chemicals continues at a rate far faster than our capacity to learn how safe extended exposures to these substances are. ...The U.S. Environmental Protection Agency was mandated to test existing pesticides—just one class of chemicals—for health risks by 1972, but the job still isn't completed today, and regulators are falling further behind."[104]

"The new residents [on grounds of former U.S. Clark Air Base, in the Philippines] dug wells, planted crops...unaware that the ground water they drank and bathed in, the soil their rice and sweet potatoes grew in, and the creeks and ponds they fished in were contaminated by toxic substances dumped during a half century of U.S. tenure. Within a few years, health workers began tracking a rise in spontaneous abortions, stillbirths, and birth defects; kidney, skin, and nervous disorders; cancers, and other conditions.... Today, the Pentagon acknowledges polluting major overseas bases, but insists that the Unites States isn't obligated to clean them up."[105]

(On the bright side: "Air-pollution emissions have dropped 7.8% since 2000 [what pollutants are measured, and where, is unstated].... Critics say the drop in water-quality complaints reflects laggard enforcement...."[106])

Anyone who wanted to search the media could go on and on citing things of this sort. And if what I've seen is any indication, he would find vastly more on the negative than on the positive side.

Perhaps the biggest pollution problem of all is global warming, which scientists now agree is due at least in part to human production of "greenhouse gasses," carbon dioxide in particular.[107] It's not just a matter of temperatures rising a few degrees; the consequences of global warming are extremely serious. They include the spread of disease,[108] extreme weather conditions such as storms, tornados, and floods,[109] possible extinction of arctic species such as the polar bear,[110] disruption of the way of life of arctic residents,[111] rising sea levels that will flood parts of the world,[112] and drought.[113] "More of the Earth is turning to dust[.] 'It's a creeping catastrophe', says a U.N. spokesman. Desertification's pace has doubled since the 1970s...."[114] However, global warming is only one of the causes of desertification.[115]

Your colleague's proposed "general pattern" doesn't work here, because you can't just turn something like global warming around when enough people become concerned about it. No matter what measures are taken now, we will be stuck with the consequences of global warming for (at least!) a matter of centuries. In fact, some scientists fear that human modification of the atmosphere may soon "throw a switch" that will trigger a dramatic, disastrous, and irreversible change in the Earth's climate.[116]

Since it is in the system's own interest to keep pollution and global warming under control, it is conceivable that solutions may be found that will prevent these problems from becoming utterly disastrous. But what will be the cost to human beings? In particular, what will be the cost to human freedom and dignity, which so often get in the way of the system's technical solutions?

Letter to David Skrbina, January 3, 2005

First point (freedom). ...I and some other people place an extremely high value on freedom; and I do so because today there is an acute shortage of freedom as I've defined it. If I had grown up in a society in which there was an abundance of freedom but an acute shortage of (for example) physical necessities, I might well have been willing to sacrifice some of my freedom for physical necessities. Poncins says that the Eskimos he knew considered it a reward and not a punishment to be imprisoned, because in prison they were fed and kept warm without having to exert themselves.[117]

Second point (autonomy/freedom). ...I wouldn't say flatly that medieval peasants (for example) had more freedom than we have today, but I think one could make a strong argument that they did have more of the kind of freedom that really counts. See my letters to J. N. (in the Labadie Collection).

Third point (surrogate activities). I've never said that surrogate activities "must be abandoned." Also, the line between surrogate activities and purposeful activities often is not easy to draw. See ISAIF, §§40, 84, 90. And surrogate activities are not peculiar to modern society. What is true is that surrogate activities have come to play an unusual, disproportionate, and exaggerated role in modern society. ...In any case, I don't see that anything would be accomplished by attacking surrogate activities. But I think that the concept of surrogate activity is important for an understanding of the psychology of modern man.

Fourth point (revolution). ...In the present historical context a successful revolution would consist in bringing about the complete dissolution of the technoindustrial system.

Fifth point (reform). Essentially I agree with this, though I wouldn't express it in exactly the same words.

Sixth point (revolution is demanded). Yes, revolution is demanded. I've never said, and I certainly do not believe, that a revolutionary movement must be peaceful and nonviolent. I have simply declined to discuss the violent aspects of revolution, because I don't want to give the authorities an excuse to cut off my communications with you on the ground that I'm "inciting violence." I do think that a revolutionary movement should have one *branch* that will avoid all violent or otherwise illegal activities in order to be able to function openly and publicly. I've never said that a revolution should be led by a "small group," which to me would mean 10, 20, 50, or at most 100 people. (The "Handful" of people I referred to in an earlier letter would be initiators, probably would not retain leadership permanently.) I do think that the active and effective part of a revolutionary movement would comprise only a small fraction of the entire population. Finally, I've never said that the revolution should be led by intellectuals. Of course, that would depend on what one means by an "intellectual." I suppose that term is most commonly taken to include college and university faculty in the humanities and social sciences, and persons in closely related occupations, such as professional writers who write on serious subjects. When the word "intellectual" is understood in that sense, it is my impression that very, very few if any present-day intellectuals are potential members of a revolutionary movement. I can imagine that some intellectuals could play a very important role in formulating, articulating, and disseminating ideas that would subsequently form part of the basis for a revolutionary movement. But in reading *The New York Review*, *The London Review*, and *The Times Literary Supplement* over the last several years I've found virtually *no* mention of the technology problem. It's as if the intellectuals were willfully avoiding what is obviously the most critical issue of our time. That's why I'm so pleased to find at least two intellectuals—yourself and your unnamed colleague— who take a serious interest in the technology problem.

Seventh point (avoidance of stress-reduction). ...I decidedly disagree with your sentence, which says: "In fact, [revolutionaries] should actively OPPOSE such actions...." Absolutely not! Let's take minority rights, for example. The big problem there is that the fuss over minority rights absorbs the rebellious energies of would-be radicals and distracts attention from the critical issue of technology. By *opposing* equal rights for non-whites, women, homosexuals, etc., revolutionaries would merely intensify the fuss over minority rights and thus distract even more attention from the issue of technology. What revolutionaries have to do is show people that the fuss over minority rights is largely irrelevant.

Further, the principle that revolutionaries should work to increase the tensions in society is merely a general rule of thumb, not a rigid law that can be applied mechanically. One has to give separate consideration to each individual case. Are the social tensions arising from discrimination against minorities useful from a revolutionary point of view? Clearly not!

For example, if black people are harassed by police, then their attention will be focused on that problem and they will have no time for the technology problem. Thus, again, problems of minority rights distract attention from the technology problem, and we would be better off if all minority problems had already been solved, because the associated tensions are *not* productive. See ISAIF, §§190-92.

For another example, suppose revolutionaries were to oppose political action designed to reduce pollution. In that case people concerned about pollution would become hostile toward the revolutionaries. Further, tension between opponents of pollution and the system would be *reduced*, because opponents of pollution would attribute continued pollution in part to the obstructive behavior of the revolutionaries. They would say, "The problem is those damned extremists! If it weren't for them, we would be able to swing the system around and reduce pollution." So, instead of *opposing* reformist efforts to reduce pollution, revolutionaries have to emphasize: (i) that such efforts can never really solve the pollution problem, but only alleviate it to a limited extent; (ii) that pollution is only one of many grave problems associated with the technoindustrial system; and (iii) that it is futile to try to attack all of these problems separately and individually—the only effective solution is to bring down the whole system.

The tensions that are useful are the tensions that pit people against the technoindustrial system. Other tensions—e.g, racial tensions, which pit different racial groups against each other rather than against the system—are counterproductive and actually relieve the tension against the system, because they serve as a distraction. See ISAIF, §§190-92.

On page 4 you write that "we should seek *optimum levels* of technology and social order." Several other people who have written to me have raised similar questions about an optimal or acceptable level of technology. My position is that we have *only two choices*. It's like flipping a light-switch. Either your light is on or your light is off, and there's nothing more to be said. Similarly, with only minor reservations and qualifications, we have only two choices at the present point in history: We can either allow the technoindustrial system to continue on its present course, or we can destroy the technoindustrial system. In the first case, technology will eventually swallow everything. In the second case, technology will find its own level as determined by circumstances over which we have no control. Consequently, it is idle to speak of finding an "optimal" level of technology. Any conclusion we might reach about an "optimal" level of technology would be useless, because we would have no means of applying that conclusion in the real world. The same is true of any "optimal" level of social order.

I've read the pieces by Jacques Ellul and Ivan Illich that you sent me. Illich wrote: "If within the very near future man cannot set limits to the interference of his tools with the environment and practice effective birth control, the next generations will experience the gruesome apocalypse predicted by many ecologists." Illich wrote that 32 years ago, and the "apocalypse" is not yet upon us. I think it's safe to say that the system will break down *eventually*—if only because every previous civilization has broken down eventually—and the breakdown when it comes will no doubt be gruesome, but I see no reason to believe that the system is now on the brink of collapse. Dire predictions made by "ecologists" 30-odd years ago have proved to be exaggerated and/or premature.

To me, a lot of what Illich writes is completely incomprehensible. E.g., on page 109 he says: "When business is normal the procedural opposition between corporations and clients usually heightens the legitimacy of the latter's dependence." Can you explain what this sentence means? I find it hopelessly obscure.

As for Ellul, "Anarchy from a Christian Standpoint, 1. What is Anarchy?," I think he's all wrong. It would take too much time to discuss all the ways in which I think he's wrong, so I'll just mention a couple of points. First, he's wrong in claiming that, in history, violence has proven to be an ineffective tactic. Actually violence has been effective or ineffective, depending on the historical circumstances of each particular case. See James F. Kirkham, Sheldon G. Levy, and William J. Crotty, *Assassination and Political Violence: A Report to the National Commission on the Causes and Prevention of Violence*, Praeger Publishers, New York, 1970, page 4. The authors concluded that, in history, systematic assassination had been "effective in achieving the long-range goals sought, although not so in advancing the short-term goals or careers of the terrorists themselves." On this subject the authors go farther than I would.

Second, Ellul writes: "[The] two great characteristics [of people], no matter what their society or education, are covetousness and a desire for power. We find these traits always and everywhere." It's not completely clear to me what Ellul means by "covetousness." But he writes that covetousness "can never be assuaged or satisfied, for once one thing is acquired it directs its attention to something else." So Ellul evidently has in mind a desire to accumulate property indefinitely. If my interpretation of his meaning is correct, then Ellul is dead wrong about covetousness. There have been many societies in which the desire to accumulate property has been absent. E.g., most if not all nomadic hunting-and-gathering societies. To take a concrete case, the Mbuti pygmies: According to Schebesta, "No urge for possession... seems to dwell in them"; "there is also the fact that among the Mbuti, any intention to pile up supplies, or at all to accumulate wealth, is lacking."[118]

The need for power undoubtedly is universal, but it does not have to take the form of a desire to dominate other people, as Ellul seems to assume. It may well be true that an impulse to dominance is innate in humans, especially in males, but I think Ellul greatly overestimates its strength. Moreover, there have existed societies in which any impulse to dominance has been kept well under control: Among the Mbuti, and among the Bushmen studied by Richard Lee, no one was allowed to set himself up above the rest.[119] Thus, these societies came surprisingly close to the anarchist ideal.

Letter to David Skrbina, March 17, 2005

I. WHY REFORM WILL FAIL

You and your colleague make a series of related assertions: We "would act...to restrict technology as it becomes necessary." "People in the future will likely act to mitigate technological advances or effects that begin to significantly undermine their wellbeing." Success in "adequately overcoming technologically-induced adversities" will be more likely through reform than through revolution. There's a "general pattern: A technical problem arises and ... [eventually]... a compromise solution is implemented that reduces the level of harm to 'a generally acceptable level.'"

In my letter of 11/23/04, I answered these claims in part. Addressing your four examples of the purported "general pattern," I argued that even assuming that the achieved solutions to the problems were adequate ones (which in three of the four cases was debatable at best): (i) The "solutions" came about largely through the operation of "objective" factors and independently of human will. (ii) In two of the four cases (political oppression, slavery) the solutions were reached, in important part, through warfare and violent revolution, hence could not fairly be characterized as reform. (iii) In the same two of the four cases, the solutions were not reached until thousands of years after the problems arose. In other words, the solutions did not happen when we needed them, but when the "objective" conditions were by chance right for them.

I.A. The most important point in the foregoing is:

1. The course of history, in the large, is generally determined not by human choice but by "objective" factors, especially by the kind of "natural selection" that I discussed in my letter of 10/12/04. Consequently, we can't achieve a long-lasting solution to a major social problem by superficial tinkering designed merely to correct particular symptoms. If a solution is possible at all, it can be reached only by finding a way to change the underlying "objective" factors that are responsible for the existing situation.

There are several other reasons why acceptable solutions[120] to the problems of the technological society will not be reached through the "general pattern" of compromise and reform that you and your colleague propose.

2. Generally speaking, reform is possible only in cases where the interests of the system coincide with the interests of human beings. Where the interests of the system conflict with those of human beings, there is no meaningful reform.[121] E.g., sanitation has improved because it is in the system's interest to avoid epidemics. But nothing has been done about the unsatisfactory nature of modern work, because if most people worked as independent artisans rather than as cogs in the system, the economic efficiency of the system would be drastically impaired.

"Natural selection" is at work here: Systems that compromise their own power and efficiency for the sake of "human values" are at a competitive disadvantage vis-à-vis systems that put power and efficiency first. Hence, the latter expand while the former fall behind.

3. You claim that people will act to mitigate problems "that begin to significantly undermine their well-being." But often, once a problem begins to significantly undermine people's well-being, it is too late to solve the problem; or even if the problem can be solved the cost of solving it may be unacceptably high.

For example, it is too late to solve the problem of the Greenhouse Effect (global warming). Whatever is done now, we will be stuck with its consequences for centuries to come. We can hope to "solve" the problem only to the extent of keeping the effect within certain limits, and it's not clear that even that much can be done without drastic cuts in energy consumption that will have unacceptable economic consequences.

Apparently the threat represented by nuclear weapons has not undermined people's well-being enough to lead to the abolition of these weapons. If there is ever a major nuclear war, people's well-being will be undermined very dramatically; but then it will be too late.

Right now biotechnicians are playing with fire. The escape from the laboratory of some artificially-created organisms or genetic material could have disastrous consequences, yet nothing is being done to restrain the biotechnicians. If there is ever a major biological disaster, people's well-being will indeed be undermined, but then it will be too late to correct the problem. For example, the so-called "killer bees" are a hybrid of South American and African bees that escaped from a research facility somewhere in South America. Once the bees had escaped, all efforts to stop them proved futile. They have spread over much of South America and into the U.S. and have killed hundreds of people. With the experimentation in biotechnology that is now going on, something much, much worse could happen. See Bill Joy's article.

4. Often a bad thing cannot be fixed because its specific cause is not known. Consider for example the steady increase in the rate of mental disorders that I discussed in my letter of 11/23/04. It seems almost certain that this increase is in some way an outgrowth of technological progress, since the entire lifestyle of modern man is essentially determined by his technology. But no one knows *specifically* why the rate of mental disorders has been increasing. My personal opinion is that the high rate of depression has a great deal to do with deprivation with respect to the power process,[122] but even if I'm right that still leaves a great deal unanswered, e.g., in regard to mania and anxiety disorders.

Again, it is believed that the rate of mortality due to cancer has increased by a factor of more than *ten* since the late 19th century,[123] and that this is not a result merely of the aging of the population. This too is almost certainly in some way an outcome of the technoindustrial lifestyle, but, while some causes of cancer are known, the reason for the overall massive increase in the incidence of this disease is still a mystery.

5. Even where a problem can be solved, the solution itself often is offensive to human dignity. For example, because the causes of depression,

mania and attention-deficit disorder either are unknown or cannot be removed without excessive cost to the system, these problems are "solved" by giving the patients drugs. So the system makes people sick by subjecting them to conditions that are not fit for human beings to live in, and then it restores their ability to function by feeding them drugs. To me, this is a colossal insult to human dignity.

6. Where a problem is of long standing, people may fail to realize even that there is a problem, because they have never known anything better. I've already suggested something like this in regard to stress. See my letter of 5/19/04.

7. Some problems are insoluble because of the very nature of modern technology. For example, the transfer of power from individuals and small groups to large organizations is inevitable in a technological society for several reasons, one of which is that many essential operations in the functioning of the technological system can be carried out only by large organizations. E.g., if petroleum were not refined on a large scale, the production of gasoline would be so costly and laborious that the automobile would not be a practical means of transportation.

8. Your formulations, as quoted on the first page of this letter, rely on such terms as "well-being," "adversities," and "generally acceptable level" of "harm." These terms may be subject to a variety of interpretations, but I assume that what you mean is that when conditions make people sufficiently uncomfortable they will act to reduce their discomfort to an acceptable level. I deny that this is consistently true, but even if it were true it would not solve the problem as I see it.

One of the most dangerous features of the technoindustrial system is precisely its power to make people comfortable (or at least reduce their discomfort to a relatively acceptable level) in circumstances under which they should *not* be comfortable, e.g., circumstances that are offensive to human dignity, or destructive of the life that evolved on Earth over hundreds of millions of years, or that may lead to disaster at some future time. Drugs (as I've just discussed, I.A.5) can alleviate the discomfort of depression and attention-deficit disorder, propaganda can reconcile the majority to environmental destruction, and the entertainment industry gives people forgetfulness so that they won't worry too much about nuclear weapons or about the fact that they may be replaced by computers a few decades from now.

So comfort is not the main issue. On the contrary, one of our most important worries should be that people may be made comfortable with almost anything, including conditions that we would consider horrifying. Perhaps you've read Aldous Huxley's *Brave New World*, a vision of a society in which nearly everyone was supremely comfortable; yet Huxley intended this vision to repel the reader, as being inconsistent with human dignity.

9. What happens is that social norms, and people themselves, change progressively over time in response to changes in society. This occurs partly through a spontaneous process of adaptation and partly through the agency of propaganda and educational techniques; in the future, biotechnology too may alter human beings. The result is that people come to accept conditions that earlier generations would have considered inconsistent with freedom or intolerably offensive to human dignity.

For example, failure or inability to retaliate for an injury was traditionally seen as intensely shameful. To the ancient Romans, it was "the lowest depth of shame to submit tamely to wrongs."[124] To the 17th-century Spanish playwright Calderón de la Barca, a man who had been subjected to a wrong was degraded but could perhaps redeem himself by seeking revenge. The same attitude—that to be wronged is a shame that can be wiped away only through revenge—persists today in the Middle East.[125] In the English-speaking world, even into the early 19th century, duels were fought over points of "honor." (We all know about the famous duel in which Aaron Burr killed Alexander Hamilton, and my recollection is that Andrew Jackson, before he became President, killed a man in a duel.)

Today, however, "revenge" is a bad word. Dueling and private retaliation not only are illegal, but by well-socialized people are seen as immoral. We are expected to submit meekly to an injury or humiliation unless a legal remedy is available through the courts. Of course, it's easy to see why modern society's need for social order makes it imperative to suppress dueling and private revenge.

Prior to the advent of the Industrial Revolution in England and America, police forces were intentionally kept weak because people saw police as a threat to their freedom. People relied for protection not primarily on the police but on themselves, their families and their friends. Effective law enforcement came to be regarded as desirable only as a result of the social changes that the Industrial Revolution brought.[126] Today, needless to say, hardly any respectable middle-class person sees the presence of strong police forces as an infringement of his freedom.

I'm not trying to persuade you to advocate the abolition of police or to approve of dueling and private revenge. My point is simply that attitudes regarding what is consistent with human dignity and freedom have changed in the past in response to the needs of the system, and will continue to change in the future, also in response to the needs of the system. Thus, even if future generations are able to "solve" social problems to the extent necessary to secure what *they* conceive of as human dignity and freedom, their solutions may be totally incompatible with what we would want for our posterity.

10. When a problem persists for a long time without substantial progress toward a solution, most people just give up and become passive with respect to it. (Note the connection with "learned helplessness.") This of course is one of the mechanisms that help bring people to accept what they formerly regarded as intolerable indignities as I described above.

For example, back in the late '50s or early '60s, Vance Packard published a book titled *The Hidden Persuaders*, which was an exposé of the manipulative techniques that advertisers used to sell products or political candidates to consumers or voters. When the book first appeared it received a great deal of attention, and my recollection is that the most common reaction among intellectuals and other thinking people was: "Isn't this scandalous? What is the world coming to when people's attitudes, voting choices, and buying habits can be manipulated by a handful of skilled professional propagandists?" At that time I was in my late teens and was naive enough to believe that, as a result of Packard's book and the attention it received, something would be done about manipulative advertising. Obviously nothing was done about it, and nowadays if anyone published a book about manipulative advertising it wouldn't get much attention. The reaction of most well-informed people would be: "Yeah, sure, we know all that. It's too bad... but what can you do?" They would then drop the unpleasant subject and talk or think about something else. They have lapsed into passive resignation.

Of course, nothing could be done about manipulative advertising because it would have cost the system too much to do anything about it. However insulting it may be to human dignity, the system needs propaganda, and as always happens when the needs of the system come into conflict with human dignity, the system's needs take precedence. (See I.A.2 above.)

11. There is the "problem of the commons": It may be to everyone's advantage that everyone should take a certain course of action, yet it may be to the advantage of each particular individual to take the *opposite* course of action. For example, in modern society, it is to everyone's advantage that everyone should pay a portion of his income to support the functions of government, but it is to the advantage of each particular individual to keep all of his income for himself. (That's why payment of taxes has to be compulsory.)

Similarly, I know people who think the technological society is horrible, that the automobile is a curse, and that we would all be better off if no one used modern technology. Yet they drive cars themselves and use all the usual technological conveniences. And why shouldn't they? If individual X refuses to drive a car, the technological system will go on as before; X's refusal to drive a car will accomplish nothing and will cost him a great deal of inconvenience. For the same reason, X in most cases will not participate in an effort to form a movement designed to remedy some problem of the technological society, because his participation would cost him time and energy, and there is at most a minimal chance that his own personal effort would make the difference between success and failure for the movement. People take action on social problems, even the most important ones, only under special circumstances. See my letter of 11/23/04, Note 101.

12. Most people, most of the time, are not particularly foresighted, and take little account of social dangers that lie decades in the future. As a result, preventive measures commonly are postponed until it is too late.

If I remember correctly, the Swedish chemist Svante Arrhenius predicted the Greenhouse Effect way back in the 19th century; certainly it was predicted at least as early as the 1960s. Yet no one tried to do anything about it until recently, when it was already too late to avoid many of its consequences.

The problem of the disposal of nuclear waste was obvious as soon as the first nuclear power-plants were set up decades ago. No one knew of a safe way to dispose of the waste, but it was simply assumed that a solution to the problem would eventually be found and the development of nuclear power-generation was pushed ahead. Worse still, nuclear power-generation was intentionally introduced to third-world countries under the "Atoms for Peace" program without any apparent consideration of the obvious question whether their often irresponsible little governments would dispose of the wastes safely or whether they would use their nuclear capacity for the development of weapons.

Today, in this country, nuclear wastes are still piling up, and there is every reason to think that they will keep piling up indefinitely. And there is still no generally accepted solution to the problem of disposing of these wastes, which will remain dangerous for many thousands of years. It is claimed that the disposal site at Yucca Mountain in Nevada is safe, but this is widely disputed. Experience has shown again and again that technological solutions, excepting only the most minor innovations, need to be tested before they can be relied on. Usually they work only after they have been corrected through trial and error. The Nevada disposal site is an experiment the result of which won't be known for thousands of years—when it will be too late. Simply on the basis of the demonstrated unreliability of *untested* technological solutions, I would guess it's more likely than not that the Nevada disposal site will prove a failure.

Of course, most people would rather stick future generations with the difficult and perhaps insoluble problem of dealing with our nuclear waste, than accept any substantial reduction in the availability of electricity now.

If the nuclear waste problem in the U.S. is worrisome, you can imagine how some of these irresponsible little third-world countries are disposing of their nuclear waste. Not to mention the fact that some of them have made or are trying to make nuclear bombs. So much for the foresight of the presumably intelligent people who promoted nuclear power-generation several decades ago.

13. The threatening aspects of technology often are balanced by temptingly attractive features. And once people have given in to the temptation of accepting an attractive but dangerous technological innovation, there is no turning back—short of a breakdown of technological civilization. See ISAIF §129. Biotechnology can increase agricultural production and provide new medicines; in the future it will probably help to eliminate genetic diseases and allow parents to give their children desired traits. As computers grow faster and more sophisticated, they give people more and more powers that they would not otherwise have. The latest electronic entertainment media give people new and exciting kicks.

Your claim that people will correct problems when these make them sufficiently uncomfortable, even if it were true, would have no clear application to such cases. Technical innovations make people comfortable in some ways and uncomfortable in other ways, and while the comforts are obvious and direct, the discomforts often are indirect and not obvious. It may be difficult or impossible even to recognize and prove the connection between the technology and the discomfort.

E.g., people directly experience the fun that they get from computers and electronic entertainment media, but it is by no means obvious that exposure of children to computers and electronic media may cause attention-deficit disorder. Some research suggests such an effect, but it remains an open question whether the effect is real. As for the possibility of correcting this problem through reform—let's watch your efforts to curtail the use of computers in the schools. If you have any great success even locally, I think you will be doing very well indeed. And I predict with 99.9% certainty that you will not succeed in curtailing the use of computers in the schools on a nationwide basis.

14. Most people, most of the time, follow the path of least resistance. That is, they do what will make them comfortable for the present and the near future. This tendency deters people from addressing the underlying causes of the discomforts of modern life.

The underlying problems are difficult to attack and can be corrected only at a certain price, so most people take the easy way out and utilize one of the avenues of escape that offer them quick alleviation of their discomfort. For those who are not satisfied simply with immersion in the pleasures provided by the entertainment industry, there are surrogate activities and there are religions, as well as ideologies that serve psychological needs in the same way that religions do. For many who suffer from a sense of powerlessness, it will be more effective to strive for a position of power within the system than to try to change the system. And for those who do struggle against the system, it will be easier and more rewarding to concentrate on one or a few limited issues in regard to which there is a reasonable chance of victory than to address the intractable problems that are the real sources of their discontent.

Consider for example the kook variety of Christianity that has become a serious political force in recent years. I'm referring to people who believe that the world will end within 40 years and that sort of thing (see enclosed article by Bill Moyers).[127] It seems fairly obvious that these people retreat into their fantasy world in order to escape from the anxieties and frustrations of modern life. Who needs to worry about nuclear war or about the environment when the world will end soon anyway, and all the true believers will go to heaven? For those who are disturbed by the decay of traditional morality, it is much easier to fight abortion and gay marriage than to recognize that rapid technological change necessarily leads to rapid changes in social values. See ISAIF §50. In ISAIF §§219-222 and in "The System's Neatest Trick," I argued that the "causes" to which leftists devote themselves similarly represent a form of escapism.

Through recourse to these various forms of escapism, people avoid the need to address the real sources of their discontent.

15. Technological progress brings too many problems too rapidly. Even if we make the extremely optimistic assumption that any one of the problems could be solved through reform, it is unrealistic to suppose that *all* of the most important problems can be solved through reform, and solved in time. Here is a partial list of problems: War (with modern weapons, not comparable to earlier warfare), nuclear weapons, accumulation of nuclear waste, other pollution problems of many different kinds, global warming, ozone depletion, exhaustion of some natural resources, overpopulation and crowding, genetic deterioration of humans due to relaxation of natural selection, abnormally high rate of extinction of species, risk of disaster from biotechnological tinkering, possible or probable replacement of humans by intelligent machines, biological engineering of humans (an insult to human dignity),[128] dominance of large organizations and powerlessness of individuals, surveillance technology that makes individuals still more subject to the power of large organizations,[129] propaganda and other manipulative psychological techniques, psychoactive medications,[130] mental problems of modern life, including, inter alia, stress, depression, mania, anxiety disorders, attention-deficit disorder, addictive disorders, domestic abuse, and generalized incompetence. If you want more, see the enclosed review of books by Jared Diamond and Richard Posner.[131]

The solution of any *one* of the foregoing problems (if possible at all) would require a long and difficult struggle. If your colleague thinks that *all* of these problems can be solved, and solved in time, by attacking each problem separately, then he's dreaming. The only way out is to attack the underlying source of all these problems, which is the technoindustrial system itself.

16. In a complex, highly-organized system like modern industrial society, you can't change just one thing. Everything is connected to everything else, and you can't make a major change in any one thing without changing the whole system. This applies not only to the physical components of the

system, but to the whole mind-set, the whole system of values and priorities that characterizes the technological society.

If you try to fix things by addressing each problem separately, your reforms can't go far enough to fix any one of the problems, because if you make changes that are far-reaching enough to fix problem X, those changes will have unacceptable consequences in other parts of the system. As pointed out in ISAIF §§121-24, you can't get rid of the bad parts of technology and still retain the good parts.

Consider for example the problem of manipulative advertising and propaganda in general. Any serious restriction on manipulative advertising would entail interference with the advertisers' First Amendment right to free expression, so a radical restructuring of our First Amendment jurisprudence would be required. The news media are supported by advertising. If there were a drastic decline in advertising, who would support the vast network that collects information around the world and funnels it to the TV-viewer and the newspaper-reader? Maybe the government would support it, but then the government could control the news we receive, and you know what that implies.

Even more important, with an end to manipulative advertising there would probably be a major drop in consumption, so the economy would go to hell. You can imagine the consequences of that as well as I can.

Since the problems can't be solved one at a time, you have to think in terms of changing the entire system, including the whole mind-set and system of values associated with it.

17. What you ask for has no precedent in history. Societies sometimes fix problems of relatively limited scope; e.g., a country that has suffered a military defeat may be able to reorganize its army on new principles and win the next battle. But historically, short of a radical transformation of the entire social fabric (i.e., revolution), it has proved impossible for societies to solve deep-lying problems of the kind we face today. I challenge you and your colleague to produce even one example from history of a society that has solved through piecemeal reform problems of the number and seriousness of those that I've listed above (see I.A.15).

I.B. If, in spite of the foregoing, you still think that reform will work, just look at our past record. To take only a few of the most conspicuous examples:

1. *Environmental destruction*. People damaged their environment to some degree even at the hunting-and-gathering stage. Forests were burned, either

through recklessness or because burned-over lands produced more food for hunter-gatherers.[132] Early hunters may have exterminated some species of large game.[133] As technology increased man's power, environmental destruction became more serious. For example, it is well known that the Mediterranean region was largely deforested by pre-modern civilizations.[134] But forests are only one part of the picture: Preindustrial societies had no radioactive waste, no chemical factories, no diesel engines, and the damage they did to their environment was minor in comparison with what is being done today. In spite of the feeble palliative measures that are now being taken, the overall picture is clear: For thousands of years, the damage that humans have done to their environment has been steadily increasing. As for reform—there *is* an environmental movement, but its successes have been very modest in relation to the magnitude of the problem.

2. *War*. War existed among nomadic hunter-gatherers, and could be nasty.[135] But as civilization and military technology advanced, war became more and more destructive. By the 20th century it was simply horrible. As Winston Churchill put it: "War, which once was glorious and cruel, has now become sordid and cruel." Private efforts to end war began at least as early as the 1790s,[136] and efforts by governments began at least as early as the end of World War I with the League of Nations. You can see how little has been accomplished.

3. *Psychological problems incident to modern life*. I discussed these in my letter of 11/23/04. But the presence of such problems was already evident early in the 20th century in the neurotic tendency of the arts. In reading a history of Spanish literature recently, I was struck by the way the neurotic made its appearance as the historian moved from the 19th to the 20th century. E.g.: "The poetry of Dámaso Alonso [born in 1898]...is a cry... of anguish and anger; *an explosion of impotent rage against his own misery and against the pain of the world around him*."[137] Artists of this type can't be dismissed simply as individuals with psychological problems peculiar to themselves, because the fact that their work has been accepted and admired among intellectuals is an indication that the neurosis is fairly widespread.

And what has been done about the psychological problems of modern times? Drugs, psychotherapy—in my view insults to human dignity. Where is the reform movement that, according to your theory, is supposed to fix things?

4. *Propaganda*. As I mentioned above (see I.A.10), the problem of propaganda was well publicized by Vance Packard ca 1960, and the problem was certainly recognized by others (e.g., Harold Lasswell) long before that. And what has been done to correct this insult to human dignity? Nothing whatsoever.

5. *Domination of our lives by large organizations*. This is a matter of fundamental importance, and nothing effective has been done to alleviate the problem. As I've pointed out (see I.A.7), nothing *can* be done about this problem in the context of a technological society.

6. *Nuclear weapons*. This is perhaps the star exhibit. Of all our technologically induced problems the problem of nuclear weapons should be the easiest to solve through reform: The danger presented by these weapons is in no way subtle—it is obvious to anyone with a normal IQ. While such things as genetic engineering and superintelligent computers promise benefits that may seem to offset their menace, nuclear weapons offer no benefits whatever—only death and destruction. With the exception only of a tiny minority of dictators, military men, and politicians who see nuclear weapons as enhancing their own power, virtually every thinking person agrees that the world would be better off without nuclear weapons.

Yet nuclear weapons have been around for 60 years, and almost no progress has been made toward eliminating them. On the contrary, they proliferate: The U.S., Russia, Britain, France; then China, Israel, India, Pakistan; now North Korea, and in a few years probably Iran...

If reform can't solve the problem of nuclear weapons, then how can it solve the far more subtle and difficult problems among those that modern technology has created?

So it's clear that reform isn't working, and there's no reason to hope that it will ever work. Obviously it's time to try something else.

II. Why Revolution May Succeed

II. A. There are several reasons why revolution may succeed where reform has made no progress.

1. Until ca 1980 I used to think the situation was hopeless, largely because of people's thoughtlessness and passivity and their tendency to take the easy way out. (See I.A.6, 8-14, above.) Up to that point I had never read much history. But then I read Thomas Carlyle's history of the French Revolution, and it opened my eyes to the fact that, in time of revolution, the usual rules do not apply: People behave differently. Subsequent reading about revolutions, especially the French and the Russian ones, confirmed that conclusion. Once a revolutionary fever has taken hold of a country, people throw off their passivity and are willing to make the greatest efforts and endure the greatest hardships for the sake of their revolution. In such cases it may be that only a minority of the population is gripped by the revolutionary fever, but that minority is sufficiently large and energetic so that it becomes the dominant force in the country. See ISAIF §142.

2. Long before that large and dominant revolutionary minority develops, that is, long before the revolution actually begins, an avowedly revolutionary movement can shake a much smaller minority out of its apathy and learned helplessness and inspire it to passionate commitment and sacrifice in a way that a moderate and "reasonable" reform effort cannot do. See ISAIF §141. This small minority may then show remarkable stamina and long-term determination in preparing the way for revolution. The Russian revolutionary movement up to 1917 provides a notable example of this.

3. The fact that revolutions are usually prepared and carried out by minorities is important, because the system's techniques of propaganda almost always enable it to keep the attitudes and behavior of the majority within such limits that they do not threaten the system's basic interests. As long as society is governed through the usual democratic processes—elections, public-opinion polls, and other numerical indices of majority choice—no reform movement that threatens the system's basic interests can succeed [138], because the system can always contrive to have the majority on its side. 51% who are just barely interested enough to cast a vote will always defeat 49%, no matter how serious and committed the latter may be. But in

revolution, a minority, if sufficiently determined and energetic, can outweigh the relatively inert majority.

4. Unlike reformers, revolutionaries are not restrained by fear of negative consequences (see I.A.16, above). Consider for example the emission of greenhouse gasses and/or creation of nuclear waste associated with the generation of electric power. Because it is unthinkable that anyone should have to do without electricity, the reformers are largely stymied; they can only hope that a technological solution will be found in time. But revolutionaries will be prepared to shut down the power plants regardless of consequences.

5. As noted above (see I.A.15), reformers have to fight a number of different battles, the loss of any one of which could lead either to physical disaster or to conditions intolerably offensive to human dignity. Revolutionaries whose goal is the overthrow of the technoindustrial system have only *one* battle to fight and win.

6. As I've argued (see I.A.1), history is guided mainly by "objective" circumstances, and if we want to change the course of history we have to change the "objective" circumstances to that end. The dominant "objective" circumstances in the world today are those created by the technoindustrial system. If a revolutionary movement could bring about the collapse of the technoindustrial system, it would indeed change the "objective" circumstances dramatically.

7. As I've pointed out (see I.A.17), your proposed solution through piecemeal reform has no historical precedents. But there are numerous precedents for the elimination through revolution of an existing form of society. Probably the precedent most apposite to our case is that of the Russian Revolution, in which a revolutionary movement systematically prepared the way for revolution over a period of decades, so that when the right moment arrived the revolutionaries were ready to strike.

8. Even if you believe that adequate reforms are possible, you should still favor the creation of an effective revolutionary movement. It's clear that the necessary reforms—if such are possible—are not currently being carried out. Often the system needs a hard kick in the pants to get it started on necessary reforms, and a revolutionary movement can provide that kick in the pants.

Further, if it is an error to attempt revolution—that is, if adequate reforms are possible—then the error should be self-correcting: As soon as the system has carried through the necessary reforms, the revolutionary movement will no longer have a valid cause, so it will lose support and peter out.

For example, in the U.S. during the early part of the 20th century, insufficient attention was paid to the problems of the working class. Labor violence ensued and provided the kick in the pants necessary to get the government to pay attention to the problems. Because adequate reforms were carried through, the violence died down;[139] this in contrast to what happened in Russia, where the Tsarist regime's stubborn resistance to reform led to revolution.

II. B. You write: "Perhaps it would be useful to focus on specific actions necessary to alter our present technological path rather than to use loaded terms like 'revolution,' which may alienate as many, or more, supporters of change as it would galvanize adherents. Or so my colleague suggests."

1. Once one has decided that the overthrow of the technoindustrial system is necessary, there is no reason to shrink from using the word "revolution." If a person is prepared to embrace a goal as radical as that of overthrowing the technoindustrial system, he is hardly likely to be alienated by the term "revolution."

Furthermore, if you want to build a movement dedicated to such a radical goal, you can't build it out of lukewarm people. You need people who are passionately committed, and you must be careful to avoid allowing your movement to be swamped by a lot of well-meaning do-gooders who may be attracted to it because they are concerned about the environment and all that, but will shrink from taking radical measures. So you *want* to alienate the lukewarm do-gooders. You need to keep them away from your movement.

A mistake that most people make is to assume that the more followers you can recruit, the better. That's true if you're trying to win an election. A vote is a vote regardless of whether the voter is deeply committed or just barely interested enough to get to the polls. But when you're building a revolutionary movement, the number of people you have is far less important than the quality of your people and the depth of their commitment. Too many lukewarm or otherwise unsuitable people will ruin the movement. As I pointed out in an earlier letter, at the outset of the Russian Revolution of 1917 the Social Revolutionary party was numerically dominant because it was a catch-all party to which anyone who was vaguely in favor of revolution could belong.[140] The more radical Bolsheviks were numerically far inferior, but they were deeply committed and had clear goals. The Social Revolutionaries proved ineffective, and it was the Bolsheviks who won out in the end.

2. This brings me to your argument that if the nomadic hunting-and-gathering (NHG) society is taken as the social ideal, the pool of potential revolutionaries would be minimal. You yourself (same page of same letter) suggested a possible answer to this, namely, that the NHG ideal might "draw in the most committed activists," and that is essentially the answer that I would give. As I've just argued, level of commitment is more important than numbers. But I would also mention that of all societies of biologically modern humans, the nomadic hunting-and-gathering ones were those that suffered least from the chief problems that modern society brings to the world, such as environmental destruction, dangerous technological powers, dominance of large organizations over individuals and small groups. This fact certainly weighs in favor of the NHG ideal. Moreover, I think you greatly underestimate the number of potential revolutionaries who would be attracted by such an ideal. I may say more about that in a later letter.

III. Necessity Of Revolution

You challenge me to present evidence that "the situation is so urgent that truly revolutionary action is demanded," and you write: "If in fact the situation is as serious as you portray, then surely there would be other rational thinkers who would come to the same conclusion. Where are the other intelligent voices that see this reality, and likewise conclude that revolution is the only option?." But there are two separate issues here: The seriousness and urgency of the situation is one question and the call for revolution is another.

III.A. I shouldn't have to offer you any evidence on the seriousness and urgency of the situation, because others have already done that. You're familiar with Bill Joy's article. Jared Diamond and Richard Posner (U.S. Circuit Judge, conservative, pro-government) have written books about the risk of catastrophe. I'm enclosing herewith a review of these two books.[141] According to a review[142] of *Our Final Century*, by the British Astronomer Royal, Sir Martin Rees estimates that "the odds are no better than fifty-fifty that our present civilization on Earth will survive to the end of the present century." (E.g.: "[E]xperiments at very high energies, perhaps a hundred times those reached by today's particle accelerators, [could create] a tiny bubble which then [would] expand[ ] at almost the speed of light, consuming our entire galaxy for a start. In 1983 Martin Rees helped to convince physicists that no all-destroying bubble could be born inside the accelerators of those days. He now stresses the need for caution as accelerator energies grow.")[143] I don't think your colleague will dismiss any of the foregoing people as "raving anarchists."

The people mentioned in the preceding paragraph warn of dangers in the hope that these can be forestalled. I think there are many others who see the situation as hopeless and believe that disaster is inevitable. Several years ago someone sent me what seemed to be a responsible article titled "Planet of Weeds."[144] I didn't actually read the article, I only glanced through it, but I think the thesis was that our civilization would cause the extinction of most life on Earth, and that when our civilization was dead—and the human race with it—the organisms that would survive would be the weed-like ones, i.e., those that could grow and reproduce quickly under adverse conditions. Many of the original members of Earth First!—before it was taken over by the leftists—were political conservatives and I don't think your colleague could reasonably dismiss them as "raving anarchists." Their view was that the collapse of industrial civilization through environmental disaster was inevitable in the relatively near future. They felt that it was impossible to prevent the disaster, and their goal was merely to save some remnants of wilderness that could serve as "seeds" for the regeneration of life after industrial society was gone.[145]

So I think there are significant numbers of intelligent and rational people who see the situation as more serious and urgent than I do. The people I've mentioned up to this point have considered mainly the risk of physical disaster. Ellul and others have addressed the issues of human dignity, and if my recollections of his book *Autopsy of Revolution* are correct, Ellul felt that there was at most a minimal chance of avoiding a complete and permanent end to human freedom and dignity. So Ellul too saw the situation as worse than I see it.

III. B. Why then is rational advocacy of revolution so rare? There are several reasons that have nothing to do with the degree of urgency or seriousness of the situation.

1. In mainstream American society today, it is socially unacceptable to advocate revolution. Anyone who does so risks being classified as a "raving anarchist" merely by virtue of the fact that he advocates revolution.

2. Many would shrink from advocating revolution simply because of the physical risk that they would run if a revolution actually occurred. Even if they survived the revolution, they would likely have to endure physical hardship. We live in a soft society in which most people are much more fearful of death and hardship than the members of earlier societies were. (The anthropologist Turnbull records the contempt that traditional Africans have for modern man's weakness in the face of pain and death.[146])

3. Most people are extremely reluctant to accept fundamental changes in the pattern of life to which they are adapted. They prefer to cling to familiar ways even if they know that those ways will lead to disaster 50 years in the future. Or even 40, 20, or 10 years. Turnbull observes that "few of us would be willing to sacrifice" modern "achievements," "even in the name of survival."[147] Instead of "achievements" he should have said "habitual patterns of living." Jared Diamond has pointed out that societies often cling stubbornly to their established ways of life even when the price of doing so is death.[148] This alone is enough to explain why calls for revolution are hardly ever heard outside of the most radical fringe.

4. Even people who might otherwise accept a radical change in their way of life may be frightened at the prospect of having to get by without the technological apparatus on which they feel themselves to be dependent. For instance, I know of a woman in the Upper Peninsula of Michigan who hates the technological system with a passion and hopes for its collapse. But in a letter to me dated August 19, 2004, she wrote: "A lightning strike on June 30 'fried' our power inverter at the cabin. For three weeks I lived without electricity. ...I realized how much I was dependent. I grew to hate the night. I think that humans will do whatever possible to preserve the electrical power grids...."

5. Many people (e.g., the original Earth First!ers whom I mentioned above, III.A) think the system will collapse soon anyway, in which case no revolution will be necessary.

6. Finally, there is hopelessness and apathy. The system seems so all-powerful and invulnerable that nothing can be done against it. There's no point in advocating a revolution that is impossible. This, rather than that revolution is unnecessary or too extreme, is the objection I've heard from some people. But it is precisely the general assumption that revolution is impossible that makes it impossible in fact. If enough people could be made to *believe* that revolution was possible, then it would *be* possible. One

of the first tasks of a nascent revolutionary movement would be to get itself taken seriously.

III. C. Your colleague insists that "the case for revolution needs to be demonstrated virtually *beyond doubt*, because it is so extreme and serious." I disagree. The possible or probable consequences of continued technological progress include the extinction of the human race or even of all of the more complex forms of life on Earth; or the replacement of humans by intelligent machines; or a transformation of the human race that will entail the permanent loss of all freedom and dignity as these have traditionally been conceived. These consequences are so much more extreme and serious than those to be expected from revolution that I don't think we need to be 100% certain, or even 90% certain, that revolution is really necessary in order to justify such action.[149]

Anyway, the standard that your colleague sets for the justification of revolution ("virtually beyond doubt") is impossibly high. Since major wars are just as dangerous and destructive as revolutions, he would have to apply the same standard to warfare. Does your colleague believe, for example, that the Western democracies acted unjustifiably in fighting World War II? If not, then how would he justify World War II under the "virtually beyond doubt" standard?

III. D. Even if we assume that it is not known at present whether revolution will ever be necessary or justifiable, the time to begin building a revolutionary movement is now. If we wait too long and it turns out that revolution *is* necessary, we may find that it is too late.

Revolutions can occur spontaneously. (For example, the way for the French Revolution was not consciously prepared in advance.) But that is a matter of chance. If we don't want to merely hope for luck, then we have to start preparing the way for revolution decades in advance as the Russian revolutionaries did, so that we will be ready when the time is ripe.

I suggest that as time goes by, the system's tools for forestalling or suppressing revolution get stronger. Suppose that revolution is delayed until after computers have surpassed humans in intelligence. Presumably the most intelligent computers will be in the hands of large organizations such as corporations and governments. At that point revolution may become impossible because the government's computers will be able to outsmart revolutionaries at every step.

Revolutions often depend for their success on the fact that the revolutionaries have enough support in the army or among the police so that at least some elements of these remain neutral or aid the revolutionaries. The revolutionary sympathies of soldiers certainly played an important part in the French and Russian Revolutions. But the armies and police forces of the future may consist of robots, which presumably will not be susceptible to subversion.

This is not science fiction. "[E]xperts said that between 2011 and 2015, every household will have a robot doing chores such as cleaning and laundering."[150] The Honda company already claims to have "an advanced robot with unprecedented humanlike abilities. ASIMO walks forward and backward, turns corners, and goes up and down stairs with ease.... The future of this exciting technology is even more promising. ASIMO has the potential to respond to simple voice commands, recognize faces.... [O]ne day, ASIMO could be quite useful in some very important tasks. Like assisting the elderly, and even helping with household chores. In essence, ASIMO might serve as another set of eyes, ears and legs for all kinds of people in need."[151] Police and military applications of robots are an obvious next step, and in fact the U.S. military is already developing robotized fighting machines for use in combat.[152]

So if we're going to have a revolution we had better have it before technology makes revolution impossible. If we wait until the need for revolution is "virtually beyond doubt," our opportunity may be gone forever.

III. E. Here's a challenge for your colleague: Outline a plausible scenario for the future of our society in which everything turns out alright, and does so *without* a collapse of the technoindustrial system, whether through revolution or otherwise. Obviously, there may be disagreement as to what is "alright." But in any case your colleague will have to explain, inter alia: (1) How he expects to prevent computers more intelligent than humans from being developed, or, if they are developed, how he expects to prevent them from supplanting humans; (2) how he expects to avoid the risk of biological disaster that biotechnological experimentation entails; (3) how he expects to prevent the progressive lowering of standards of human dignity that we've been seeing at least since the early stages of the Industrial Revolution; and (4) how he expects nuclear weapons to be brought under control. As I pointed out above (see I.B.6), of all our technology-related problems, the problem of nuclear weapons should be by far the easiest to solve, so if your colleague can't give a good and convincing answer to question (4)—something better than just a pious hope that mankind will see the light and dismantle all the nukes in a spirit of brotherhood and reconciliation—then I suggest it's time to give up the idea of reform.

Letter to David Skrbina, April 5, 2005

First, as to the likelihood that computers will catch up with humans in intelligence by the year 2029, which I think is the date predicted by Ray Kurzweil: My guess is that this will not happen until significantly later than 2029. I have no technical expertise that qualifies me to offer an opinion on this subject. My guess is based mainly on the fact that technical experts tend to underestimate the time it will take to achieve fundamental breakthroughs. In 1970, computer experts predicted that computers would surpass humans in intelligence within 15 years,[153] and obviously that didn't happen.

I do think it's highly probable that machines will *eventually* surpass humans in intelligence. I'm enough of a materialist to believe that the human brain functions solely according to the laws of physics and chemistry. In other words, the brain is in a sense a machine, so it should be possible to duplicate it artificially. And if the brain can be duplicated artificially, it can certainly be improved upon.

Second, while I think it's highly probable that the technosystem is headed for *eventual* physical disaster, I don't think the risk of a massive, worldwide physical disaster within the next few decades is as high as some people seem to believe. Again, I have no technical expertise on which to base such an opinion. But back in the late 1960s there were supposedly qualified people who made dire predictions for the near future—e.g., Paul Ehrlich in his book *The Population Bomb*. Their predictions were not entirely without substance. They predicted the Greenhouse Effect, for example;[154] they predicted epidemics, and we have AIDS. But on the whole the consequences of overpopulation and reckless consumption of natural resources have been nowhere near as severe as these people predicted.

On the other hand, there is a difference between the doomsday prophets of the 1960s and people like Bill Joy and Martin Rees. Certainly Paul Ehrlich and probably many of the other 1960s doomsdayers were leftish types, and leftish types, as we know, look for any excuse to rail against the existing society; hence, their criticisms tend to be wildly exaggerated. But Bill Joy and Martin Rees are not leftish types as far as I know; in fact, they are dedicated technophiles. And dedicated technophiles are not likely to be motivated to exaggerate the dangers of technology. So maybe I'm naive in feeling that the risk of physical disaster is less imminent than Joy and Rees seem to think.

The foregoing remarks are intended to clarify matters that I discussed in my letter of 3/17/05. Now I'd like to address specifically some points raised in your letters.

I. You write: "Art, music, literature, and (for the most part) religion are considered by most people to be true and important achievements of humanity.... You seem to undervalue any such accomplishments, and in fact virtually advocate throwing them away...; art and literature are nothing more than 'a harmless outlet for rebellious impulses.'"

I A. I did write in "Morality and Revolution": "Art, literature and the like provide a harmless outlet for rebellious impulses...." (I think Ellul somewhere says much the same thing.) But I've never said that art and literature were *nothing more* than that. In any case, I don't *advocate* "throwing away" art and literature. I do recognize that the loss of much art and literature would be a consequence of the downfall of the technoindustrial system, but getting rid of art and literature is not a *goal*.

I. B. It could be argued that the arts actually are in poor health in modern society and have been in much better health in many primitive societies. You claim that in our society the arts "are considered by most people to be true and important achievements of humanity." But how often do most people visit an art museum, listen to classical music, or read serious literature? Very seldom, I think. Furthermore, even if we include commercial graphic art, television, light novels, and the like among the arts, only a small minority of people today participate *actively* in the arts, whether as professionals or as amateurs. Most people participate only as spectators or consumers of art.

Primitives too may have specialists in certain arts, but active participation tends to be much more widespread among them than it is in the modern world. For instance, among the African pygmies, *everyone* participated in song and dance. After describing the dances of the Mbuti pygmies, their "angeborene Schauspielkunst" (inborn dramatic art), and their music, Schebesta writes: "Here I will go into no further detail about Mbuti art, of whatever kind, for I only wanted to show what significance all of this has for their daily life. Here opens a source that feeds the life-energies of the primitives, that brightens and pleasantly adorns their forest life, which is otherwise so hard. That is probably why the Mbuti are so devoted to these pleasures."[155]

Compare industrial society, in which most people participate in the arts only to the extent of watching Hollywood movies, reading popular magazines or light novels, and having a radio blaring in their ears without actually listening to it.

Admittedly, much primitive art is crude, but this is by no means true of all of it. You must have seen reproductions of the magnificent paintings found on the walls of caves in Western Europe, and the polyphony of the African pygmies is much admired by serious students of music.[156] Of course, no premodern society had a body of art that matched in range and elaborate development the arts of present-day industrial society, and much of the latter would undoubtedly be lost with the collapse of the system. But the argument I would use here is that of...

I. C. The monkey and the peanut. When I was a little kid, my father told me of a trick for catching monkeys that he had read about somewhere. You take a glass bottle the neck of which is narrow enough so that a monkey's clenched fist will not pass through it, but wide enough so that a monkey can squeeze his open hand into the bottle. You put a piece of bait—say, a peanut—into the bottle. A monkey reaches into the bottle, clutches the peanut in his little fist, and then finds that he can't pull his hand out of the bottle. He's too greedy to let go of the peanut, so you can just walk over and pick him up. Thus, because the monkey refuses to accept the loss of the peanut, he loses everything.

If we continue on our present course, we'll probably be replaced by computers sooner or later. What use do you think the machines will have for art, literature, and music? If we aren't replaced by computers, we'll certainly be changed profoundly. See ISAIF §178. What reason do you have to believe that people of the future will still be responsive to the art, music, and literature of the past? Already the arts of the past have been largely superseded by the popular entertainment media, which offer intense kicks that make the old-time stuff seem boring. Shakespeare and Cervantes wrote, Vermeer and Frans Hals painted[157] for ordinary people, not for an elite minority of intellectuals. But how many people still read Shakespeare and Cervantes when they're not required to do so as part of a college course? How many hang reproductions of the Old Masters' paintings on their walls? Even if the human race still exists 200 years from now, will *anyone* still appreciate the classics of art, music, and literature? I seriously doubt it. So if we continue on our present course we'll probably lose the Western artistic tradition anyway, and we'll certainly lose a great deal more besides.

So maybe it's better to let go of the peanut than to lose everything by trying to hang onto it. Especially since we don't have to give up the whole peanut. If the system collapses before it's too late, we'll retain our humanity and our capacity to appreciate art, literature and music. It's safe to assume then that people will continue to create art, literature, and music as they always have in the past, and that works of high quality will occasionally appear.

I. D. Along with art, literature, and music you mention religion. I'm rather surprised that you regard religion as something that would be lost with the collapse of modern civilization, since modern civilization is notorious for its secularity. The explorer and ethnographer Vilhjalmur Stefansson wrote: "One frequently hears the remark that no people in the world have yet been found who are so low that they do not have a religion. This is absolutely true, but the inference one is likely to draw is misleading. It is not only true that no people are so low that they do not have a religion, but it is equally true that the lower you go in the scale of human culture the more religion you find...."[158]

Actually Stefansson's observation is not strictly accurate, but it is true that in most primitive societies religion played a more important role than it does in modern society. Colin Turnbull makes clear how much religious feeling was integrated into the daily lives of the Mbuti pygmies,[159] and the North American Indians had a similarly rich religious life, which was intimately interwoven with their day-to-day existence.[160] Compare this with the religious life of most modern people: Their theological sophistication is virtually zero; they may go to church on Sundays, but the rest of the week they govern their behavior almost exclusively according to secular mores.

However, a reservation is called for: It's possible that a resurgence of religion may occur in the modern world. See the article by Bill Moyers[161] that I enclosed with my last letter. But I certainly *hope* that the kind of kook religion described by Moyers is not the kind of religion of which your colleague would regret the loss if the system collapsed. Among other things, that brand of religion is irrational, intolerant, and even hate-filled. It's worth noting that a similar current has developed within Hinduism (see enclosed article);[162] and of course we all know what's going on in Islam. None of this should surprise us. Each of the great world religions claims to have exclusive possession of the truth, and ever since their advent religion has been a source and/or instrument of conflict, often very deadly conflict. Primitive religions, in contrast, are generally tolerant, syncretistic, or both.[163] I know of no religious wars among primitives.

So if your colleague believes that modern religions would be lost with the collapse of the system (a proposition which unfortunately I think is very doubtful), it's not clear to me why he should regret it.

II. You read me as holding that "we have now passed...the point at which reform was a viable option." But that is not my view. I don't think that reform was ever a viable option. The Industrial Revolution and succeeding developments have resulted from the operation of "objective" historical forces (see my letter of 10/12/04), and neither reform nor (counter)revolution could have prevented them. However, we may now be approaching a window of opportunity during which it may be possible to "kill" the technoindustrial system.

A simple, *decentralized* organism like an earthworm is hard to kill. You can cut it up into pieces and each piece will grow into a whole new worm. A complex and *centralized* organism like a mammal is easy to kill. A blow or a stab to a vital organ, a sufficient lowering of body temperature, or any one of many other factors can kill a mammal.

Northwestern Europe in the 18th century was poised for the Industrial Revolution. However, its economy was still relatively simple and decentralized, like an earthworm. Even in the unlikely event that war or revolution had wiped out half the population and destroyed half the infrastructure, the survivors would have been able to pick up the pieces and get their economy functioning again. So the Industrial Revolution probably would have been delayed only by a few decades.

Today, on the other hand, the technoindustrial system is growing more and more to resemble a single, centralized, worldwide organism in which every part is dependent on the functioning of the whole. In other words, the system increasingly resembles a complex, easy-to-kill organism like a mammal. If the system once broke down badly enough it would "die," and its reconstruction would be extraordinarily difficult. See ISAIF §§207-212. Some believe that its reconstruction would even be impossible. This was the opinion of (for example) the distinguished astronomer Fred Hoyle.[164]

So only now, in my opinion, is there a realistic possibility of altering the course of technoindustrial development.

Letter to David Skrbina, July 10, 2005

Regarding the material about monkey genes—yes, it's not uncommon to read reports of new ways of monkeying with the brain (no pun intended), and there is plenty of reason to worry about this stuff, not so much because employers might force their employees to take gene treatments to turn them into workaholics (which I think is unlikely), as because increased understanding of the brain leads to solutions that are, at the least, insulting to human dignity.See ISAIF §§143-45, 149-156.

Regarding Ray Kurzweil's "Promise and Peril," you write, "I'm not sure which disturb me more, his 'promises' or his 'perils'." I feel the same way. To me they are all just perils. I'm skeptical about Kurzweil's predictions, though. I'll bet that a lot of them will turn out to be just pie in the sky. In the past there have been too many confident predictions about the future of technology that have not been fulfilled. It's certainly not that I would want to downplay the power or the danger of technology. However, I do question Kurzweil's ability to predict the future. I'll be very surprised if everything that he predicts actually materializes, but I won't be a bit surprised if a lot of scary stuff happens that neither Kurzweil nor anyone else can now anticipate.

To address a few specific points from Kurzweil's article:

He asks: "Should we tell the millions of people afflicted with cancer and other devastating conditions that we are canceling the development of all bioengineered treatments because there is a risk that these same technologies may someday be used for malevolent purposes?" Kurzweil fails to note that cancer results largely from the modern way of life (see my letter of 3/17/05), and the same is true of many other "devastating conditions," e.g., AIDS, which, assuming that it occurred at all, would probably have remained localized if it had not been for modern transportation facilities, which spread the disease everywhere. In any case, what is at stake now are the most fundamental aspects of the fate of the whole world. It would be senseless to risk a disastrous outcome in order to prolong artificially the lives of people suffering from "devastating conditions."

Throughout his essay Kurzweil romanticizes the technological way of life, while he paints a misleading and grim picture of preindustrial life. In my letter of 11/23/04, I pointed out some reasons for considering primitive life better than modern life. To address specifically Kurzweil's point about life-expectancy— he mentions an expectancy of 35 years for preindustrial Swedish females and 33 for males. Let's split the difference and make it 34 years overall. Assuming this figure is correct, it is misleading because it gives the impression that few people lived beyond their mid-30s. I've more than once read statements by demographers to the effect that the low life-expectancies of preindustrial times largely reflected the high rate of infant and early-childhood mortality. Once the vulnerable first few years were past, people's lives were not so very much shorter than they are today. I'm depending on memory here and can't cite my sources. But information for which I *can* cite sources is consistent with what I've just said. According to Rousseau, in mid-18th-century France 50% of children died before reaching the age of eight.[165] Since mortality must have been highest in the earliest years, let's suppose that the average age of these children at death was 3 years. Assuming that this is applicable to Sweden, accepting the above figure of 34 years for average age at death, and setting A = average age at death of all people who survived beyond the age of eight, we have 0.5 x 3 + 0.5 x A = 34. Solving for A gives an average age at death of 65 for those who survived beyond the age of eight. This of course is only a crude estimate, and I'm not suggesting that the high child mortality rate should be discounted as a triviality, but we do see here how misleading it is to cite the 34-year life-expectancy without further explanation. It's worth noting that about 8% of a population of Kalahari Bushmen (hunter-gatherers) was said to consist of persons from 60 to more than 80 years old.[166] My recollection is that according to the 1970 census, 10% of the American population was then aged 65 or older. This figure has stuck in my mind because I read it not long after reading the foregoing figure for the Bushmen.

Kurzweil states not only that technological progress proceeds exponentially but that biological evolution has always done so. This statement is almost meaningless. To say that something grows exponentially means that it follows a curve of the form: y equals e to the ax power, where "a" is a constant. So, before you can meaningfully say that a thing grows exponentially, you have to have a quantitative measure of that thing. Where is Kurzweil's quantitative measure of evolutionary progress? How would he assign numerical values to fishes, amphibians, reptiles, mammals, etc., that would show the rate of evolution in quantitative terms?

It's easy to establish quantitative measures of progress in specific aspects of technology. E.g., one can speak of the number of operations that a computer performs in one second. But on what quantitative measure does Kurzweil rely in stating that *overall* technological progress is and always has been exponential? I don't doubt that technological progress has been "exponential" in some vague subjective sense, at least for the last few centuries. A responsible commentator might say just that, or he might say that as measured by some specified numerical index progress has been exponential. But Kurzweil just says flatly and without qualification: "Exponential growth is a feature of any evolutionary process...." This kind of overconfidence is apparent also in other parts of the article, and it reinforces my suspicion (which I mentioned in an earlier letter) that Kurzweil is more of a showman than a serious thinker.

Again, I myself believe that technology is carrying us forward at an accelerating and extremely dangerous rate; on that point I fully agree with Kurzweil. But I question whether he is a responsible, balanced, and reliable commentator.

Kurzweil admits that we can't "absolutely ensure" the survival of human ethics and values, but he does seem to believe we can do a lot to promote their survival. And throughout his article generally he shows his belief that humans can to a significant degree control the path that technological progress will take. I maintain that he is dead wrong. History shows the futility of human efforts to guide the development of societies, and, given that the pace of change—as Kurzweil himself says—will keep accelerating indefinitely, the futility of such efforts in the future will be even more certain. So Kurzweil's ideas for limiting the dangerous aspects of technological progress are completely unrealistic. Relevant here are my remarks about "natural selection" (see my letter of 10/12/04). For example, "human values" in the long run will survive only if they are the "fittest" values in terms of natural selection. And it is highly unlikely that they will continue to be the fittest values in the world of the future, which will be utterly unlike the world that has existed heretofore.

What Kurzweil says about "distributed technologies" makes me uneasy. He may be right in claiming that the system will tend toward the development of decentralized facilities, thus decreasing its dependence on centralized facilities such as power-plants, oil refineries, and so forth. The more decentralized the system becomes, the more difficult it will be to eliminate it. This is one reason why I oppose decentralization.

A question has to be raised about the people who are promoting all this mad technological growth—those who do the research and those who provide the funds for research. Are they criminals? Should they be punished?

# # #

Concerning the recent terrorist action in Britain: Quite apart from any humanitarian considerations, the radical Islamics' approach seems senseless. They take a hostile stance toward whole nations, such as the U.S. or Britain, and they indisciminately kill ordinary citizens of those countries. In doing so they only strengthen the countries in question, because they provide the politicians with what they most need: a feared external enemy to unite the people behind their leaders. The Islamics seem to have forgotten the principle of "divide and conquer": Their best policy would have been to profess friendship for the American, British, etc. *people* and limit their expressed hostility to the elite groups of those countries, while portraying the ordinary people as victims or dupes of their leaders. (Notice that this is the position that the U.S. usually adopts toward hostile countries.)

So the terrorists' acts of mass slaughter seem stupid. But there may be an explanation other than stupidity for their actions: The radical Islamic leaders may be less interested in the effect that the bombings have on the U.S. or the U.K. than in their effect within the Islamic world. The leaders' main goal may be to build a strong and fanatical Islamic movement, and for this purpose they may feel that spectacular acts of mass destruction are more effective than assassinations of single individuals, however important the latter may be. I've found some support for this hypothesis:

"[A] radical remake of the faith is indeed the underlying intention of bin Laden and his followers. Attacking America and its allies is merely a tactic, intended to provoke a backlash strong enough to alert Muslims to the supposed truth of their predicament, and so rally them to purge their faith of all that is alien to its essence. Promoting a clash of civilizations is merely stage one. The more difficult part, as the radicals see it, is convincing fellow Muslims to reject the modern world absolutely (including such aberrations as democracy), topple their own insidiously secularizing quisling governments, and return to the pure path."[167] •

ENDNOTES

[1] *Encyclopædia Britannica*, 15th Ed., 2003, Vol. 28, article "Technology," p. 451.

[2] Or something to that effect. This is probably from Ellul's *Autopsy of Revolution*. Here, and in any letter I may write you, please bear in mind the caveat about the unreliability of memory that I mentioned in an earlier letter. Whenever I fail to cite a source, down to the page number, for any fact I state, you can assume that I'm relying for that fact on my (possibly wrong) memory of something I've read (possibly many years ago), unless the fact is common knowledge or can be looked up in readily available sources such as encyclopedias or standard textbooks.

[3] Leon Trotsky, *History of the Russian Revolution*, trans. by Max Eastman, 1980 ed., Vol. One, pp. xviii-xix.

[4] E.g., Elizabeth Cashdan, "Hunters and Gatherers: Economic Behavior in Bands," in S. Plattner (editor), *Economic Anthropology*, 1989, pp. 22-23.

[5] "In every well-documented instance, cases of hardship [=starvation] may be traced to the intervention of modern intruders." Carleton S. Coon, *The Hunting Peoples*, 1971, pp. 388-89.

[6] I take this to be "common knowledge" among anthropologists. However, I have little specific information on this subject.

[7] *Encyclopædia Britannica*, 15th ed., 1997, Vol. 10, article "Slave," p. 873.

[8] *Ibid.*, Vol. 26, article "Propaganda," pp. 175-76 ("The propagandist must realize that neither rational arguments nor catchy slogans can, by themselves, do much to influence human behavior.")

[9] *Ibid.*, p. 176.

[10] *Ibid.*, p. 174.

[11] *Encyclopædia Britannica*, 15th ed., 2003, Vol. 16, article "Christianity," p. 261.

[12] Trotsky, *op. cit.*, Vol. One, p. 223.

[13] *Ibid.*, p. 324. On this subject generally, see *Ibid.*, pp. 223-331.

[14] Trotsky, *op. cit.*, Vol. One, Chapter VIII, pp. 136-152.

[15] See Trotsky, *op. cit.*, or any history of Russia during the relevant period.

[16] Admittedly, one would have to stretch a point to say that (II) here is identical with the second objective for a revolutionary movement that I listed in my letter of 8/29/04: "to increase the tensions within the social order until those tensions reach the breaking point." But one thing I've learned about expository writing is that too much precision is counterproductive. In order to be understood one has to simplify as much as possible, even at the cost of precision. For the purposes of my letter of 8/29/04, the point I needed to emphasize was that a revolutionary movement has to increase social tensions rather than relieving them through reform. If I had given a more detailed and precise account of the task of a revolutionary movement, as in the present letter, it would only have distracted attention from the point that I needed to make in my letter of 8/29/04. So I beg your indulgence for my failure to be perfectly consistent in this instance.

[17] The suggestion that a biotechnological accident could provide a trigger for revolution is in tension with my earlier suggestion (letter of 8/29/04, page 12) that it might be desirable to slow the progress of biotechnology in order to postpone any biotechnological catastrophe. On the one hand, such a catastrophe might be so severe that afterward there would be nothing left to save; on the other hand, a lesser catastrophe might provide the occasion for revolution. It's arguable which consideration should be given more weight. But on the whole I think it would be best to try to slow the progress of biotechnology.

[18] *The New Encyclopædia Britannica*, 15th ed., 2003, Vol. 28, article "Union of Soviet Socialist Republics," p. 1000.

[19] *Ibid.*, Vol. 26, article "Propaganda," p. 176 ("the most effective media as a rule... are not the impersonal mass media but rather those few associations or organizations [reference groups] with which the individual feels identified.... Quite often the ordinary man not only avoids but actively distrusts the mass media...but in the warmth of his reference group he feels at home....").

[20] Here, the usual caveat about the unreliability of memory.

[21] Martin E. P. Seligman, *Helplessness: On Depression, Development, and Death*, W. H. Freeman and Company, New York, 1975, p. 55.

[22] Nathan Keyfitz, reviewing Gerard Piel's *Only One World: Our Own to Make and to Keep*, in *Scientific American*, February, 1993, p. 116.

[23] See, e.g., Tacitus, *Germania* 46 (hunter-gatherers present in the Baltic area < 2,000 years ago); *Encyclopædia Britannica*, 15th ed., 2003, Vol. 28, article "Spain," p. 18 (hunter-gatherers present in Spain up to 5,500 years ago).

[24] "Ten thousand years ago all men were hunters, including the ancestors of everyone reading this book. The span of ten millennia encompasses about four hundred generations, too few to allow for any notable genetic changes." Carleton S. Coon, *The Hunting Peoples*, 1971, p. xvii. Admittedly, it may be open to argument whether 400 generations allow for any "notable genetic changes."

[25] Robert Wright, "The Evolution of Despair," *Time* magazine, August 28, 1995.

[26] There is no claim here that this is an exhaustive list of the ways in which human intentions for a society can be realized on a historical scale. If you can identify any additional ways that are relevant for the purposes of the present discussion, I'll be interested to hear of them.

[27] Rafiq Zakaria, *The Struggle Within Islam*, Penguin Books, 1989, p. 59.

[28] *Encycl. Britannica*, 15th ed., 2003, Vol. 27, article "Social Structure and Change," p. 369.

[29] "[E]ach territorial clan had its own headman and council, and there was also a paramount chief for the entire tribe. The council members of each clan were elected in a meeting between the middle-aged and elderly men, and a few of the outstanding younger ones as well." Coon, *op. cit.*, p. 253.

[30] *Encycl. Britannica*, 15th ed., 2003; Vol. 28, article "Ukraine," p. 985.

[31] Buccaneers elected their own captains: *Encycl. Britannica*, Vol. 2, article "buccaneers," p. 592. For deposition of captains I'm relying on my memory of books read 40 years ago.

[32] *Encycl. Britannica*, 15th ed., 2003, Vol. 19, article "Geneva," p. 743.

[33] *Ibid.*, Vol. 20, article "Greek and Roman Civilizations," p. 294.

[34] The Russian armies played a much greater role in the defeat of Germany in World War II than the Western armies did, but the Russians received massive quantities of military aid—trucks, for example—that were produced by American industry. Moreover, British and American factories produced the thousands of bombers—not to mention bombs—that shattered German cities, though admittedly the military utility of World War ii strategic bombing is a matter of controversy. see *Encycl. Britannica*, 15th ed., 2003, Vol. 29, article "World Wars," pp. 997, 999, 1019; John Keegan, *The Second World War*, Penguin Books, 1990, pp. 44 (photo caption), 215, 218, 219, 416, 430, 432; Freeman Dyson, "The Bitter End," *The New York Review*, April 28, 2005, p. 4 ("German soldiers consistently fought better than Britons or Americans. Whenever they were fighting against equal numbers, the Germans always won....").

[35] Jeffrey Kaplan and Leonard Weinberg, *The Emergence of a Euro-American Radical Right*, Rutgers University Press, 1998, Chapter II. William E. Leuchtenburg,

Franklin D. Roosevelt and the New Deal, 1932-1940, Harper & Row, New York, 1963, pages 26, 27, 30 & footnote 43, 102 & footnote 22, 182-83, 221 & footnote 78, 224, 275-77, 279, 288.

[36] Warren Angus Ferris, *Life in the Rocky Mountains*, edited by Paul C. Phillips, pp. 40-41.

Concerning Notes 36, 41, and 43: These citations are from notes that I made many years ago, at a time when I was often careless about the completeness (though not about the accuracy) of bibliographical information that I recorded. I neglected to write down the dates of publication of the books cited here. So if you should consult different editions of these books than the ones I used, you may not find the words I've quoted on the pages that I've cited.

[37] *Ibid.*, p. 289.

[38] Gontran de Poncins, *Kabloona*, Time-Life Books, 1980, p. 78.

[39] *Ibid.*, p. 111.

[40] Domingo Faustino Sarmiento, *Civilización y Barbarie*. Regrettably, I can't give the page number. But the quotation should be accurate, since I copied it (i.e., I copied the Spanish original of it) years ago out of a book that quoted Sarmiento. However, I neglected to record the author or the title of the latter book.

[41] Osborne Russell, *Journal of a Trapper*, Bison Books, p. 26.

[42] Colin M. Turnbull, *The Forest People* and *Wayward Servants*, passim.

[43] Colin M. Turnbull, *The Mountain People*, p. 21.

[44] Colin M. Turnbull, *The Forest People*, Simon & Schuster, 1962, p. 26.

[45] Paul Schebesta, *Die Bambuti-Pygmäen vom Ituri*, Vol. I, Institut Royal Colonial Belge, 1938, p. 73. I have not had an opportunity to examine Vols. II and III of this work, which contain most of the ethnographic information.

[46] *Ibid.*, p. 205.

[47] Gontran de Poncins, *op. cit.*, pp. 212-213.

[48] *Ibid.*, p. 292.

[49] *Ibid.*, p. 273.

[50] James Axtell, *The Invasion Within*, Oxford University Press, 1985, pp. 326-27.

[51] *Ibid.* also at various other places in the same book.

[52] E.g., Francis Parkman, *The Conspiracy of Pontiac*, Little, Brown and Company, 1917, Vol. II, p. 237; *The Old Regime in Canada*, same publisher, 1882, pp. 375-76.

[53] Robert Wright, "The Evolution of Despair," *Time* magazine, August 18, 1995.

[54] Paul Schebesta, *op. cit.*, p. 228.

[55] *Ibid.*, p. 213.

[56] Catherine Edwards, "Look Back at Anger" (book review), *Times Literary Supplement*, August 23, 2002, p. 25. However, it seems to me that I recall stories from Ovid's *Metamorphoses* that could be understood as portraying depression.

[57] Gontran de Poncins, *op. cit.*, pp. 169-175, 237.

[58] Coon, *op. cit.*, pp. 72, 184.

[59] *Ibid.*, pp. 372-373.

[60] *The Denver Post*, December 30, 2003, p. 5A, reporting on a paper by Daniel Hamermesh and Jungmin Lee published during December 2003 by the National Bureau of Economic Research.

[61] *Time* magazine, June 10, 2002, p. 48.

[62] *U.S. News & World Report*, March 6, 2000, p. 45.

[63] *Ibid.*, February 18, 2002, p. 56.

[64] Elliot S. Gershon and Ronald O. Rieder, "Major Disorders of Mind and Brain," *Scientific American*, September 1992, p. 129.

[65] *Funk & Wagnalls New Encyclopedia*, 1996, Vol. 24, article "Suicide," p. 423.

[66] *Los Angeles Times*, September 15, 1998, p. A1. The study was reported at about that date in the *Archives of General Psychiatry*, according to the *L.A. Times* article.

[67] *The New Encyclopædia Britannica*, 15th ed., 2003, Vol. 29, article "United Kingdom," p. 38.

[68] *Ibid.*, pp. 61-66.

[69] *Ibid.*, Vol. 27, article "Southern Africa," section "South Africa," p. 920.

[70] *Ibid.*, p. 925.

[71] *Ibid.*, pp. 928-929.

[72] *Ibid.*, p. 929.

[73] *Ibid.*, p. 925.

[74] *Newsweek*, September 27, 2004, p. 36.

[75] *The Denver Post*, October 19, 2004, p. 15A.

[76] *Ibid.*, October 18, 2004, p. 15A.

[77] As of 2008.

[78] Here I am not including Finland among the Scandinavian countries.

[79] *Encycl. Britannica*, 15th ed., 2003, Vol. 24, article "Netherlands," pp. 891-94.

[80] *Ibid.*, p. 894 ("When the crisis of the 1848 revolutions broke... [a] new constitution was written...").

[81] *Ibid.*, Vol. 28, article "Sweden," pp. 335-38.

[82] *Ibid.*, Vol. 24, article "Norway," pp. 1092-94.

[83] *Ibid.*, Vol. 17, article "Denmark," pp. 240-41.

[84] *Ibid.*, Vol. 28, article "Switzerland," pp. 352-56.

[85] *Ibid.*, p. 354 ("a new constitution, modeled after that of the United States, was established in 1848...").

[86] Sometimes a country can be intentionally and calculatedly assimilated to the technoindustrial system and the culture thereof. This falls under one of the exceptions (exception [iii], my letter of 10/12/04) that I noted, in which human intentions for the future of a society can be successfully realized.

[87] "[A]ntislavery groups estimated that 27 million people were enslaved at the beginning of the 21st century, more than in any previous historical period," *Encycl. Britannica*, 15th ed., 2003, Vol. 27, article "Slavery," p. 293. I assume, however, that the *percentage* of the world's population that lives in slavery today is smaller than in earlier times, and that the elimination of slavery from fully modernized countries is very nearly complete.

[88] "The fate of slavery [in most of the world outside the British Isles] depended on the British abolition movement...," *Encycl. Britannica*, 15th edition, 2003, Vol. 27, article "Slavery," p. 293.

[89] *Ibid.*, p. 299.

[90] *Ibid.*, pp. 298-99.

[91] G. A. Zimmermann, *Das Neunzehnte Jahrhundert*, Zweite Hälfte, Zweiter Teil, Milwaukee, 1902, pp. 30-31.

[92] *Encycl. Britannica*, 15th ed., 2003, Vol. 20, article "Greek and Roman Civilizations," p. 277.

[93] Simón Bolívar, letter to the editor of the *Gaceta Real de Jamaica*, September 1815; in Graciela Soriano (ed.), *Simón Bolívar: Escritos políticos*, Madrid, 1975, p. 86.

[94] *Ibid.*, p. 87.

[95] *Encycl. Britannica*, 15th ed., 2003, Vol. 27, article "Slavery," p. 288.

[96] Theodore H. von Laue, *Why Lenin? Why Stalin?*, J. B. Lippencott Co., New York, 1971, p. 202.

[97] I don't mean to suggest that discipline as such is necessarily bad. I suspect that any successful revolutionary movement directed against the technoindustrial system will have to be well disciplined.

[98] Friedrich Nietzsche, *Twilight of the Idols/The Antichrist*, trans. by R. J. Hollingdale, Penguin Books, 1990, p. 103.

[99] *U.S. News & World Report*, May 8, 2000, pp. 47-49. *National Geographic*, May, 2006, pp. 127, 129.

[100] *Encycl. Britannica*, 15th ed., 2003, Vol. 20, article "Hitler," p. 628.

[101] See, e.g., *Encycl. Britannica*, 15th ed., 1997, Vol. 26, article "Propaganda," p. 175 ("The rank and file of any group, especially a big one, have been shown to be remarkably passive until aroused by quasi-parental leaders whom they admire and trust."); Trotsky, *op. cit.*, Vol. Two, p. vii ("[T]he mere existence of privations is not enough to cause an insurrection....It is necessary that...new conditions and new ideas should open the prospect of a revolutionary way out.").

[102] *Vegetarian Times*, May, 2004, p. 13 (quoting *Los Angeles Times* of January 13, 2004).

[103] *Science News*, February 1, 2003, Vol. 163, p. 72.

[104] "Kids Need more Protection From chemicals," *Los Angeles Times*, January 28, 1999, page number not available.

[105] *U.S. News & World Report*, January 24, 2000, pp. 30-31.

[106] *Time* magazine, October 18, 2004, p. 29.

[107] E.g., Bill McKibben, "Acquaintance of the Earth" (book review), *New York Review*, May 25, 2000, p. 49. *U.S. News & World Report*, February 5, 2001, p. 44.

[108] *Time* magazine, July 1, 2002, p. 57. *U.S News & World Report*, February 5, 2001, pp. 46, 48, 50.

[109] *U.S. News & World Report*, February 5, 2001, pp. 44-46.

[110] *Time* magazine, November 22, 2004, pp. 72-73.

[111] *Ibid.*, October 4, 2004, pp. 68-70.

[112] *U.S. News & World Report*, February 5, 2001, pp. 48, 50.

[113] *Ibid.*, p. 50.

[114] *The Denver Post*, June 16, 2004, p. 2A.

[115] *Ibid.*

[116] *Christian Science Monitor*, March 8, 2001, p. 20. Elizabeth Kolbert, "Ice Memory," *The New Yorker*, January 7, 2002, pp. 30-37.

[117] Poncins, *op. cit.,* pp. 164-65.

[118] Paul Schebesta, *Die Bambuti-Pygmäen vom Ituri*, II. Band, I. Teil, Institut Royal Colonial Belge, Brussels, 1941, pp. 8, 18. I received Schebesta's Vol. II as a Christmas gift this year from my beloved lady, the schoolteacher whom I mentioned to you in an earlier letter. Two years ago I received Schebesta's Vol. I from her as a Christmas gift.

[119] Colin Turnbull, *Wayward Servants*, Natural History Press, 1965, pages 27, 28, 42, 178-181, 183, 187, 228, 256, 274, 294, 300; *The Forest People*, pages 110, 125. Nancy Bonvillain, *Women and Men*, Prentice-Hall, 1998, pages 20-21.

[120] Obviously, there may be disagreement as to what constitutes an "acceptable" solution. I suspect that you and I may not be too far apart as to what we would consider acceptable, but I have no idea where your colleague stands in that respect.

[121] Admittedly there is a gray area: Sometimes a reform is in the interest of the system only because conditions are so hard on human beings that they will rebel if there is no alleviation. E.g., the government acted to solve the labor problems of the early 20th century only after violence by workers made clear that it was in the interest of the system to solve the problems. I think there is a chapter on these labor problems in Hugh Davis Graham and Ted Robert Gurr (editors), *Violence in America: Historical and Comparative Perspectives.*

[122] See my letter of 10/12/04; ISAIF §§44, 58, 145.

[123] Mel Greaves, *Cancer: The Evolutionary Legacy*, Oxford University Press, 2000, p. 16. Greaves actually writes, "overall *age-related* mortality from the major types of cancer in Western society at the end of the twentieth century was probably more than ten times that at the end of the nineteenth century." I assume this means that cancer mortality *in any given age group* has increased by a factor of more than ten. For balance: "overall rates of new cancer cases and deaths from cancer in the U.S. have been declining gradually since 1991... .," *University of California, Berkeley, Wellness Letter*, September 2004, p. 8.

[124] From speech attributed to Gaius Memmius by Sallust, *Jugurthine War*, Book 31, somewhere around Chapt. 16. Roman historians commonly invented the speeches that they put into the mouths of their protagonists, but the quotation reflects Roman attitudes even if it was invented by Sallust rather than spoken by Memmius.

[125] Pedro Calderón de la Barca, *La vida es sueño*, Jornada primera, Escena cuarta (Edilux Ediciones, Medellin, Colombia, 1989, p. 25): "hombre que está agraviado es infame...," etc. Mark Danner, "Torture and Truth," *The New York Review*, 6/10/04, p. 45.

[126] Hugh Davis Graham and Ted Robert Gurr, *op. cit.*; Chapter 12, by Roger Lane. *The New Encyclopædia Britannica*, 15th ed., 2003, Vol. 25, article "Police," pp. 959-960.

[127] Bill Moyers, "Welcome to Doomsday," *New York Review*, 3/24/05, pp. 8, 10.

[128] Some of us would add: biological engineering of other organisms (an insult to the dignity of all life).

[129] *National Geographic*, November 2003, pp. 4-29, had a surprisingly vigorous article on surveillance technology (e.g., p. 9: "Cameras are becoming so omnipresent that all Britons should assume that their behavior outside the home is monitored.... Machines will recognize our faces and our fingerprints. They will watch out for... red-light runners and highway speeders."). For other scary stuff on surveillance, see, e.g., *Denver Post,* 7/13/04, p. 2A ("Mexico has required some prosecutors to have tiny computer chips implanted in their skin as a security measure...."); *Time Bonus Section,* Oct. 2003, pp. A8-A16; *Time*, 1/12/04, "Beyond the Sixth Sense."

[130] The claim here is not that governments or corporations will directly use psychoactive medications to control people, but that people will "voluntarily" medicate themselves (e.g., for depression) or their children (e.g., for hyperactivity or attention-deficit disorder) in order to enable them to meet the system's demands.

[131] Clifford Geertz, "Very Bad News," *New York Review*, 3/24/05, pp. 4-6.

[132] Julio Mercader (ed.), *Under the Canopy: The Archaeology of Tropical Rain Forests*, Rutgers University Press, 2003, pp. 235, 238, 239, 241, 282. Carleton S. Coon, *The Hunting Peoples*, Little, Brown and Co., 1971, p. 6.

[133] E.g., Mercader, *op. cit.*, p. 233.

[134] *Encyclopædia Britannica*, 15th ed., 2003, Vol. 14, article "Biosphere," pp. 1190, 1202. *Ibid.*, Vol. 19, article "Forestry and Wood Production," p. 410.

[135] E.g., Coon, *op. cit.*, pp. 243-44.

[136] Neil J. Smelser, *Theory of Collective Behavior*, The Macmillan Company, New York, 1971, p. 273 ("The peace movement is a general social movement which has been in existence since its beginning in england during the revolutionary and Napoleonic Wars").

[137] J. García López, *Historia de la literatura española*, 5th ed., Las Americas Publishing Co., New York, 1959, p. 567.

[138] Unless it is rich enough to undertake a massive, long-term propaganda campaign on a national scale—a possibility too far-fetched to be considered here.

[139] See Note 121.

[140] Trotsky, *op. cit.,* Vol. One, p. 223. See my letter of 8/29/04.

[141] Clifford Geertz, *op. cit.* (see Note 131).

[142] *Times Literary Supplement*, 8/1/03, pp. 6-7.

[143] *Ibid.* This danger was also discussed by Russell Ruthen in "Science and the Citizen," *Scientific American*, August, 1993.

[144] David Quammen, "Planet of Weeds," *Harper's Magazine*, October 1998, pp. 57-69.

[145] These statements about Earth First! are based mainly on my recollection of Martha F. Lee, *Earth First!: Environmental Apocalypse.*

[146] Colin Turnbull, *The Mbuti Pygmies: Change and Adaptation*, Harcourt Brace College Publishers, 1983, pp. 89-90, 92.

[147] *Ibid.*, p. 11.

[148] Malcolm Gladwell, *The New Yorker*, 1/3/05, p. 72. (reviewing Jared Diamond's *Collapse: How Societies Choose to Fail or Succeed).*

[149] In terms of freedom and dignity I personally feel that the situation is *already* bad enough to justify revolution, but I don't need to rely on that.

[150] *Denver Post*, 1/25/05, p. 11A.

[151] Advertisement by Honda in *National Geographic*, February 2005 (unnumbered page).

[152] *Denver Post*, 2/18/05, pp. 28A-29A.

[153] *Chicago Daily News*, November 16, 1970. I don't have a record of the page number.

[154] *Encyclopædia Britannica*, 15th ed., 2003, Vol. 16, article "Climate and Weather," has a good section on the Greenhouse Effect, pp. 508-511.

[155] Paul Schebesta, *Die Bambuti-Pygmäen vom Ituri*, Institut Royal Colonial Belge, Brussels, II. Band, I. Teil, 1941, p. 261.

[156] See Louis Sarno, *The Song from the Forest.*

[157] *Encyclopædia Britannica*, Vol. 24, article "The Netherlands," p. 891.

[158] Vilhjalmur Stefansson, *My Life with the Eskimo*, Macmillan, 1951, p. 38.

[159] Colin Turnbull, *The Forest People*, Simon and Schuster, 1962, pp. 92-93, 145. *Wayward Servants*, The Natural History Press, 1965, pp. 19, 234, 252-53, 271, 278. I have not seen the volume of Schebesta's work that deals specifically with Mbuti religion, but in his II. Band, I. Teil, I find some incidental remarks that seem inconsistent with Turnbull's account of Mbuti religion. The inconsistency is perhaps explained by the fact that Turnbull and Schebesta focused their main studies on different groups of Mbuti.

[160] E.g., Clark Wissler, *Indians of the United States*, Revised Edition, Anchor Books, 1989, pp. 179-182, 304-09.

[161] Bill Moyers, "Welcome to Doomsday," *New York Review*, 3/24/05, pp. 8, 10.

[162] William Dalrymple, "India: The War Over History," *New York Review*, April 7, 2005, pp. 62-65.

[163] If I remember correctly, James Axtell, *The Invasion Within: The Contest of Cultures in Colonial North America*, Oxford University Press, 1985, discusses the tolerant and syncretistic character of American Indian religion in the eastern U.S.

[164] Fred Hoyle was quoted to this effect by Richard C. Duncan in an Internet article. The quote is probably from Hoyle's book *Of Men and Galaxies*, University of Washington Press, Seattle, 1964.

[165] Jean-Jacques Rousseau, *Emile or On Education*, trans. by Allan Bloom, Basic Books, HarperCollins, 1979, p. 47.

[166] John E. Pfeiffer, *The Emergence of Man*, Harper & Row, 1969, p. 344. I question whether Pfeiffer is reliable, but it should be possible to check this information by consulting Pfeiffer's sources.

[167] Max Rodenbeck, "Islam Confronts its Demons," *New York Review*, April 29, 2004, p. 16.

10

Excerpts from Letters to a German

Written by TJK During 2006

There are two difficulties connected with the characteristic victimization issues of the left, such as the alleged oppression of women, homosexuals, racial or ethnic minorities, and animals.

First, these issues distract attention from the technology problem. Rebellious energies that might have been directed against the technological system are expended instead on the irrelevant problems of racism, sexism, etc. Therefore it would have been better if these problems had been completely solved. In that case they could not have distracted attention from the technology problem.

But revolutionists should not attempt to solve the problems of racism, sexism, and so forth, because, in addressing these problems, they would further distract attention from the problem of technology. Furthermore, revolutionists could contribute very little to the solution of the problems of women, minorities, etc., because technological society itself is already working to solve these problems. Every day (at least in the United States) the media teach us that women are equal to men, that homosexuals should be respected, that all races should receive equal treatment, and so forth. Hence, any efforts in this direction by revolutionists would be superfluous.

Through their obsessive concentration on victimization issues such as the alleged oppression of women, homosexuals, and racial minorities, leftists vastly increase the extent to which these issues distract attention from the technology problem. But it would be counterproductive for revolutionists to try to obstruct leftists' efforts to solve the problems of women, minorities, and so forth, because such obstruction would intensify the controversy over these issues and therefore would distract even more attention from the technology problem.

Instead, revolutionists must repeatedly point out and emphasize that the energy expended on the leftists' victimization issues is wasted, and that that energy should be expended on the technological problem.

A second difficulty connected with victimization issues is that any group that concerns itself which such issues will attract leftists.

As the Manifesto argues, leftists are useless as revolutionists because most of them don't really want to overthrow the existing form of society.

They are interested only in satisfying their own psychological needs through vehement advocacy of "causes." Any cause will do as long as it is not specifically right-wing.

Thus, when any movement (other than a right-wing movement) arises that aspires to be revolutionary, leftists come swarming to it like flies to honey until they outnumber the original members of the movement, take it over, and transform it into a leftist movement. Thereafter the movement is useless for revolutionary purposes. The case of the movement called Earth First! provides a neat example of this process. (See Martha F. Lee, *Earth First!: Environmental Apocalypse*, Syracuse University Press, Syracuse, New York, 1995.) Thus, the left serves as a mechanism for emasculating nascent revolutionary movements and rendering them harmless.

Therefore, in order to form an effective movement, revolutionists must take pains to exclude leftists from the movement. In order to drive away leftists, revolutionists should not only avoid involvement in efforts to help women, homosexuals, or racial minorities; they should specifically disavow any interest in such issues, and they should emphasize again and again that women, homosexuals, racial minorities, and so forth should consider themselves lucky because our society treats them better than most earlier societies have done. By adopting this position, revolutionists will separate themselves from the left and discourage leftists from attempting to join them.

# # #

You seem to think that increasing the pressure to which people are subject in modern society will be sufficient to produce a revolution. But this is not correct. Certainly a serious grievance must be present in order for a revolution to occur, but a serious grievance, or even the greatest suffering, by itself is not sufficient to bring about a revolution. People who have studied the process of revolution are agreed that in addition to a grievance, some precipitating factor is necessary. The precipitating factor might be a dynamic leader, some extraordinary event, or anything that arouses new hope that rebellion can bring relief from the grievance.

Thus Trotsky wrote:

"In reality the mere existence of privations is not enough to cause an insurrection.... It is necessary that...new conditions and new ideas should open the prospect of a revolutionary way out."[1]

In the opinion of the philosopher-sociologist Eric Hoffer: "[T]he presence of an outstanding leader is indispensable. Without him there will be no movement. The ripeness of the times does not automatically produce a mass movement... ."[2]

Similarly the Encyclopædia Britannica: "The rank and file of any group; especially a big one, have been shown to be remarkably passive until aroused by quasi-parental leaders whom they admire and trust."[3]

Of course, the prerequisites for revolution are much more complex than the mere presence of dynamic leaders or of "new conditions and new ideas" that arouse hope. For an extended discussion, see Neil J. Smelser, *Theory of Collective Behavior,* Macmillan Company, New York, 1971, pages 313-384. The point is, however, that revolutionists cannot simply wait passively for hard conditions to produce a revolution. Instead, revolutionists must actively prepare the way for revolution.

I should add that the remarks about leftism, here and in the Manifesto, are based on observation of the American left. I do not know whether the remarks can be applied without modification to the European left.

# # #

You write: "Let us not deceive ourselves about the real role of women." If you mean that motherhood is the *only* suitable role for women, then I disagree. Quite apart from child-rearing, women have always done very important, even indispensable work, and work that was often very hard physically or required great skill. To mention only a few examples: Among the Mbuti pygmies of Africa and exclusive of child-rearing, the women worked far more than the men, they provided the greater part of the food, they built the huts, and their work was often very hard. Among other things, they carried huge stacks of firewood into camp on their backs.[4] The women of hunting-and-gathering societies of warm climates usually provided the greater part of the food, whereas in cold countries the men provided the greater part through hunting.[5] But in cold countries the women produced the clothing,[6] which in such climates was indispensable, and in doing so the women of certain hunting-and-gathering societies showed extraordinary skill.[7]

Thus, without denying the importance of their role as mothers, we must also acknowledge the importance of the role of women as laborers and skilled handworkers. And moreover I maintain that women, just as much as men, need work, that is, activities directed toward a goal (the "power process").[8] And I suspect that the reason why today's women want to take up masculine occupations is that their role as mother is not enough to satisfy them now that technology has reduced other traditional feminine occupations to triviality. The modern woman doesn't need to make clothes, because she can buy them; she doesn't need to weave baskets, because she has at her disposal any number of good containers; she doesn't need to look for fruits, nuts, and roots in the forest, because she can purchase good food; and so forth.

# # #

You write: "The system operates so insidiously that it talks ethnic minorities into believing that the loss of their identity is a good thing. Minorities are manipulated to their own disadvantage, and entirely without any perceptible compulsion." Yes, I agree with this, except that in some countries the system is more cunning: Instead of telling ethnic minorities that the loss of their identity is a good thing it tells them to maintain their ethnic identity, but at the same time the system knows very well how to drain ethnic identity of its real content and reduce it to empty external forms. This has happened both in the United States[9] and in the Soviet Union.

# # #

Of course, I know very little about German universities, but American university intellectuals, apart from rare exceptions, are not at all suited to be members of an effective revolutionary movement. The majority belong to the left. Some of these intellectuals might make themselves useful by spreading ideas about the technology problem, but most of them are frightened at the idea of the overthrow of the system and cannot be active revolutionaries. They are the "men of words" of whom Eric Hoffer has spoken:

"The preliminary work of undermining existing institutions, of familiarizing the masses with the idea of change, and of creating a receptivity to a new faith, can be done only by men who are, first and foremost, talkers or writers.... Thus imperceptibly the man of words undermines the established institutions, discredits those in power, weakens prevailing beliefs and loyalties, and sets the stage for the rise of a mass movement."[10]

"When the old order begins to fall apart, many of the vociferous men of words, who prayed so long for the day, are in a funk. The first glimpse of the face of anarchy frightens them out of their wits."[11]

"The creative man of words, no matter how bitterly he may criticize and deride the existing order, is actually attached to the present. His passion is to reform and not to destroy. When the mass movement remains wholly in his keeping, he turns it into a mild affair. The reforms he initiates are of the surface, and life flows on without a sudden break."[12]

# # #

You write: "The movement should be a completely new beginning, beyond all positions of the left and of the right." Yes indeed! I agree completely!

# # #

You're right: We need to worry about the time factor. But we also have to take into consideration the possibility that the struggle will last a very long time, perhaps many decades. We should overthrow the system as soon as possible, but we must nevertheless prepare ourselves for a long-term revolutionary effort, because it may turn out that no quick overthrow of the system will be feasible.

You point out that technological progress proceeds at lightning speed; that it will take perhaps twenty years to develop the first computers that will surpass every human brain in computing power; that genetic engineering will inevitably be applied for the "improvement" of human beings; that new drugs will be developed. All of this may be true. But the future may be different from what we expect. For example:

"A scientist at the Massachusetts Institute of Technology believes that within eight years a machine with more intelligence than the genius level will be developed.... Other scientists...disagreed only on the timetable. They suggested 15 years...."

This is from the newspaper *The Chicago Daily News,* November 16, 1970. Obviously, what the scientists predicted has not happened. Similarly, attempts to cure certain human diseases by means of genetic technology have run into difficulties: Gene therapy can cause cancer. Thus it is possible that computers may not surpass human beings in intelligence as soon as is believed; genetic engineering may not be so easily applied to humans; and so forth. On the other hand, it is also possible that these developments will proceed even faster than anyone now suspects. In any case the social consequences of the new technology are unforeseeable and may be different from what we expect. The social consequences of the technological progress that has occurred up to the present time are different from what I expected when I was young. Therefore we have to prepare ourselves for all possibilities, including the possibility that our struggle may last a very long time.

# # #

There are two mistakes that almost all people, with the exception of experienced politicians and social scientists, make when they devise a plan for changing society.

The first mistake is that one works out a plan through pure reason, as if one were designing a bridge or a machine, and then one expects the plan to succeed.

One can successfully design a bridge or the like because material objects reliably obey precise rules. Thus one can predict how material objects will react under given circumstances. But in the realm of social phenomena we have at our disposal very few reliable, exact rules; therefore, in general, we cannot reliably predict social phenomena.

Among the few reliable predictions that we can make is the prediction that a plan will not succeed. If you let an automobile without a driver roll down a rough slope, you can't predict the route that the automobile will take, but you can predict that it will not follow a previously selected route. If you release a group of mice from a cage, you can't predict which way each mouse will run, but you can predict that the mice will not march in accord with a previously specified plan. So it goes, in general, in the domain of social phenomena.

Social scientists understand how difficult it is to carry out any long-term plan:

"History has no lessons for the future except one: that nothing ever works out as the participants quite intended or expected."[13]

"World War I...ended in various plans for peace as illusory as the plans for war had been. As the historian William McNeill wrote, 'The irrationality of rational, professionalized planning could not have been made more patently manifest.'"[14]

"Most social planning is short-term...; the goals of planning are often not attained, and, even if the plan is successful in terms of the stated goals, it often has unforeseen consequences. The wider the scope and the longer the time span of planning, the more difficult it is to attain the goals and to avoid unforeseen and undesired consequences.... Large-scale and long-term social developments in any society are still largely unplanned."[15]

The foregoing is indisputably true, and moreover it refers to the plan of the State. The State has power, vast quantities of information, and the capacity to analyze and utilize such quantities of information. We have no power and relatively little capacity to gather and analyze information. If it is impossible for the State to carry out a long-term social plan successfully, then all the more is it impossible for us.

Therefore I maintain that revolutionaries should not commit themselves to any predetermined, long-term or comprehensive plan. Instead, they should as far as possible rely on experience and proceed by trial and error, and commit themselves only to simple, short-term plans. Of course, revolutionaries should also have a comprehensive, long-term plan, but this must always be provisional, and the revolutionaries must always be ready to modify the comprehensive plan or even abandon it altogether, provided that they never forget the final goal, which is to overthrow the system. In other words, the movement must be flexible and prepared for all eventualities.

The second of the above-mentioned errors is that one proposes a plan (let us assume that it is a very good plan) and then believes that a sufficient number of people will follow the plan merely because it is a good one. But if the goal of a plan is to change society, then, however excellent the plan may be, its excellence is not what will move people to follow it. We have to take human motivations into consideration.

In private life pure reason may often move a person to follow a good plan. For example, if through the use of reason we can convince a person that one doctor is more skillful than another, then the person will probably consult the more skillful doctor, because he knows that in this way he will recover better from his ailment.

On the other hand, if we can convince a person that a certain plan will be useful to society provided that a sufficient number of people follow the plan, this provides the person with at most a very weak motive to follow the plan, for he knows that it is very unlikely, or even impossible, that his own individual participation will by itself have any perceptible effect on society. For example: Many people know that it would be better for the world if everyone refused to use automobiles. Nevertheless, apart from rare exceptions, each one of these people has his automobile, because he says to himself that if he refuses to drive he will suffer great inconvenience without doing any perceptible good for the world; for the world will derive no perceptible advantage unless many millions of people refuse to use automobiles.

So we must always bear in mind that, with only rare exceptions, a person joins a revolutionary movement not primarily in order to achieve the movement's objective, but in order to fulfill his own psychological or physical needs or to experience some form of pleasure. However loyal and sincerely devoted he may later be to the revolutionary goal, his devotion has in some way grown out of his own needs or out of the pleasures he has experienced. Of course, the attainment of a movement's goal can fulfill the needs of a member, but in general only the actions of a few leaders can perceptibly increase the likelihood that the goal will be attained. As previously indicated, the rank-and-file member knows that his own individual participation will have at most only an imperceptible effect on the progress toward the goal. Therefore the goal by itself, and through cold reason alone, cannot motivate the rank-and-file member.

Since enthusiasm produces great pleasure, enthusiasm for a strongly desired goal can be enough to move a person to revolutionary action, but only when the attainment of the goal is very near. When the attainment of the goal appears to be improbable or distant in time, the goal by itself cannot arouse much enthusiasm.

When the attainment of the goal is not near, then the following satisfactions, for example, can motivate the rank-and-file member of a revolutionary movement: (i) Sense of purpose, the feeling that one has a goal around which to organize one's life. (ii) Sense of power. (iii) Sense of belonging, the feeling of being part of a cohesive social group. (iv) Status or prestige within the movement; the approval of other members of the movement. (v) Anger, revenge; the opportunity to retaliate against the system.

Of course, one can also find satisfaction in one's contribution to the future attainment of the revolutionary goal, even if one's own individual contribution has only an imperceptible effect, but in that case the satisfaction is too weak to move anyone to make significant revolutionary efforts—apart from rare, exceptional cases. Therefore a revolutionary movement must be based chiefly on other motivations.

# # #

As for the sense of power—a cell consisting of ten people cannot afford a member much sense of power. The member will gain a sense of power only when he joins the power-holding circles of society, and then the member receives his sense of power not from the revolutionary movement but from his position within the system. He has perhaps one chance in a hundred of gaining a position of power, and he can reach such a position only through efforts extending over a long period.

A person will undertake such efforts and persist in them only if he finds satisfaction in his career. Let us assume, then, that a member of a revolutionary cell has had a successful career and after twenty years of effort has joined the power-holding circles. He likes his career, he now has power, and he has achieved these satisfactions through long years of effort. Will he want to lose all this through the destruction of the system? In rare, exceptional cases he will, but usually he will not. History offers countless examples of the young, hot-blooded rebel who swears to resist the system forever, but who then has a successful career, and when he is older and richer and has status and prestige, he comes to the conclusion that the system is not so bad after all, and that it is better to adapt himself to it.

There are further reasons to believe that your plan cannot succeed. The plan requires that the movement should remain secret and unknown to the public. But that is impossible. One can be quite sure that some member of the movement will change his mind or make a mistake, so that the existence of the movement will become publicly known. Then there will be official investigations and so forth. In history one finds examples of sophisticated spy networks the secrecy of which was carefully guarded, but which nevertheless became known, though some of their cells may have succeeded in remaining secret. The existence of the movement that you propose likewise would surely become known.

In the fourth section of your letter you propose that leaders and agitators from the ranks of the leftists should be "instructed" by members of the movement. But, apart from exceptional cases, it is impossible to believe that members of the movement could have so much control over people who have the ability to become successful leaders and agitators.

If you succeeded in infiltrating into the power-holding circles just three or four revolutionaries who, moreover, did not subsequently betray the revolution in order to keep their power and their prestige, that would be an amazing success. Such infiltrators could perhaps play a role in the revolution, but their role probably would not be decisive.

# # #

You say that revolutions are never planned on a drawing-board, and you are right. But I wouldn't say that revolutions have always been attributable to the dissatisfactions of some large segment of a society. Dissatisfaction is a precondition for revolution, but dissatisfaction by itself is not enough to bring about a revolution. I've emphasized that previously. Among other things a revolutionary myth is needed, and on this subject you write that revolutions have never chosen their ideals and myths freely, which is quite true. But then you write: "The circumstances under which people live leave them no other choice than to adopt exactly these myths and ideals and no others." I do not entirely agree with this. A myth can't be chosen arbitrarily. A myth can succeed only if it responds to the prevailing (perhaps in part unconscious) dissatisfactions and yearnings. But I'm not convinced that the circumstances under which people live always must precisely determine a single myth. For example: The Prophet Mohammed created an extraordinarily successful myth when he wrote the Koran. Would you venture to say that nothing other than precisely the Koran could have responded to the yearnings of the Arabs?

Even if you were right and for each revolution only a single myth were possible, still we would not be entitled to assume that people would develop the right myth on their own, and develop it in time. The myths of the French and Russian revolutions were not developed by the people at large, but by a small number of intellectuals. Maybe the work of the intellectuals consisted only in giving form and structure to the formless or unconscious dissatisfactions and yearnings of the nation; nevertheless, this work was indispensable for the success of the revolution.

So I maintain that the task of revolutionaries is not to increase or intensify the objective grounds for dissatisfaction. There are already plenty enough grounds for dissatisfaction. Instead, revolutionaries should do the following:

(a) There are certain counterfeit grounds for dissatisfaction (e.g., the alleged problems of women, ethnic minorities, homosexuals, cruelty to animals, etc.), that serve to divert attention from the real grounds for dissatisfaction. Revolutionaries must somehow circumvent or negate these diversionary tactics.

(b) Revolutionaries must bring into effective operation the genuine but as yet poorly perceived grounds for dissatisfaction.

(c) To this end revolutionaries must (among other things) develop a revolutionary myth. This doesn't mean that they should invent a myth arbitrarily. Instead, they must discover and bring to light the real myth that already exists in inchoate form, and give it a definite structure.

# # #

You are right in saying that the role of the revolutionaries is only that of a catalyst. Revolutionaries can't create a revolution from nothing. All they can do is realize those possibilities that are offered by the conditions under which people live, just as a catalyst can bring about a chemical reaction only if all of the necessary reagents are available. You seem to believe that one can best play the role of a catalyst by intensifying the objective grounds for dissatisfaction. But I am convinced that the objective grounds for dissatisfaction are already sufficient. In order to play the role of a catalyst one must achieve a psychological effect; for example, by discovering and utilizing the right myth.

# # #

There are many young people who recognize that the technological system is destroying our world and our freedom; they want to resist it, but they know that they can't achieve anything alone, therefore they look for a group or a movement that they can join. Under the circumstances existing today, they can find no groups or movements other than the leftist or similar ones. So a young person joins one of these groups and either is converted to its ideology or else gets discouraged, leaves the group, gives up, and becomes apathetic. What is needed is a real revolutionary movement that such young people could join before they are lured by some leftist group and ruined by it.

# # #

Speeding up the system. It is not always safer to proceed on the assumption that the worst case will occur. For example: We are on a ship that is sinking. The "worst case" is that the ship will sink within two minutes. So we immediately throw the boat into the water, jump into the boat and row hurriedly away from the ship. Then we notice that we are going to die because we haven't taken any food or water with us. It would have been better to provide ourselves with food and water instead of rowing away in such a hurry, for the ship has not sunk as fast as we feared. But now it's too late....

So we should not prepare ourselves for the worst case only but, as far as possible, for all cases.

You maintain that we should speed up the action of "the machine" (that is, of the system) so that the machine will destroy itself. But in destroying itself the machine will also destroy us and our world, and perhaps all higher forms of life. Remember that not all of the destructive processes initiated by the system will stop as soon as the system falls apart. Consider for example the greenhouse effect.

"[G]lobal climate systems are booby-trapped with tipping points and feedback loops, thresholds past which the slow creep of environmental decay gives way to sudden and self-perpetuating collapse. Pump enough CO_2

into the sky, and that last part per million of greenhouse gas behaves like the 212th degree Fahrenheit [212° Fahrenheit = 100° Celsius] that turns a pot of hot water into a plume of billowing steam.... 'Things are happening a lot faster than anyone predicted,' says Bill Chameides, chief scientist for the advocacy group Environmental Defense and a former professor of atmospheric chemistry. 'The last 12 months have been alarming,' adds Ruth Curry of the Woods Hole Oceanographic Institute in Massachusetts. 'The ripple through the scientific community is palpable.'...Is it too late to reverse the changes global warming has wrought? That's still not clear...." *Time* magazine, April 3, 2006, pages 35, 36.

By releasing so much carbon dioxide into the atmosphere, the system has already disrupted the Earth's climate to such an extent that even specialists in the field can't predict the consequences. Even if the system immediately stopped releasing carbon dioxide, the Earth's climate probably would not revert to its previous condition. No one knows where our climate will go. We don't even know for certain whether the Earth will still be inhabitable at the end of this century. Of course, the more carbon dioxide the system releases, the greater the danger is. Yes, the system could destroy itself by progressing faster and releasing greater quantities of carbon dioxide, but in the process it would destroy everything else, too.

I have already emphasized that what could lead to a revolution would not be the worsening of living conditions, but a psychological situation conducive to revolution. And one of the indispensable psychological preconditions for revolution is that people should have hope. If there's no hope, there will be no revolution. A serious problem is the fact that many of the most intelligent people have already lost hope. They think that it's too late, the Earth can't be saved. If we speeded up the destructive action of the system , we would only spread and deepen this hopelessness. •

ENDNOTES

[1] Leon Trotsky, *The History of the Russian Revolution,* translated by Max Eastman (three volumes in one), Pathfinder, New York, 1980, Vol. Two, page vii.

[2] Eric Hoffer, *The True Believer,* §90.

[3] *The New Encyclopædia Britannica*, 15th edition, 2003, Vol. 26, article "Propaganda," page 175.

[4] Paul Schebesta, *Die Bambuti-Pygmäen vom Ituri,* II. Band, I. Teil, Institut Royal Colonial Belge, Brussels, 1941, pages 11-21, 31, 142, 170.

[5] Carleton S. Coon, *The Hunting Peoples*, Little, Brown and Company, Boston and Toronto, 1971, pages 72-73. Elizabeth Cashdan, "Hunters and Gatherers: Economic Behavior in Bands," in S. Plattner, *Economic Anthropology,* Stanford University Press, 1989, page 28.

[6] Coon, *op. cit.*, page 48.

[7] Gontran de Poncins, *Kabloona*, Time-Life Books, Alexandria, Virginia, USA, 1980, pages 14, 15, 124.

[8] *Industrial Society and its Future,* paragraphs 33-37.

[9] See *Industrial Society and its Future*, paragraph 29.

[10] Hoffer, *op. cit.*, section 104.

[11] Hoffer, *op. cit.*, section 110.

[12] Hoffer, *op. cit.*, section 111.

[13] Gordon S. Wood, "The Making of a Disaster," *The New York Review*, April 28, 2005, page 34.

[14] *The New Encyclopædia Britannica*, 15th edition, 2003, Volume 21, article "International Relations," page 807.

[15] *Ibid.*, Volume 27, article "Social Structure and Change," page 370.

Additional Letters

Extract from a Letter to A.O.

You write: "Even some primitive people from Mexico join the values of modern society (because of TV). What could make them go back to the forest?"

What could "make them go back to the forest" would be an end to the functioning of the world's industrial centers. The Mexican Indians couldn't use their TV sets if the TV stations were no longer broadcasting. They couldn't use motor vehicles or any internal combustion engines if the refineries were no longer producing fuel. They couldn't use any electrical appliances if the electrical power-plants were no longer producing electricity. Or, even if the Indians relied on small, local, water-powered generators, these would become useless when parts of the generators or of the appliances wore out and could not be replaced with new parts produced in factories. For example, could a group of Mexican Indians make a light bulb? I think it would be impossible, but even if it were possible it would be so difficult that it would not be worth the trouble. Thus, if the world's industrial centers stopped functioning, the Mexican Indians would have no choice but to revert to simple, preindustrial methods.

But what could make the TV stations stop broadcasting, the power-plants stop generating electricity, the refineries stop producing fuel, and the factories stop making parts? If the power-plants stopped producing electricity, then the TV stations would no longer be able to broadcast, the refineries would no longer be able to produce fuel, and the factories would no longer be able to make things. If the refineries stopped producing fuel, then the transportation of goods and people would have to cease, and therefore the factories would no longer be able to make things. If the factories were no longer able to make things, then there would be no more replacement parts to keep the TV stations, power-plants, and petroleum refineries functioning. Moreover, every factory needs things produced by other factories in order to keep operating.

Thus, modern industrial society can be compared to a complex organism in which every important part is dependent on every other important part. If any one important part of the system stops functioning, then the whole system stops functioning. Or even if the complex and finely-tuned relationship between the various parts of the system is severely disrupted, the system must stop functioning. Consequently, like any other highly complex organism, the modern industrial system is much easier to kill than a simple organism.[1] Compare a human being with an earthworm: You can cut an earthworm into many pieces, and each piece will grow into a whole new worm. But a human being can be killed by a blow to the head, a stab to the heart or the kidney, the cutting of a major artery—even a psychological condition such as severe depression can kill a human being. Like a human being, the industrial system is vulnerable because of its complexity and the interdependence of its parts. And the more the system comes to resemble a single, highly organized worldwide entity, the more vulnerable it becomes.

Thus, to your question about what could make Mexican Indians give up modernity, the answer is: the death of the industrial system. Is it possible for revolutionary action to kill the industrial system? Of course, I can't answer that question with any certainty, but I think it may be possible to kill the industrial system. I suggest that the movement that led to the Russian Revolution of 1917, and the Bolsheviks in particular, could provide a model for action today. I don't mean that anyone should look at the Bolsheviks and say, "The Bolsheviks did such-and-such and so-and-so, therefore we should do the same." What I do mean is that the Russian example shows what a revolutionary movement might be able to accomplish today.

Throughout its history up to 1917, the Bolshevik party remained small in relation to the size of Russia. Yet when the time of crisis arrived the Bolsheviks were able to assume control of the country, and they were able to inspire millions of Russians to heroic efforts that enabled them against all odds to triumph over enormous difficulties.

Of course, the Russian Revolution is accounted a failure because the ideal socialist society of which the Bolsheviks dreamed never materialized. Revolutions never succeed in creating the new social order of which the revolutionaries dream. But destruction is usually easier than construction, and revolutions often do succeed in destroying the old social order against which they are directed. If revolutionaries today were to abandon all illusions about the possibility of creating a new and better society and take as their goal merely the death of the industrial system, they might well succeed in reaching that goal. •

ENDNOTE

[1] I don't mean to say that modern industrial society is literally an organism in the same sense in which an earthworm or a human being is an organism. But the analogy with an organism is instructive for some purposes.

Letter from FC to *Scientific American*, 1995.

We write in reference to a piece by Russell Ruthen, "Strange Matters: Can Advanced Accelerators Initiate Runaway Reactions?," Science and the Citizen, *Scientific American*, August, 1993.

It seems that physicists have long kept behind closed doors their concern that experiments with particle accelerators might lead to a world-swallowing catastrophe. This is a good example of the arrogance of scientists, who routinely take risks affecting the public. The public commonly is not aware that risks are being taken, and often the scientists do not even admit to themselves that there are risks. Most scientists have a deep emotional commitment to their work and are not in a position to be objective about its negative aspects.

We are not so much concerned about the danger of experiments with accelerated particles. Since the physicists are not fools, we assume that the risk is small (though probably not as small as the physicists claim). But scientists and engineers constantly gamble with human welfare, and we see today the effects of some of their lost gambles: ozone depletion, the greenhouse effect, cancer-causing chemicals to which we cannot avoid exposure, accumulating nuclear waste for which a sure method of disposal has not yet been found, the crowding, noise and pollution that have followed industrialization, massive extinction of species and so forth. For the future, what will be the consequences of genetic engineering? Of the development of superintelligent computers (if this occurs)? Of understanding of the human brain and the resulting inevitable temptation to "improve" it? No one knows.

We emphasize that negative PHYSICAL consequences of scientific advances often are completely unforeseeable. (It probably never occurred to the chemists who developed early pesticides that they might be causing many cases of disease in humans.) But far more difficult to foresee are the negative SOCIAL consequences of technological progress. The engineers who began the industrial revolution never dreamed that their work would result in the creation of an industrial proletariat or the economic boom and bust cycle. The wiser ones may have guessed that contact with industrial society would disrupt other cultures around the world, but they probably never imagined the extent of the damage that these other cultures would suffer. Nor did it occur to them that in the West itself technological progress would lead to a society tormented by a variety of social and psychological problems.

EVERY MAJOR TECHNICAL ADVANCE IS ALSO A SOCIAL EXPERIMENT. These experiments are performed on the public by the scientists and by the corporations and government agencies that pay for their research. The elite groups get the fulfillment, the exhilaration, the sense of power involved in bringing about technological progress while the average man gets only the consequences of their social experiments. It could be argued that in a purely physical sense the consequences are positive, since life expectancy has increased. But the acceptability of risks cannot be assessed in purely actuarial terms. "[P]eople also rank risks based on...how equitably the danger is distributed, how well individuals can control their exposure and whether risk is assumed voluntarily." (M. Granger Morgan, "Risk Analysis and Management," *Scientific American*, July, 1993, page 35.) The elite groups who create technological progress share in control of the process and assume the risks voluntarily, whereas the role of the average individual is necessarily passive and involuntary. Moreover, it is possible that at some time in the future the population explosion, environmental disaster or the breakdown of an increasingly troubled society may lead to a sudden, drastic lowering of life expectancy.

However it may be with the PHYSICAL risks, there are good reasons to consider the SOCIAL consequences of technological progress as highly negative. This matter is discussed at length in a manuscript that we are sending to the *New York Times*.

The engineers who initiated the industrial revolution can be forgiven for not having anticipated its negative consequences. But the harm caused by technological progress is by this time sufficiently apparent so that to continue to promote it is grossly irresponsible. •

Letter to M. K., Dated October 4, 2003

Up to the time when I entered Harvard University at the age of sixteen, I used to dream of escaping from civilization and going to live in some wild place. During the same period, my distaste for modern life grew as I became increasingly aware that people in industrial society were reduced to the status of gears in a machine, that they lacked freedom and were at the mercy of the large organizations that controlled the conditions under which they lived.

After I entered Harvard University I took some courses in anthropology, which taught me more about primitive peoples and gave me an appetite to acquire some of the knowledge that enabled them to live in the wild. For example, I wished to have their knowledge of edible plants. But I had no idea where to get such knowledge until a couple of years later, when I discovered to my surprise that there were books about edible wild plants. The first such book that I bought was *Stalking the Wild Asparagus*, by Euell Gibbons, and after that when I was home from college and graduate school during the summers, I went several times each week to the Cook County Forest Preserves near Chicago to look for edible plants. At first it seemed eerie and strange to go all alone into the forest, away from all roads and paths. But as I came to know the forest and many of the plants and animals that lived in it, the feeling of strangeness disappeared and I grew more and more comfortable in the woodland. I also became more and more certain that I did not want to spend my whole life in civilization, and that I wanted to go and live in some wild place.

Meanwhile, I was doing well in mathematics. It was fun to solve mathematical problems, but in a deeper sense mathematics was boring and empty because for me it had no purpose. If I had worked on applied mathematics I would have contributed to the development of the technological society that I hated, so I worked only on pure mathematics. But pure mathematics was only a game. I did not understand then, and I still do not understand, why mathematicians are content to fritter away their whole lives in a mere game. I myself was completely dissatisfied with such a life.

I knew what I wanted: To go and live in some wild place. But I didn't know how to do so. In those days there were no primitivist movements, no survivalists, and anyone who left a promising career in mathematics to go live among forests or mountains would have been regarded as foolish or crazy. I did not know even one person who would have understood why I wanted to do such a thing. So, deep in my heart, I felt convinced that I would never be able to escape from civilization.

Because I found modern life absolutely unacceptable, I grew increasingly hopeless until, at the age of 24, I arrived at a kind of crisis: I felt so miserable that I didn't care whether I lived or died. But when I reached that point, a sudden change took place: I realized that if I didn't care whether I lived or died, then I didn't need to fear the consequences of anything I might do. Therefore I could do anything I wanted. I was free! That was the great turning-point in my life because it was then that I acquired courage, which has remained with me ever since. It was at that time, too, that I became certain that I would soon go to live in the wild, no matter what the consequences. I spent two years teaching at the University of California in order to save some money, then I resigned my position and went to look for a place to live in the forest.

# # #

I wrote for my journal on August 14, 1983: "The fifth of August I began a hike to the east. I got to my hidden camp that I have in a gulch beyond what I call "Diagonal Gulch." I stayed there through the following day, August 6. I felt the peace of the forest there. But there are few huckleberries there, and though there are deer, there is very little small game. Furthermore, it had been a long time since I had seen the beautiful and isolated plateau where the various branches of Trout Creek originate. So I decided to take off for that area on the 7th of August. A little after crossing the roads in the neighborhood of Crater Mountain I began to hear chain saws; the sound seemed to be coming from the upper reaches of Rooster Bill Creek. I assumed they were cutting trees; I didn't like it but I thought I would be able to avoid such things when I got onto the plateau. Walking across the hillsides on my way there, I saw down below me a new road that had not been there previously, and that appeared to cross one of the ridges that close in Stemple Creek. This made me feel a little sick. Nevertheless, I went on to the plateau. What I found there broke my heart. The plateau was criss-crossed with new roads, broad and well-made for roads of that kind. The plateau is ruined forever. The only thing that could save it now would be the collapse of the technological society. I couldn't bear it. That was the best and most beautiful and isolated place around here and I have wonderful memories of it.

"One road passed within a couple of hundred feet of a lovely spot where I camped for a long time a few years ago and passed many happy hours. Full of grief and rage I went back and camped by South Fork Humbug Creek..."

The next day I started for my home cabin. My route took me past a beautiful spot, a favorite place of mine where there was a spring of pure water that could safely be drunk without boiling. I stopped and said a kind of prayer to the spirit of the spring. It was a prayer in which I swore that I would take revenge for what was being done to the forest.

My journal continues: "...and then I returned home as quickly as I could because—I have something to do!" You can guess what it was that I had to do.

# # #

The problem of civilization is identical with the problem of technology. Let me first explain that when I speak of technology I do not refer only to physical apparatus such as tools and machines. I include also techniques, such as the techniques of chemistry, civil engineering, or biotechnology. Included too are human techniques such as those of propaganda or of educational psychology, as well as organizational techniques could not exist at an advanced level without the physical apparatus—the tools, machines, and structures—on which the whole technological system depends.

However, technology in the broader sense of the word includes not only modern technology but also the techniques and physical apparatus that existed at earlier stages of society. For example, plows, harness for animals, blacksmith's tools, domesticated breeds of plants and animals, and the techniques of agriculture, animal husbandry, and metalworking. Early civilizations depended on these technologies, as well as on the human and organizational techniques needed to govern large numbers of people. Civilizations cannot exist without the technology on which they are based.

Conversely, where the technology is available civilization is likely to develop sooner or later.

Thus, the problem of civilization can be equated with the problem of technology. The farther back we can push technology, the farther back we will push civilization. If we could push technology all the way back to the stone age, there would be no more civilization.

# # #

In reference to my alleged actions you ask, "Don't you think violence is violence?" Of course, violence is violence. And violence is also a necessary part of nature. If predators did not kill members of prey species, then the prey species would multiply to the point where they would destroy their environment by consuming everything edible. Many kinds of animals are violent even against members their own species. For example, chimpanzees often kills other chimpanzees. In some regions, fights are common among wild bears. The magazine *Bears and Other Top Predators*, Volume 1, Issue 2, pages 28-29, shows a photograph of bears fighting and a photograph of a bear wounded in a fight, and mentions that such wounds can be deadly. See article "Sibling Desperado," *Science News*, Volume 163, February 15, 2003.

Human beings in the wild constitute one of the more violent species. A good general survey of the cultures of hunting-andgathering peoples is *The Hunting Peoples*, by Carleton S. Coon, published by Little, Brown and Company, Boston and Toronto, 1971, and in this book you will find numerous examples in hunting-and-gathering societies of violence by human beings against other human beings. Professor Coon makes clear (pages XIX, 3, 4, 9, 10) that he admires hunting-and-gathering peoples and regards them as more fortunate than civilized ones. But he is an honest man and does not censor out those aspects of primitive life, such as violence, that appear disagreeable to modern people. Thus, it is clear that a significant amount of violence is a natural part of human life. There is nothing wrong with violence in itself. In any particular case, whether violence is good or bad depends on how it is used and the purpose for which it is used.

So why do modern people regard violence as evil in itself? They do so for one reason only: They have been brainwashed by propaganda. Modern society uses various forms of propaganda to teach people to be frightened and horrified by violence because the technoindustrial system needs a population that is timid, docile, and afraid to assert itself, a population that will not make trouble or disrupt the orderly functioning of the system. Power depends ultimately on physical force. By teaching people that violence is wrong (except, of course, when the system itself uses violence via the police or the military), the system maintains its monopoly on physical force and thus keeps all power in its own hands.

Whatever philosophical or moral rationalizations people may invent to explain their belief that violence is wrong, the real reason for that belief is that they have unconsciously absorbed the system's propaganda.

# # #

All of the groups you mention here are part of a single movement. (Let's call it the "GA [Green Anarchist] Movement.") Of course, these people are right to the extent that they oppose civilization and the technology on which it is based. But, because of the form in which this movement is developing, it may actually help to protect the technoindustrial system and may serve as an obstacle to revolution. I will explain:

It is difficult to suppress rebellion directly. When rebellion is put down by force, it very often breaks out again later in some new form in which the authorities find it more difficult to control. For example, in 1878 the German Reichstag enacted harsh and repressive laws against the Social-Democratic movement, as a result of which the movement was crushed and its members were scattered, confused, and discouraged. But only for a short time. The movement soon reunited itself, became more energetic, and found new ways of spreading its ideas, so that by 1884 it was stronger than ever. G. A. Zimmermann, *Das Neunzehnte Jahrhundert*, Zweite Hälfte, Zweiter Teil, Druck und Verlag von Geo. Brumder, Milwaukee, 1902, page 23.

Thus, astute observers of human affairs know that the powerful classes of a society can most effectively defend themselves against rebellion by using force and direct repression only to a limited extent, and relying mainly on manipulation to deflect rebellion. One of the most effective devices used is that of providing channels through which rebellious impulses can be expressed in ways that are harmless to the system. For example, it is well known that in the Soviet Union the satirical magazine *Krokodil* was designed to provide an outlet for complaints and for resentment of the authorities in a way that would lead no one to question the legitimacy of the Soviet system or rebel against it in any serious way. But the "democratic" system of the West has evolved mechanisms for deflecting rebellion that are far more sophisticated and effective than any that existed in the Soviet Union. It is a truly remarkable fact that in modern Western society people "rebel" in favor of the values of the very system against which they imagine themselves to be rebelling. The left "rebels" in favor of racial and religious equality, equality for women and homosexuals, humane treatment of animals, and so forth. But these are the values that the American mass media teach us over and over again every day. Leftists have been so thoroughly brainwashed by media propaganda that they are able to "rebel" only in terms of these values, which are values of the technoindustrial system itself. In this way the system has successfully deflected the rebellious impulses of the left into channels that are harmless to the system.

Rebellion against technology and civilization is real rebellion, a real attack on the values of the existing system. But the green anarchists, anarcho-primitivists, and so forth (the "GA Movement") have fallen under such heavy influence from the left that their rebellion against civilization has to a great extent been neutralized. Instead of rebelling against the values of civilization, they have adopted many civilized values themselves and have constructed an imaginary picture of primitive societies that embodies these civilized values.

# # #

[At this point the letter to M. K. contained a long section debunking the anarcho-primitivist myth. That section is omitted here because it only duplicates some of the material found in "The Truth About Primitive Life," above, pages 126-189.]

# # #

I don't mean to say that the hunting-and-gathering way of life was no better than modern life. On the contrary, I believe it was better beyond comparison. Many, perhaps most investigators who have studied hunter-gatherers have expressed their respect, their admiration, or even their envy of them.

But obviously the reasons why primitive life was better than civilized life had nothing to do with gender equality, kindness to animals, non-competitiveness, or nonviolence. Those values are the soft values of modern civilization. By projecting those values onto hunting-and-gathering societies, the GA Movement has created a myth of a primitive utopia that never existed in reality. Thus, even though the GA Movement claims to reject civilization and modernity, it remains enslaved to some of the most important values of modern society. For this reason, the GA Movement cannot be an effective revolutionary movement.

In the first place, part of the GA Movement's energy is deflected away from the real revolutionary objective—to eliminate modern technology and civilization in general—in favor of the pseudo-revolutionary issues of racism, sexism, animal rights, homosexual rights, and so forth. In the second place, because of its commitment to these pseudo-revolutionary issues, the GA Movement may attract too many leftists—people who are less interested in getting rid of modern civilization than they are in the leftist issues of racism, sexism, etc. This would cause a further deflection of the movement's energy away from the issues of technology and civilization. In the third place, the objective of securing the rights of women, homosexuals, animals, and so forth, is incompatible with the objective of eliminating civilization, because women and homosexuals in primitive societies often do not have equality, and such societies are usually cruel to animals. If one's goal is to secure the rights of these groups, then one's best policy is to stick with modern civilization. In the fourth place, the GA Movement's adoption of many of the soft values of modern civilization , as well as its myth of a soft primitive utopia, attracts too many soft, dreamy, lazy, impractical people who are more inclined to retreat into utopian fantasies than to take effective, realistic action to get rid of the technoindustrial system.

The GA Movement may be not only useless, but worse than useless, because it may be an obstacle to the development of an effective revolutionary movement. Since opposition to technology and civilization is an important part of the GA Movement's program, young people who are concerned about what technological civilization is doing to the world are drawn into that movement. Certainly not all of these young people are leftists or soft, dreamy, ineffectual types; some of them have the potential to become real revolutionaries. But in the GA Movement they are outnumbered by leftists and other useless people, so they are neutralized, they become corrupted, and their revolutionary potential is wasted. In this sense, the GA Movement could be called a destroyer of potential revolutionaries.

It will be necessary to build a new revolutionary movement that will keep itself strictly separate from the GA Movement and its soft, civilized values. I don't mean that there is anything wrong with gender equality, kindness to animals, tolerance of homosexuality, or the like. But these values have no relevance to the effort to eliminate technological civilization. They are not revolutionary values. An effective revolutionary movement will have to adopt instead the hard values of primitive societies, such as skill, self-discipline, honesty, physical and mental stamina, intolerance of externally-imposed restraints, capacity to endure physical pain, and, above all, courage. •

Letter to J.N., Dated April 29, 2001

The text of the following extract has been altered only minimally, but the notes have been greatly expanded beyond those of the original.

You write, "Watching a documentary on a tribe of Amazon Indians, I found that their life was as ordered as any modern man's... their day seemed as regimented as an office worker's."

You reached this conclusion on the basis of one documentary that you watched. I would say you were a bit hasty. I can't comment on that particular tribe because I know nothing about it. You didn't even say what tribe it was.

I wouldn't necessarily say that the life of every primitive people is less regimented than ours is. Among the Aino (a sedentary hunting-and-gathering people who formerly occupied part of Japan), ritual obligations were so elaborate and pervasive that they imposed a heavy psychological burden, often leading to serious disorders.[1]

But unquestionably many primitive societies were far less regimented than ours is. Regarding the African Pygmies, see Colin Turnbull's books on that subject,[2] or Louis Sarno's *Song from the Forest*. One who lived among the North American Indians early in the 19th century wrote that they consisted of "individuals who had been educated to prefer almost any sacrifice to that of personal liberty.... The Indians individually acknowledge no superior, nor are they subordinate to any government.... [I]n general, the warriors while in their villages are unyielding, exceedingly tenacious of their freedom, and live together in a state of equality, closely approximated to natural rights... [A]lthough [their governments] somewhat resemble the democratic form, still a majority cannot bind a minority to a compliance with any acts of its own."[3]

Of course, you have to understand that prior to the modern era freedom was not conceived, as it often is today, as the freedom to just fritter away one's time in aimless, hedonistic pursuits. It was taken for granted that survival required effort and self-discipline. But there is a world of difference between the discipline that a small band of people imposes on itself in order to meet practical necessities, and discipline that is imposed from the outside by large organizations.

You write, "High infant- and child-mortality must affect women in these cultures with a level of angst about their children and their own lives that we can't imagine."

This is a good point. The anarcho-primitivists find it convenient to overlook the high infant- and child-mortality rate (typically around 50%) of most preindustrial societies, including Western society up to the 18th century. The basic answer to this is simply that you can't have it both ways: If you want to escape the evils of industrial society, then you have to pay a price for it. However, it's likely that the high infant-mortality rate was necessary to preserve the health of the species. Today, weak and sickly babies survive to pass on their defective genes.

How do primitive women feel about it? I don't know whether anyone has ever taken the trouble to ask them. It's presumably very painful to them (and their husbands) when one of their babies dies. But I doubt that they feel the extreme anxiety that you suggest. A study of the Kalahari Bushmen found that they had very low levels of psychological stress,[4] and I assume this included the women. When people see it as normal and expected that half their children should die during the first few years of life, they probably take it in stride and don't worry about it unduly.[5] The human race doubtless has had that high infant- and child-mortality rate for the last million years and is presumably adapted to it. For a woman to be tormented by constant anxiety about her children would be maladaptive, hence a tendency to such anxiety would probably be eliminated by natural selection.

Still, a 50% infant-mortality rate is no joke. It's one of the hard aspects of forgoing industrial civilization.

You ask, "Is it not possible that our culture's unhappiness stems from our lack of strong religious beliefs, not our industrial lifestyle?"

Undoubtedly *some* people are happier for having strong religious beliefs. On the other hand, I don't think that strong religious belief is a prerequisite for happiness. Whether religion is *usually* conducive to happiness is open to argument.

But the point I want to make here is that the decline of religion in modern society is not an *accident*. It is a *necessary result* of technical progress. There are several reasons for this, of which I will mention three.

First, as page 42 of *Mean*,[6] April 2001, puts it, "Every curtain science pulls away is another that God cannot hide behind." In other words, as science advances, it disproves more and more traditional religious beliefs and therefore undermines faith.

Second, the need for toleration is antagonistic to strong religious belief. Various features of modern society, such as easy long-distance transportation, make mixing of populations inevitable. Today, people of different ethnic groups and different religions have to live and work side by side. In order to avoid the disruptive conflicts to which religious hatred would give rise, society has to teach us to be tolerant.

But toleration entails a weakening of religious faith. If you unquestioningly believed that your own creed was absolutely right, then you would also have to believe that every creed that disagreed with it was absolutely wrong, and this would imply a certain level of intolerance. In order to believe that all religions are just as good as yours is, you have to have, deep in your heart, considerable uncertainty about the truth of your own religion.

Third, all of the great world religions teach us such virtues as reverence and self-restraint. But the economists tell us that our economic health depends on a high level of consumption. To get us to consume, advertisers must offer us endless pleasure, they must encourage unbridled hedonism, and this undermines religious qualities like reverence and self-restraint.

\# \# \#

Regarding your question, there is so much to say in reply to it that I find it impossible to keep my answer brief. I'll confine myself to three points of the many that could be made.

(a) It's true that in many societies the extended family, the clan, or the village could be very confining. The paterfamilias (the "old man" who headed the extended family), or the council of village elders, kept people on a leash.

But when the paterfamilias and the village elders lost their grip on the leash as a result of modernization, it was picked up by "the system," which now holds it much more tightly than the old-timers ever did.

The family or the village was small enough so that individuals within it were not powerless. Even where all authority was theoretically vested in the paterfamilias, in practice he could not retain his power unless he listened and responded to the grievances and problems of the individual members of his family.[7]

Today, however, we are at the mercy of organizations, such as corporations, governments and political parties, that are too large to be responsive to single individuals. These organizations leave us a great deal of latitude where harmless recreational activities are concerned, but they keep under their own control the life-and-death issues on which our existence depends. With respect to these issues, individuals are powerless.

(b) In former times, for those who were willing to take serious risks, it was often possible to escape the bonds of the family, of the village, or of feudal structures. In medieval Western Europe, serfs ran away to become peddlers, robbers, or town-dwellers. Later, Russian peasants ran away to become Cossacks, black slaves ran away to live in the wilderness as "Maroons," and indentured servants in the West Indies ran away to become buccaneers.[8]

But in the modern world there is nowhere left to run. Wherever you go, you can be traced by your credit card, your social-security number, your fingerprints. You, Mr. N., live in California. Can you get a hotel or motel room there without showing your picture I.D.? You can't survive unless you fit into a slot in the system, otherwise known as a "job." And it is becoming increasingly difficult to get a job without making your whole past history accessible to prospective employers. So how can you defend your statement that "[m]odern urban society allows one to escape into an anonymity that family and clan based cultures couldn't"?

Granted, there are still corners of the world where one can find wilderness, or governments so disorganized that one can escape from the system there. But these are relics of the past, and they will disappear as the system continues to grow.

(c) "Today," you write, "one can...adopt whatever beliefs or lifestyle one wants. One can also easily travel, experiencing other cultures...."

But to what end? What, in practical terms, does one accomplish by changing one's beliefs or lifestyle, or by experiencing other cultures? Essentially nothing—except whatever fun one gets from it.

People don't need only fun, they need purposeful work, and they need to have control not only over the pleasure-oriented aspects of their lives but over the serious, practical, purposeful, life-and-death aspects. That kind of control is not possible in modern society because we are all at the mercy of large organizations.

Up to a point, having fun is good for you. But it's not an adequate substitute for serious, purposeful activity. For lack of this kind of activity people in our society get bored. They try to relieve their boredom by having fun. They seek new kicks, new thrills, new adventures. They masturbate their emotions by experimenting with new religions, new art-forms, travel, new cultures, new philosophies, new technologies. But still they are never satisfied, they always want more, because all of these activities are *purposeless*. People don't realize that what they really lack is serious, practical, purposeful work—work that is under their own control and is directed to the satisfaction of their own most essential, practical needs.

You ask, "How do we know that the breakdown of technological society won't lead to a simpler but more oppressive system?"

We don't know it. If the technological system should break down completely, then in areas unsuitable for agriculture—such as rugged mountains, arid plains, or the subarctic—people would probably be nomadic, supporting themselves as pastoralists or by hunting and gathering. Historically, nomadic peoples have tended to have a high degree of personal freedom.

But in areas suitable for large-scale, sedentary, intensive agriculture, people would probably support themselves by that kind of agriculture. And under those conditions it's likely that an oppressive landlord-class would tend to develop, like the feudal nobility of medieval Europe or the *latifundistas* of modern Latin America.

But even under the most oppressive conditions of the past, people were not as powerless as they are today. Russian serfs, for example, had means of resisting their landlords. They engaged in deception, theft, poaching, evasion of work, arson. If a peasant got angry enough, he would kill his landlord. If many peasants got angry at the same time, there would be a bloody revolt, a "jacquerie."[9]

It's not a pretty picture. But it is at least arguable that Russian serfs had more freedom—the kind of freedom that really counts—than does the average well-trained, modern middle-class person, who has almost unlimited freedom in regard to recreational activities but is completely impotent vis-à-vis the large organizations that control the conditions under which he lives and the life-and-death issues on which his existence depends.

If the technoindustrial system collapses the probable result will be a reversion to a situation roughly equivalent to that which existed several hundred years ago, in the sense that people will live under widely varying conditions in different parts of the world. There will be sickness and health, full bellies and starvation, hatred and love, brotherhood and ethnic bitterness, war and peace, justice and oppression, violence and kindliness, freedom and servitude, misery and contentment. But it will be a world in which such a thing as freedom will at least be *possible*, even though everyone might not have it.

If this were all that were involved, one might reasonably argue that it would be better to maintain the existing system rather than encourage it to collapse. If the collapse is rapid—as I think it probably will have to be—there is bound to be bloodshed, starvation, and death for many people. Though our society is a generally unhappy one, most people are not sufficiently dissatisfied to want to undergo great risks and hardships in order to achieve an outcome that will by no means be universally idyllic.

But there is much more at stake than the relative advantages of a collapse versus the *currently existing* conditions of life. We also have to ask where so-called "progress" will take us in the future. What kinds of monstrous crimes will be committed with the godlike powers of the new technology? Will human behavior be so regulated through biological and psychological techniques that the concept of freedom becomes meaningless? Will there be environmental disasters, even disasters that will make the world uninhabitable? Will we be replaced by machines or by bioengineered freaks? The future is impossible to predict. But two things are certain:

First, all of the deepest human values, and the qualities that have been most respected and admired since prehistoric times, will become meaningless or obsolete in the techno-world of the future. What is the meaning of personal identity if you are someone else's clone? What is the meaning of achievement if your innate abilities have been planned for you by biotechnicians? What is the meaning of free will if your behavior can be predicted and guided by psychologists, or explained in mechanistic terms by neurophysiologists? Without free will, what is the meaning of freedom or of moral choice? What is the meaning of nature when wild organisms are allowed to survive only where and as the system chooses, and when they are altered by genes introduced, accidentally or intentionally, by human beings?

Already we can see that the prevailing concepts of traditional values like loyalty, friendship, honesty, and morality have been seriously altered under modern conditions. Courage has been devalued, personal honor has practically disappeared. In the future, with intelligent machines, human manipulation of other humans' genetic endowment, and the fact of living in a wholly artificial environment, conditions of life will be so radically different, so far outside the range of anything that the human race has experienced in the past, that all traditional values will become irrelevant and will die. The human race itself will be transformed into something entirely different from what it has been in the past.

Second, whatever may happen with technology in the future, it will *not* be rationally planned. Technology will *not* be used "wisely." In view of our society's past record, anyone who thinks that technology will be used wisely is completely out of touch with reality. Technology will take us on a course that we can neither predict nor control. All of history, as well as understanding of complex systems in general, supports this conclusion. No society can plan and control its own development.

The changes that technology will bring will be a hundred times more radical, and more unpredictable, than any that have occurred in the past. The technological adventure is wildly reckless and utterly mad, and the people who are responsible for it are the worst criminals who have ever lived. They are worse than Hitler, worse than Stalin. Neither Stalin nor Hitler ever dreamed of anything so horrible.

# # #

Who says I love to read and write? Of course, when you're stuck in prison you have to have some sort of entertainment, and reading and writing are better than watching television (which I do not do). But when you're living out in the mountains you don't *need* entertainment. During my best time in the mountains I did very little reading, and what writing I did was mostly in my diary and was not for pleasure but for the purpose of recording my experiences so that I would never lose the memory of them.

Later, beginning roughly around 1980, I did embark on a program of reading. But that was purposeful reading, mostly in the social sciences. My goal was to understand more about human nature and about history, especially about the way societies develop and change.

# # #

I've never had anything but contempt for the so-called "'60s kids," the radicals of the Vietnam-War era. (The Black Panthers and other black activists are possible exceptions, since black people had then, and still have today, more genuine grievances on the score of discrimination than anyone else does.) I was a supporter of the Vietnam War. I've changed my mind about that, but not for the reasons you might expect.

I knew all along that our political and military leaders were fighting the war for despicable reasons—for their own political advantage and for the so-called "national interest." I supported the war because I thought it was necessary to stop the spread of communism, which I believed was even more dangerous to freedom, and even more committed to technology, than the system we have in this country is.

I've changed my mind about the war because I've concluded that I vastly overestimated the danger of communism. I overestimated its danger partly as a result of my own naivety and partly because I was influenced by media propaganda. (At the time, I was under the mistaken impression that most journalists were reasonably honest and conscientious.)

As it turned out, communism broke down because of its own inefficiency, hence no war was needed to prevent its spread. Despite its ideological commitment to technology, communism showed itself to be less effective than capitalism in bringing about technological progress. Finally—again because of its own inefficiency—communism was far less successful than it would have liked to be in strangling individual freedom. Thirty years ago I accepted the image of communist countries that the media projected. I believed that they were tightly regulated societies in which virtually the individual's every move was supervised by the Party or the State. Undoubtedly this was the way the communist leaders would have *liked* to run their countries. But it now seems that because of corruption and inefficiency in communist systems the average man in those countries had a great deal more wiggle-room than was commonly assumed in the West. Very instructive is Robert W. Thurston's study, *Life and Terror in Stalin's Russia, 1934-1941* (Yale University Press, 1996).

On the basis of Thurston's information, one could plausibly argue that the average Russian worker under Stalin actually had more personal freedom than the average American worker has had at most times during the 20th century. This certainly was not because the communist leaders *wanted* the workers to have any freedom, but because there wasn't much they could do to prevent it.

# # #

You write that you "could go on-line and learn all about" me. Yes, and to judge from the Internet postings that people have sent me, probably most of what you learned was nonsense. Leaving aside the question of the accuracy of the information you get from the Internet and assuming for the sake of argument that the Internet is a wholly beneficial source of information, still it weighs very little when balanced against the negative aspects of technology.

ENDNOTES

[1] Carleton S. Coon, *The Hunting Peoples*, Little, Brown And Company, Boston, 1971, pages 372-73.

[2] *The Forest People*, and *Wayward Servants*.

[3] John D. Hunter, *Manners and Customs of Several Indian Tribes Located West of the Mississippi*, Ross and Haines, Minneapolis, 1957, pages 52, 319-320. The authenticity of Hunter's account has been questioned, but has been persuasively defended by Richard K. Drinnon, *White Savage: The Case of John Dunn Hunter*, Schocken Books, 1972. There are in any event plenty of other sources that refer to the freedom of primitive and barbarian peoples, e.g., E. E. Evans-Pritchard, *The Nuer*, Oxford University Press, 1972, pages 5-6, 181-83.

[4] Here I'm relying on my memory of something I read many years ago. I can't cite the source, and my memory is not infallible.

[5] "Only with difficulty could [Mbuti] mothers remember the number of their deceased children." Paul Schebesta, *Die Bambuti-Pygmäen vom ituri*, I. Band, Institut Royal Colonial Belge, Brussels, 1938, page 112. This suggests that the loss of a child was less than a devastating experience for Mbuti women.

[6] *Mean* was an obscure magazine (now no longer published) for which J. N. was a writer.

[7] I think W. I. Thomas and F. Znaniecki, in the one-volume, abridged edition of *The Polish Peasant in Europe and America*, make this point in regard to the paterfamilias of Polish peasant families, but I'm relying on memory and can't cite the page.

[8] It may have been commonplace for slaves and medieval peasants to escape their servitude by running away. See Richard C. Hoffmann, *Land, Liberties, and Lordship in a Late Medieval Countryside*, University of Pennsylvania Press, Philadelphia, 1989, pages 51-52; William H. TeBrake, *A Plague of Insurrection*, University of Pennsylvania Press, Philadelphia, 1993, page 8; Andreas Dorpalen, *German History in Marxist Perspective*, Wayne State University Press, Detroit, 1988, pages 90, 158; *Encyclopædia Britannica*, 15th edition, 2003, Volume 18, article "European History and Culture," pages 618, 629; Volume 20, article "Germany," pages 75-76, 81; Volume 27, article "Slavery," pages 298-99.

[9] For these forms of resistance by slaves and serfs generally (not just Russian ones), see, e.g., Wayne S. Vucinich (editor), *The Peasant in Nineteenth Century Russia*, Stanford, California, 1968; Hoffmann, op.cit., pages 144, 305, 356, 358; TeBrake, op.cit., pages 8-9; Dorpalen, op.cit., pages 90, 92, 123, 129, 158-59; Geir Kjetsaa, *Fyodor Dostoyevsky: A Writer's Life*, translated by Siri Hustvedt and David McDuff, Fawcett Columbine, New York, 1989, pages 32, 33; Barbara Tuchman, *A Distant Mirror*, Ballantine Books, New York, 1978, page 41; *Encycl. Brit.*, 2003, Volume 27, article "Slavery," pages 298-99. Landlords or slave-owners who abused peasant or slave women sexually may have run a grave risk of being killed by the women themselves or by their menfolk. See *Ibid.*, page 299. My recollection is that sexual abuse of their women was the most common reason for which Russian peasants killed their landlords, according to Mosse, *Alexander II and the Modernization of Russia*. (Since I'm relying on memory, I can't give the page number or the author's full name.) Some time around the end of the 19th century an adolescent Mexican peon named Doroteo Arango killed one of the owners of the estate where he worked, in revenge for an assault on his sister. He fled to the mountains, where he lived for some years as a fugitive. Subsequently he acquired a certain notoriety as a revolutionary under the nom de guerre of Pancho Villa. *Encycl. Brit.*, 2003, Volume 12, article "Villa, Pancho," page 369.

(Blackfoot Valley Dispatch, Lincoln, Montana)

12

An Interview with Ted

BY J. ALIENUS RYCHALSKI, SPECIAL CORRESPONDENT FOR THE *BVD*.

1st of Four Parts

Vol. 19, No. 01, Wednesday, January 3, 2001

In 1999 I requested an interview with Theodore J. Kaczynski for the *Blackfoot Valley Dispatch* which he kindly granted. The interview took place that same year at the United States Penitentiary, Administrative Maximum, Florence, Colorado.

BVD: Well…

TJK: Well.

BVD: Well, why did you leave your job at Berkeley and your career in mathematics?

TJK: At the time I accepted the job at Berkeley, I had already decided that I would keep it for at most two years before leaving it to go live in the woods. The fact is that I never at any time felt satisfied with the idea of spending my life as just a mathematician and nothing more. Ever since my early teens I had dreamed of escaping from civilization—as in going to live on an uninhabited island or in some other wild place.

The trouble was that I didn't know how to go about it, and it was extremely difficult to work up the nerve to cut loose from my civilized moorings and take off to the woods. It's very difficult because sometimes we don't know how much the choices we make are governed by the expectations of people around us, and the fact that we go and do something other people would regard as mad—it's very difficult to do. Furthermore, I didn't know where to go really.

But at about the beginning of my last year at the University of Michigan I went through a kind of crisis. You could say that the psychological chains with which society binds us sort of broke for me. After that I was sure that I had the courage to break away from the system, to take off and just go into some wild place and try to live there. When I went to Berkeley, I never went there with the intention of continuing there indefinitely. I took the job at Berkeley only to earn some money to get started with, to buy a piece of land.

BVD: You said that when you were in your early teens you had dreams of going to live in an uninhabited place. Do you recall anything that led you to have those dreams? Something you saw or experienced?

TJK: Certainly things I read led me in that direction. Robinson Crusoe, for one thing. And then when I was maybe 11 or 12, somewhere in around there, I read some anthropology books about Neanderthal man and speculations about the way they lived and so forth. I became very interested in reading about that stuff and at some point asked myself why I wanted to read more about this material. At some point it dawned an me that what I really wanted was not to read more about these things but to actually live that way.

BVD: It's interesting that these things impacted you so strongly that you actually acted on them. What do you think it was about the lives or lifestyles of Crusoe and Neanderthal man that appealed to you?

TJK: At the time I don't think I knew why I was attracted to those ways of life. I now think it had a great deal to do with freedom and personal autonomy.

BVD: Those things must appeal to many people. So, why not everyone who…?

TJK: I think a lot of people are attracted to these things, but they aren't especially determined to actually break away from their ties and actually go and do something like that. Robinson Crusoe is supposed to be one of the most widely read books that's ever been written. So it's obviously attractive to many people. [An investigator for my case] said that she herself was very interested in the way of life I adopted in Montana and that many other people to whom she talked about my case were also very interested in it.

And many people that her investigators talked to thought that they envied me. As a matter of fact, one of the FBI agents who arrested me said "I really envy your way of life up here." So, there are a lot of people who react that way, but they just sort of drift with the tide and don't come to a point where they break away.

BVD: When you broke away, you went to Lincoln, Montana. Why Lincoln?

TJK: Well, first of all I applied for a lease on a piece of crown land in British Columbia. After, I think, over a year, they turned it down. I spent the next winter, the winter of 1970-1971, at my parents' home in Lombard, Illinois. Meanwhile my brother had gone to live in Great Falls, Montana, where he eventually got a job at the Anaconda Company smelter. At some point during that winter he mentioned in a letter to my mother that if I wanted to buy a piece of land in his part of the country, he would be interested in going 50-50 with me on it. So during the spring I drove out to Great Falls, showed up at his apartment, and took him up on his offer. With characteristic passivity, he left it up to me to find a piece of land.

Not knowing what else to do, I just took off toward the west on Highway 200, which at the time I think was called Highway 20, to see what I could see. As I passed through Lincoln I saw a little cabin, almost just a kiosk by the side of the road, with a sign advertising real estate. I stopped and asked the realtor, an old man named Ray Jensen, whether he could show me a secluded plot of land. He showed me a place up Stemple Pass Road. I liked it. I took my brother to see it and he liked it too, so we bought it. We paid $2,100 in cash—in twenty dollar bills—to the owner, Cliff Gehring, Senior.

BVD: So it could have been almost anywhere, actually.

TJK: Yeah.

2nd of Four Parts

Vol. 19, No. 02, Wednesday, January 10, 2001

BVD: What was Lincoln like when you first moved there?

TJK: The town itself to me doesn't seem that much different. I don't notice that much change. But there has been some, like the new school, the library, and a few new businesses. Maybe I would notice the changes in the town more if I were interested in it, but since I'm not, I don't notice much of those changes.

I am interested in the surrounding countryside, and that has changed a lot because aside from logging and road building, an awful lot of people have moved in there. For example, Stemple Pass Road. There were far fewer places along Stemple Pass Road, and most of them were just log cabins. Not modern log cabins, but ones that must have been built decades and decades ago, and the few year-round residents were real old-timers, another culture, not modern people. Stemple Pass Road at that time looked like a bit left over from the old frontier days.

If you go down Stemple Pass Road today, you'll see these fancy, pretentious, modern things that really look out of place in the woods. But the very few cabins that existed before were not pretentious. They weren't modern. In fact, once when my parents came to visit me in the early 1970s, we drove along Stemple Pass Road and my mother, who is bourgeois to the core in spite of her background, asked in a sneering tone "Who are these people who live in these places? Are they just drifters or what?" They weren't drifters, but stable old-timers, retirees. But they weren't concerned about status and the appearance of their homes. They were old-fashioned enough so that they didn't care whether their houses had an appearance of middle-class respectability. So, by my mother's standard their homes looked shabby.

You can see how Stemple Pass Road has changed and similar changes, I think, characterize a lot of the country around Lincoln, because a lot of places where there are cabins now, there were no cabins when I got there.

BVD: Your cabin looked right at home—harmonious—with its surroundings in the woods. Did you use plans to help you with the building of it or did you plan the building yourself?

TJK: I just planned it myself.

BVD: And you built the cabin yourself?

TJK: I had a little help from my brother, but very little. The amount of help he gave me was insignificant. Mostly I did it myself.

BVD: How long did it take you to build it?

TJK: It took me from the beginning of July 1971 until I think late November. But the work was interrupted by some trips I made to Great Falls for various purposes. Much more important, it was interrupted when I scalded my foot. On August 1, 1971, I was so clumsy as to knock over a pot of boiling soup. It poured right down into my sneaker and scalded my foot so badly that, on doctor's orders, I remained inactive for about 5 or 5 1/2 weeks.

BVD: I'm curious. Did you have enough light in your cabin? Was it light enough in there?

TJK: In the winter?

BVD: Anytime.

TJK: Yeah. It was light enough. Except for when it got dark outside, of course.

BVD: Who were the people you first met when you came to Lincoln, and who were your neighbors?

TJK: Well, obviously, the realtor. But, the first people whom I knew socially when I moved onto my property were Glen and Dolores Williams, who still own the cabin next to mine. They never lived there permanently. It was only a vacation home for them. I was always on friendly terms with them, but I never became at all close to them. And, Irene Preston and Kenny Lee. They were, what we call, colorful characters. He used to have some interesting stories...

BVD: And when did you meet the Lundbergs?

TJK: I think I first met Dick Lundberg around 1975, because until that time I had a car, later an old pickup truck. But after about 1975 I had no functioning motor vehicle, and that was when I started riding to Helena occasionally with Dick. I think I met Eileen in the late 1970s or early '80s.

BVD: So, these people you met were the people living in close proximity to you.

TJK: Yeah. Glen and his wife, as you know, were living just below me, and I also met Bill Hull and some members of his family. Aside from clerks in stores and so forth, those were the only people I got to know until, oh, probably into the '80s. When Sherri (Wood) took over the library, I started to get to know her. Eventually I got to know Theresa and the Garlands. I got to know them by going into their store. So, I didn't really get to know people there to any significant extent for the first 10 years I was there, or more.

BVD: What about Chris Waits?

TJK: The first I met him would probably be somewhere around the mid '80s. I don't remember. He used to sometimes pass me on the road. I may have taken a ride from him once or twice—I'm not sure if I ever did at all. But I know he used to pass me on the road and say hello, and that's the only acquaintance I ever had with him, except once I was at this yard sale at Leora Hall's, and I talked briefly to him there. See, I pretty much spent my time in the woods and kept to myself, and so, really, had no occasion to meet anyone except the people living in the immediate area.

BVD: I see. He didn't really live in the immediate area. About Leora Hall's yard sale, where you briefly talked to him: in his book, Waits claims that you bought silver or silver-plated flatware there. But Leora Hall has said that you positively did not buy any silver or silver-plated flatware, because she didn't have any for sale. She does, however, remember seeing you there and even remembers the specific items bought. Any comment?

TJK: I've never bought any silver-plated or silver flatware from Leora Hall or anyone else.

BVD: Well, let's move on then. Did you follow routines in your life?

TJK: I didn't really have routines, but certain activities—such as cooking meals or fetching sticks for kindling—tended to fall into routine patterns.

BVD: What was an average day like for you in Lincoln?

TJK: That's a very difficult question to answer because I don't know that there was an average day. My activities varied so much according to the season and according to the tasks I had before me on a given day. But I will describe a representative day...

3rd of Four Parts

vol. 19, No. 03, Wednesday, January 17, 2001

TJK: ...Well, let's take a day in January, and let's suppose I wake up about 3:00 a.m. to find that snow is falling. I start a fire in my stove and put a pot of water on. When the water comes to a boil I dump a certain quantity of rolled oats into it and stir them for a few minutes until they are cooked. The I take the pot off the stove, add a couple of spoonfuls of sugar and some milk— made from powdered milk. While the oats are cooling I eat a piece of cold boiled rabbit meat. Afterward I eat the oats. I sit for a few minutes before the open door of the stove watching the fire burn down, then I take my clothes off again, get back into bed, and go to sleep. When I wake up, the sky is just starting to get light. I get out of bed and dress myself quickly because it's cold in the cabin. By the time I'm dressed there's a little more light and I can see that it's no longer snowing and the sky is clear. Because of the fresh snow, it should be a good day for rabbit hunting. So I take my old, beat-up, single-shot 22 down from the hooks on the wall. I put my little wooden cartridge-box, containing 16 cartridges, in my pocket, with a couple of books of matches wrapped in plastic bags and a sheath knife on my belt in case I have to build a fire in an emergency. Then I put on my snowshoes and take off. First there's a hard climb to get up on top of the ridge, and then a level walk of a mile or so to get to the open forest of lodgepole pines where I want to hunt. A little way into the pines I find the tracks of a snowshoe hare. I follow the trail around and around through its tangled meanderings for about an hour. Then suddenly I see the black eye and the black-tipped ears of an otherwise white snowshoe hare. It's usually the eye and the black-tipped ears that you notice first. The bunny is watching me from behind the tangled branches and green needles of a recently-fallen pine tree. The rabbit is about 40 feet away, but it's alert and watching me, so I won't try to get closer. However, I have to maneuver for an angle to shoot from, so that I can have a clear shot through the tangle of branches—even a slender twig can deflect a 22 bullet enough to cause a miss. To get that clear shot I have to lie down in the snow in an odd position and use my knee as a rest for the rifle barrel. I line up the sights on the rabbit's head, at a point just behind the eye... hold steady...ping! The rabbit is clipped through the head. Such a shot ordinarily kills the rabbit instantly, but the animal's hind legs usually kick violently for a few seconds so that it bounces around in the snow. When the rabbit stops kicking I walk up to it and see that it's quite dead. I say aloud "Thank you, Grandfather Rabbit"—Grandfather Rabbit is a kind of demigod I've invented who is the tutelary spirit of all the snowshoe rabbits. I stand for a few minutes looking around at the pure-white snow and the sunlight filtering through the pine trees. I take in the silence and the solitude. It's good to be here. Occasionally I've found snowmobile tracks along the crest of the main ridge, but in these woods where I am now, once the big-game hunting season is over, in all my years in this country I've never seen a human footprint other than my own. I take one of the noosed cords out of my pocket. For convenience in carrying I put the noose around the rabbit's neck and wrap the other end of the cord around my mittened hand. Then I go looking for the trail of another rabbit. When I have three rabbits I head home. On arriving there I've been out some six or seven hours. My first task is to peel off the skins of the rabbits and remove their guts. Their livers, hearts, kidneys, brains and some assorted scraps I put in a tin can. I hang the carcasses up under shelter, then run down to my root cellar to fetch some potatoes and a couple of parsnips. When these have been washed and some other chores performed— splitting some wood maybe, or collecting snow to melt for drinking water—I put the pot on to boil, and at the appropriate time add some dried wild greens, the parsnips, the potatoes, and the livers and other internal organs of the rabbits. By the time it's all cooked, the sky is getting dark. I eat my stew by the light of my kerosene lamp. Or, if I want to economize, maybe I open the door of the stove and eat by the light of the fire. I finish off with half a handful of raisins. I'm tired but at peace. I sit for a while in front of the open door of the stove gazing at the fire. I may read a little. More likely I'll just lie on my bed for a time watching the firelight flicker on the walls. When I get sleepy I take off my clothes, get under the blankets, and go to sleep.

BVD: I envy you, too ...While work, that does sound wonderful. Freedom and autonomy. No time clock to punch, whether literal or figurative. But let me shift topic. You just mentioned sleep. Was your bed, or bunk, comfortable?

TJK: Well, it was comfortable enough for me.

BVD: I respect and appreciate your thanking Grandfather Rabbit. I'm reminded of the real origins of the ritual or custom of saying grace before a meal: A solemn awareness of sacrifice, that all life gives itself so that other life may live...Do you believe in fate?

TJK: No.

BVD: Do you believe in God?

TJK: No. Do you?

BVD: Fate or God?

TJK: Both.

BVD: Maybe… I remember reading that your parents were atheists, that you were raised in an atheistic home.

TJK: True.

BVD: Do you remember your parents ever talking about God? Did they ever say anything like "This is what some people believe…"?

TJK: Oh, they did a little bit. For example, if my mother were reading a book to me and something about God were in there, she would explain "Well, some people believe so-and-so, but we don't believe it." That sort of thing.

BVD: I see…. Well, back on your representative day—you mentioned some of what you might eat. What was your diet like in general? What would you eat on a typical day?

TJK: This varied so much with the season…. Between 1975 and 1983 I would buy flour, rice, rolled oats, sugar, cornmeal, cooking oil and powdered milk, and a modest amount of canned fruit and, or, tomatoes for the winter. I would eat maybe one can every other day through the cold season. I would eat a small amount of canned fish and dried fruit. Other than that almost everything I ate was wild or grown in my garden. I ate deer, elk, snowshoe hare, pine squirrel, three kinds of grouse, and porcupines, and occasionally ducks, rockchucks, muskrats, packrats, weasels, coyotes, an owl killed by accident—I would never kill an owl intentionally—deer mice, and grasshoppers, huckleberries, soapberries, red twinberries, black twinberries, gooseberries, two kinds of black currents, raspberries, strawberries, Oregon grapes, choke cherries, and rose hips. Starchy roots I ate were camas, yampa, bitterroot and Lomatium, also spring beauty…. I also ate a few minor kinds of roots and a couple of dozen kinds of wild greens. During May and June, before each meal I would eat a salad, often quite a large salad, by just strolling around my property, picking a bit of this and that, and popping it into my mouth. In a few cases I ground up edible seeds and used them for bread. But grinding them was excessively time-consuming. I had no hand-mill, and ground them on a rock. In my garden I grew potatoes, parsnips, beets, onions, two kinds of carrots, spinach, radishes, broccoli, and on occasion orach, Jerusalem artichoke, and turnips.

I would dry wild greens and garden vegetables, and sometimes berries, for use in the winter. But for my starchy foods I relied mainly on potatoes and on store-bought staples such as flour, rice, et cetera. Wild starchy roots are scanty up in the high country. Bitterroot and camas are abundant in places in the lower, flat areas, but these are mostly private land and presumably the ranchers wouldn't want me digging up their meadows to get these foods. In the winters I used to use a tea made from the needles of Douglas fir as a source of vitamin C.

My last winter in Montana, 1995-1996, I was hard up. But when you have to dispense with the things that the system provides, it's surprising how well you can do by improvising on your own. I had no commercial fruits or vegetables, whether fresh, dried, or canned, but I had plenty of my own dried vegetables. I had some dried black currents and rhubarb, and I had squirrels and rabbits for meat. The commercial stuff I had was just flour—whole wheat and white—cooking oil, sugar, and I think I had a scanty supply of rice. I don't recall whether I had any oats or cornmeal. I do know that the little powdered milk that I had soon ran out and I was using plaster of Paris—dental—as a source of calcium. When that ran out I was planning to use either burnt, pulverized rabbit bones or pulverized lime-stone. But I did alright, I enjoyed my meals, and it was a good winter.

BVD: What was your favorite wild food?

TJK: Probably the tastiest wild food in the Lincoln area is partridge berries, a tiny species of Vaccinium—the blueberry genus—that grows at high altitudes. The berries are so tiny that it may take an hour to pick a cupful, but the flavor is superb. Apart from those, my favorite foods are huckleberries, yampa, and the livers of deer, snowshoe rabbit, and porcupines.

BVD: Did you have any favorite meals that you prepared?

TJK: I didn't have any standard meals, since I just ate what was available at a given time. Generally speaking, my best meals were the stews that contained meat, vegetables, and some starchy food such as potatoes, rice, noodles, or roots such as yampa.

BVD: Would you eat your meals outdoors?

TJK: I seldom did that. I usually ate indoors, at my table in the cabin… When I was done eating, I would sometimes sit back in my chair with my feet up on the table and just gaze out the window for a while…

BVD: Could you see out the window?

TJK: Pardon me?

BVD: Could you see out the window?

TJK: Yes. That's what windows are for...

4th of Four Parts

Vol. 19, No. 04, Wednesday, January 24, 2001

BVD: How did you learn which plants were edible, and their preparation, if any was needed?

TJK: For years before I left Berkeley I'd been interested in the outdoors, and I had been learning skills such as how to recognize edible wild plants and so forth. I learned how to recognize them from books on the subject, such as Edible Wild Plants of Eastern North America, by Fernald and Kinsey, and Wild Edible Plants of the Western Unites States, by Donald Kirk. The books give some information about preparation of these plants, but mostly I learned to prepare them by trial and error. I learned some edible plants by experiment. It would be dangerous to experiment with certain families of plants, such as the carrot family and the lily family, because they contain some species that are deadly poisonous. But it's safe to experiment with the mustard family; and the composite family and the beet family, as far as I know, contain no deadly species, though they do contain some that are more or less poisonous. There were a couple of members of the mustard family that I used as greens without ever learning the names of the plants. There was a member of the composite family that I ate for years before I learned that it was a species of false dandelion. And there was a member of the beet family that I often ate but never did identify.

BVD: Were you self-sufficient?

TJK: By no means wholly self-sufficient. I needed store-bought staples such as flour, rice, rolled oats, and cooking oil. I bought most of my clothing, though I also made some. Originally, complete self-sufficiency was a goal that I wanted to attain eventually, but with the shrinking of the wild country and the crowding-in of people around me, I got to feeling that there wasn't any point in it anymore, and my interests turned in other directions.

BVD: How did the way you chose to live fulfill your dreams, desires, or original motivations? That is, your dreams as a youth, and your plan and decision to leave Berkeley. And what was the most satisfying thing about your life in Lincoln?

TJK: In my life in the woods I found certain satisfactions that I had expected, such as personal freedom, independence, a certain element of adventure, and a low-stress way of life.

I also achieved certain satisfactions that I hadn't fully understood or anticipated, or that even came as complete surprises to me.

The more intimate you become with nature, the more you appreciate its beauty. It's a beauty that consists not only in sights and sounds but in an appreciation of...the whole thing. I don't know how to express it. What is significant is that when you live in the woods, rather than just visiting them, the beauty becomes part of your life rather than something you just look at from the outside.

Related to this, part of the intimacy with nature that you acquire, is the sharpening of your senses. Not that your hearing or eyesight become more acute, but you notice things more. In city life you tend to be turned inward, in a way. Your environment is crowded with irrelevant sights and sounds, and you get conditioned to block most of them out of your consciousness. In the woods you get so that your awareness is turned outward, toward your environment, hence you are much more conscious of what goes on around you. For example, you'll notice inconspicuous things on the ground, such as edible plants or animal tracks. If a human being has passed through and has left even just a small part of a footprint, you'll probably notice it. You know what the sounds are that come to your ears: This is a birdcall, that is the buzzing of a horsefly, this is a startled deer running off, this is the thump of a pine cone that has been cut down by a squirrel and has landed on a log. If you hear a sound that you can't identify, it immediately catches your attention, even if it's so faint that it's barely audible. To me this alertness, or openness of one's senses, is one of the greatest luxuries of living close to nature. You can't understand this unless you've experienced it yourself.

Another thing I learned was the importance of having purposeful work to do. I mean really purposeful work—life-and-death stuff. I didn't truly realize what life in the woods was all about until my economic situation was such that I had to hunt, gather plants, and cultivate a garden in order to eat. During part of my time in Lincoln, especially 1975 through 1978, if I didn't have success in hunting, then I didn't get any meat to eat. I didn't get any vegetables unless I gathered or grew them myself. There is nothing more satisfying than the fulfillment and self-confidence that this kind of self-reliance brings. In connection with this, one loses most of one's fear of death.

In living close to nature, one discovers that happiness does not consist in maximizing pleasure. It consists in tranquility. Once you have enjoyed tranquility long enough, you acquire actually an aversion to the thought of any very strong pleasure—excessive pleasure would disrupt your tranquility.

Finally, one learns that boredom is a disease of civilization. It seems to me that what boredom mostly is is that people have to keep themselves entertained or occupied, because if they aren't, then certain anxieties, frustrations, discontents, and so forth, start coming to the surface, and it makes them uncomfortable. Boredom is almost nonexistent once you've become adapted to life in the woods. If you don't have any work that needs to be done, you can sit for hours at a time just doing nothing, just listening to the birds or the wind or the silence, watching the shadows move as the sun travels, or simply looking at familiar objects. And you don't get bored. You're just at peace.

BVD: What was the hardest part or thing about your life in Lincoln?

TJK: The worst thing about my life in the woods was the inexorable closing-in of modern civilization. There were always more houses along Stemple Pass Road and elsewhere. More roads put through the woods, more areas logged off, more aircraft flying over. Radio collars on the elk, spraying of herbicides, et cetera, et cetera.

BVD: What are some of your fondest memories of your life in the woods?

TJK: ...Early in the springtime, when the winter's snow was melted off enough to make it possible, I would take long rambles over the hills, enjoying the new physical freedom made possible by the fact that I no longer had to wear snowshoes, and coming home with a load of fresh, young wild vegetables such as wild onions, dandelions, bitterroot, and Lomatium, with a grouse or two—killed illegally, I'll admit. Working on my garden early in the morning. Hunting snowshoe rabbits in the winter. Times spent at my hidden shack during the winter. Certain places where I camped out during spring, summer, or autumn. Autumn stews of deer meat with potatoes and other vegetables from my garden. Any number of occasions when I just sat or lay still doing nothing, not even thinking much, just soaking in the peace.

BVD: Thank you, very much …

TJK: You're welcome.

(Interviewer's note: Contrary to a published claim that purports Kaczynski's hidden shack was found, it was not found.)

Explanation of Judicial Opinions,

Afterthoughts, Bibliography & Index

EXPLANATION OF THE JUDICIAL OPINIONS

PUBLISHER'S NOTE

The judicial opinions referred to in this text are official U.S. documents published under the following URLs:

1) 262 F.3d 1034 (9th Cir. 2001)
a) http://bulk.resource.org/courts.gov/c/F3/262/262.F3d.1034.99-16531.html
b) http://caselaw.findlaw.com/data2/circs/9th/9916531op.pdf
2) 239 F.3d 1108 (9th Cir. 2001)
http://bulk.resource.org/courts.gov/c/F3/239/239.F3d.1108.99-16531.html

---

Under American law, property seized without a valid search warrant cannot be used as evidence at the trial of the person from whom the property is seized. In searching my cabin in 1996, the United States Government relied in bad faith on a warrant issued without what is called "probable cause." Quin Denvir and Judy Clarke, the lawyers appointed to represent me at my trial, told me that if my case had been an ordinary one the courts would probably have declared the warrant invalid. In that event, all the evidence seized from my cabin would have been excluded from my trial, and I could not have been convicted. I would have been a free man. But, said Denvir and Clarke, because of the political implications of my case it would be very difficult to persuade the courts to declare the warrant invalid.

After hearings that preceded my trial by several months, Judge Burrell, of the United States District Court for the Eastern District of California, refused to declare the search warrant invalid or exclude the evidence seized from my cabin. That was not the end of the matter, however, for if I had been tried and convicted I could have appealed to the United States Court of Appeals for the Ninth Circuit. My lawyers, Denvir and Clarke, estimated that there was something like a twenty-percent chance that the Court of Appeals would declare the warrant invalid, in which case I would go free. Denvir and Clarke, however, were not very interested in securing my freedom. They would have preferred to negotiate a "plea agreement" with the government; that is, an agreement that I would plead guilty on condition that the government should drop its demand for the death penalty. Because the government refused to accept any plea agreement that would allow me to appeal to the Ninth Circuit, a plea agreement would have eliminated my last chance of avoiding life imprisonment, even though it would have saved me from the death penalty. I was not interested in escaping the death penalty if the alternative were life in prison. My objective was to appeal to the Ninth Circuit in an effort to have the search warrant declared invalid.

Meanwhile, my lawyers Denvir and Clarke were preparing a defense that would have portrayed me as insane. Such a defense might have saved me from the death penalty but could not have saved me from spending the rest of my life in prison or in an insane asylum. I knew that my lawyers wanted to use a defense of that type, but until shortly before the trial they dishonestly led me to believe that they could not or would not use a defense based on a claim of insanity unless I consented to such a defense. When I learned that my lawyers could use such a defense without my consent and intended to do so, there followed a series of angry disagreements between my lawyers and me. To make a long story short, I asked Judge Burrell to let me dismiss Denvir and Clarke and be represented instead by J. Tony Serra, a lawyer who had agreed not to use a mental-illness defense. When Judge Burrell denied that request, I asked permission to dispense with representation by a lawyer and represent myself before the court. The Judge denied that request too, so I was left with only two alternatives: I could either undergo a trial in which my lawyers would portray me as insane, or I could accept a plea agreement, thus sacrificing my chance to appeal to the Ninth Circuit.

In order to persuade me to accept the plea agreement, Denvir and Clarke told me that even with a plea agreement I could challenge my conviction by way of what is called a "collateral action": Under a statute labeled 28 United States Code, §2255, I could file a motion in which I would contend that my guilty plea was involuntary. Denvir and Clarke said that if I filed such a motion my chances of eventually having the search warrant declared invalid would be almost as good as they would have been with a direct appeal. Denvir and Clarke also promised to find lawyers to file a motion for me under 28 U.S.C. §2255. Several months later, other lawyers told me that in reality my chances of succeeding with a §2255 motion were very slight. Moreover, Denvir and Clarke broke their promise to find lawyers to file a §2255 motion for me; in the end I had to file the §2255 motion and litigate the entire action myself without the help of a lawyer.

The United States Constitution, as interpreted by the Supreme Court, guarantees to every defendant in a criminal trial the right to dispense with an attorney and represent himself before the court. There are, however, certain reservations; for example, a court is not required to allow a defendant to represent himself if he has requested self-representation for the purpose of delaying the trial. Judge Burrell had justified his denial of my self-representation request by claiming that I had made that request for the purpose of delay.

My §2255 motion therefore was based primarily on the contention that there was no evidence that I had intended to delay the trial, that I therefore had been improperly deprived of my constitutional right to self-representation, and that this rendered my guilty plea involuntary in the constitutional sense. Legally my argument was air-tight except at one point: In claiming that my motive for requesting self-representation was to delay the trial, Judge Burrell was making an assertion about what I was thinking at the time I made the request, and an assertion about what a person is thinking at a given time almost never can be proved or disproved conclusively. Thus, if a judge wants to decide that a defendant's motive is to delay his trial, no one can force the judge to do otherwise, however implausible his decision may seem to an objective observer.

As the first step in challenging my conviction I was required to submit my §2255 motion to Judge Burrell himself. Needless to say, he denied the motion. The next step was to take the §2255 motion to the Court of Appeals for the Ninth Circuit. An appeal to a United States Court of Appeals ordinarily is heard by a panel of three randomly-selected judges; my appeal was heard by Judges Brunetti, Reinhardt, and Rymer. Brunetti and Rymer voted to deny my appeal. In an opinion written by Judge Rymer, they claimed to agree with Judge Burrell's conclusion that I had requested self-representation for the purpose of delaying the trial.

Judge Reinhardt disagreed with Brunetti and Rymer and wrote a dissenting opinion in which he explained that there was *no* evidence that I had intended to delay the trial. I do not appreciate Judge Reinhardt's insulting comments about me and my "twisted theories," but Reinhardt is a thoroughly conscientious and widely respected jurist of unquestioned integrity, and in his dissenting opinion he did a fine job of explicating the dispute between my lawyers, me, and Judge Burrell. However, I do have to correct Judge Reinhardt on one point: Reinhardt was mistaken in assuming that if my appeal of my §2255 motion had been successful and I had won a new trial in which I would represent myself, I would then have used the trial as an opportunity to expound my "twisted theories." Actually, if I had represented myself in a new trial I probably would have said little or nothing in court. I would have gone through the trial only so that, after being convicted, I could appeal to the Ninth Circuit on the issue of the validity of the search warrant.

After my appeal of my §2255 motion was denied by Judges Brunetti and Rymer I petitioned for a rehearing by the same three-judge panel, and simultaneously for a rehearing en banc. (When the Ninth Circuit hears a case "en banc," the case is decided by a panel of eleven judges.) Judge Reinhardt voted for a rehearing by the original three-judge panel but Brunetti and Rymer voted to the contrary, therefore there was no rehearing by the original panel. The petition for rehearing en banc was voted upon by all of the active judges of the Ninth Circuit and was denied. Reinhardt, interestingly, was one of those who voted against en banc rehearing.

When the decision to deny the petition for rehearing was published, Judge Kozinski issued a dissenting opinion in which he suggested that Judge Burrell's action in my case might have been an episode from George Orwell's novel *1984*.

In case the foregoing account leaves the reader with any doubt about my sanity, I mention the following: For about four years beginning on May 5, 1998, the date on which I first arrived at the prison where I am now held, I was visited almost every day by one or both of the two prison psychologists,

Dr. James Watterson and Dr. Michael Morrison. Drs. Watterson and Morrison did not believe these visits were necessary, but their superiors in the Bureau of Prisons had ordered them to visit me every day. In the course of four years we got to know each other rather well, and Drs. Watterson and Morrison told me repeatedly that they saw no indication that I suffered from any serious mental illness. Dr. Morrison said that the diagnosis of paranoid schizophrenia (offered by the psychologists and psychiatrists whom Denvir and Clarke had hired for that purpose) was "ridiculous" and "wildly improbable"; and on more than one occasion Morrison made caustic remarks about psychologists and psychiatrists who, he said, would provide any desired diagnosis if they were well paid for doing so.

TJK, May 4, 2007

AFTERTHOUGHTS

1. Último Reducto has recently called attention to some flaws in my work. For example, in ISAIF, paragraph 69, I wrote that primitive man could accept the risk of disease stoically because "it is no one's fault, unless it is the fault of some imaginary, impersonal demon." Último Reducto pointed out that this often is not true, because in many primitive societies people believe that diseases are caused by witchcraft. When someone becomes sick the people will try to identify and punish the witch—a specific person—who supposedly caused the illness.

Again, in paragraph 208 I wrote, "We are aware of no significant cases of regression in small-scale technology," but Último Reducto has pointed out some examples of regression of small-scale technology in primitive societies.

The foregoing flaws are not very important, because they do not significantly affect the main lines of my argument. But other problems pointed out by Último Reducto are more serious. Thus, in the second and third sentences of paragraph 94 of ISAIF I wrote: "Freedom means being in control...of the life-and-death issues of one's existence: food, clothing, shelter and defense against whatever threats there may be in one's environment. Freedom means having power...to control the circumstances of one's own life." But obviously people have never had such control to more than a limited extent. They have not, for example, been able to control bad weather, which in certain circumstances can lead to starvation. So what kind and degree of control do people really need? At a minimum they need to be free of "interference, manipulation or supervision ... from any large organization," as stated in the first sentence of paragraph 94. But if the second and third sentences meant no more than that, they would be redundant.

So there is a problem here in need of a solution. I'm not going to try to solve it now, however. For the present let it suffice to say that ISAIF is by no means a final and definitive statement in the field that it covers. Maybe some day I or someone else will be able to offer a clearer and more accurate treatment of the same topics.

2. In "The Truth About Primitive Life" and in "The System's Neatest Trick" I referred to the "politicization" of American anthropology, and I came down hard on politically correct anthropologists. See pages [144–149*] and [202–203] of this book. My views on the politicization of anthropology were based on a number of books and articles I had seen and on some materials sent to me by a person who was doing graduate work in anthropology. My views were by no means based on a systematic survey or a thorough knowledge of recent anthropological literature.

One of my Spanish correspondents, the editor of *Isumatag*, argued that I was being unfair to anthropologists, and he backed up his argument by sending me copies of articles from anthropological journals; for example, Michael J. Shott, "On Recent Trends in the Anthropology of Foragers," *Man* (N.S.), Vol. 27, No. 4, Dec., 1992, pages 843–871; and Raymond Hames, "The Ecologically Noble Savage Debate," *Annual Review of Anthropology*, Vol. 36, 2007, pages 177–190.

The editor of *Isumatag* was right. As he showed me, I had greatly underestimated the number of American anthropologists who made a conscientious effort to present facts evenhandedly and without ideological bias. But even if my point about the politicization of anthropology was overstated, it still contained a significant element of truth. First, there *are* some anthropologists whose work is heavily politicized. (I discussed the case of Haviland on pages [145, 202–203] of this book.) Second, some of the anthropologists' debates seem clearly to be politically motivated, even if the participants in these debates do strive to be honest and objective. Consider for example the article by Raymond Hames cited above, which reviews the anthropological controversy over whether primitive peoples were or were not good conservationists. Why should this question be the subject of so much debate among anthropologists? The reason, obviously, is that nowadays the problem of controlling the environmental damage caused by industrial society is a hot political issue. Some anthropologists are tempted to cite primitive peoples as moral examples from whom we should learn to treat our environment with respect; other anthropologists perhaps would prefer to use primitives as negative examples in order to convince us that we should rely on modern methods to regulate our environment.

Until roughly the middle of the 20th century, industrial society was extremely self-confident. Apart from a very few dissenting voices, everyone assumed that "progress" was taking us all to a better and brighter future. Even the most rebellious members of society—the Marxists—believed that the injustices of capitalism represented only a temporary phase that we had to pass through in order to arrive at a world in which the benefits of "progress" would be shared equally by everyone. Because the superiority of modern society was taken for granted, it seldom occurred to anyone to draw comparisons between modern society and primitive ones, whether for the purpose of exalting modernity or for the purpose of denigrating it.

But since the mid-20th century, industrial society has been losing its self-confidence. Thinking people are increasingly affected by doubts about whether we are on the right road, and this has led many to question the value of modernity and to react against it by idealizing primitive societies. Other people, whose sense of security is threatened by the attack on modernity, defensively exaggerate the unattractive traits of primitive cultures while denying or ignoring their attractive traits. That is why some anthropological questions that once were purely academic are now politically loaded.

I realize that the foregoing two paragraphs greatly simplify a complex situation, but I nevertheless insist that industrial society's loss of self-confidence in the course of the 20th century is a real event.

3. *Disposal of Radioactive Waste.* In a letter to David Skrbina dated March 17, 2005, I expressed the opinion, based on "the demonstrated unreliability of *untested* technological solutions," that the nuclear-waste disposal site at Yucca Mountain, Nevada likely would prove to be a failure. See page [315] of this book. It may be of interest to trace the subsequent history of the Yucca Mountain site as reported in the media.

On March 18, 2005, *The Denver Post*, page 4A, carried an Associated Press report by Erica Werner according to which then-recent studies had found that water seepage through the Yucca Mountain site was faster than what earlier studies had reported. The more-rapid movement of water implied a greater risk of escape of radioactive materials from the site, and there were reasons to suspect that the earlier studies had been intentionally falsified.

The Week, January 26, 2007, page 24, reported a new study: "Special new containers designed to hold nuclear waste for tens of thousands of years may begin to fall apart in just 210 years," the study found. "Researchers... had pinned their hopes on zircon, a material they thought was stable enough to store the waste...." The scientists had based this belief on computer simulations, but they were "startled" when they discovered how alpha radiation affected the "zircon" in reality.

Zircon is a gemstone. The substance referred to in the article presumably is a ceramic called zirconia. See *The New Encyclopædia Britannica*, 15th ed., 2003, Vol. 21, article "Industrial Ceramics," pages 262–63.

On September 25, 2007, *The Denver Post*, page 2A, reported: "Engineers moved some planned structures at the Yucca Mountain nuclear waste dump after rock samples indicated a fault line unexpectedly ran beneath their original location...."

On March 6, 2009, *The Denver Post*, page 14A, carried an Associated Press report by H. Jośef Hebert according to which the U.S. Government had abandoned the plan to store reactor waste at Yucca Mountain. This after having spent 13.5 billion dollars on the project.

So it appears that the problem of safe disposal of radioactive waste is no closer to a solution than it ever was.

4. *Why is Democracy the Dominant Political Form of the Modern World?*
The argument about democracy set forth in my letters to David Skrbina of October 12 and November 23, 2004 (pages [283–285] and [292–296] of this book) is incomplete and insufficiently clear, so I want to supplement that argument here.

The most important point that I wanted to make was that democracy became the dominant political form of the modern world not as the result of a decision by human beings to adopt a freer or a more humane form of government, but because of an "objective" fact, namely, the fact that in modern times democracy has been associated with the highest level of economic and technological success.

To summarize the argument of my letters to Dr. Skrbina, democratic forms of government have been tried at many times and places at least since the days of ancient Athens, but democracy did not thrive sufficiently to displace authoritarian systems, which remained the dominant political forms through the 17th century. But from the advent of the Industrial Revolution the (relatively) democratic countries, above all the English-speaking ones, were also the most successful countries economically and technologically. Because they were economically and technologically successful, they were also successful militarily. The economic, technological, and military superiority of the democracies enabled them to spread democracy forcibly at the expense of authoritarian systems. In addition, many nations voluntarily attempted to adopt democratic institutions because they believed that these institutions were the source of the economic and technological success of the democracies.

As part of my argument, I maintained that the two great military contests between the democracies and the authoritarian regimes—World Wars I and II—were decided in favor of the democracies because of the democracies' economic and technological vigor. The astute reader, however, may object that the democracies could have won World Wars I and II simply by virtue of their great preponderance in resources and in numbers of soldiers, with or without any putative superiority in economic and technological vigor.

My answer is that the democracies' preponderance in resources and numbers of soldiers was only one more expression of their economic and technological vigor. The democracies had vast manpower, territory, industrial capacity, and sources of raw material at their disposal because they—especially the British—had built great colonial empires and had spread their language, culture, and technology, as well as their economic and political systems, over a large part of the world. The English-speaking peoples moreover had powerful navies and therefore, generally speaking, command of the sea, which enabled them to assist one another in war by transporting troops and supplies to wherever they might be needed.

Authoritarian systems either had failed to build empires of comparable size, as in the case of Germany and Japan, or else they had indeed built huge empires but had left them relatively backward and undeveloped, as in the case of Spain, Portugal, and Russia. It was during the 18th century, as the Industrial Revolution was gathering force, that authoritarian France lost to semidemocratic Britain in the struggle for colonization of North America and India. France did not achieve stable democracy until 1871, when it was too late to catch up with the British.

Germany as a whole was politically fragmented until 1871, but the most important state in Germany—authoritarian Prussia—was already a great power by 1740[1] and had access to the sea,[2] yet failed to build an overseas empire. Even after the unification of their country in 1871, the Germans' efforts at colonization were half-hearted at best.

Like the English-speaking peoples, the Spanish- and Portuguese-speaking peoples colonized vast territories and populated them thickly, but the manpower of their territories could not have been used very effectively in a European war, because these peoples lacked the economic, technical, and organizational resources to assemble, train, and equip large armies, transport them to Europe, and keep them supplied with munitions while they were there. Moreover, they lacked the necessary command of the sea. The Russians did not need command of the sea in order to transport their men to a European battlefield, but, as pointed out on page [340] of this book, note 34, the Russians during World War II did need massive aid from the West, without which they could not have properly equipped and supplied their troops.

Thus the Allies' preponderance in resources and numbers of troops, at least during World War II, was clearly an expression of the democracies' economic and technological vigor. The democracies' superiority was a consequence not only of the *size* of their economies, but also of their *efficiency*. Notwithstanding the vaunted technical efficiency of the Germans, it is said that during World War II German productivity per man-hour was only *half* that of the United States, while the corresponding figure for Japan was only *one fifth* that of the U.S.[3]

Though the case may not have been as clear-cut in World War I, it does appear that there too the Allies' superiority in resources and in numbers of troops was largely an expression of the democracies' economic and technological vigor. "In munitions and other war material Britain's industrial power was greatest of all.... Britain...was to prove that the strength of her banking system and the wealth distributed among a great commercial people furnished the 'sinews of war'...."[4] Authoritarian Russia was not a critical factor in World War I, since the Germans defeated the Russians with relative ease.

Thus it seems beyond argument that democracy became the dominant political form of the modern world as a result of the democracies' superior economic and technological vigor. It may nevertheless be questioned whether democratic government was the *cause* of the economic and technological vigor of the democracies. In the foregoing discussion I've relied mainly on the example of the English-speaking peoples. In fact, France, following its democratization in 1871 and even before the devastation wrought by World War I, was *not* economically vigorous.[5] Was the economic and technological vigor of the English-speaking peoples perhaps the result, not of their democratic political systems, but of some other cultural trait?

For present purposes the answer to this question is not important. The objective fact is that since the advent of the Industrial Revolution democracy has been generally associated with economic and technological vigor. Whether this association has been merely a matter of chance, or whether there is a causative relation between democracy and economic and technological vigor, the fact remains that the association has existed. It is this objective fact, and not a human desire for a freer or a more humane society, that has made democracy the world's dominant political form.

It is true that some peoples have made a conscious decision to adopt democracy, but it can be shown that in modern times (at least since, say, 1800) such decisions have usually been based on a belief (correct or not) that democracy would help the peoples in question to achieve economic and technological success. But even assuming that democracy had been chosen because of a belief that it would provide a freer or a more humane form of government, and even assuming that such a belief were correct, democracy could not have thriven under conditions of industrialization in competition with authoritarian systems if it had not equalled or surpassed the latter in economic and technological vigor.

Thus we are left with the inescapable conclusion that democracy became the dominant political form of the modern world not through human choice but because of an objective fact, namely, the association of democracy, since the beginning of the Industrial Revolution, with economic and technological success.

It is my opinion that we have now reached the end of the era in which democratic systems were the most vigorous ones economically and technologically. If that is true, then we can expect democracy to be gradually replaced by systems of a more authoritarian type, though the external forms of democratic government will probably be retained because of their utility for propaganda purposes.

5. *Popular Rebellion as a Force for Reform.* On pages [345 note 121, 322–323] of this book I stated that in the early 20th century labor violence in the United States impelled the government to carry out reforms that alleviated the problems of the working class. This statement was based on my memory of things read many years earlier. Recent reading and rereading lead me to doubt that the statement is accurate.

It's true that labor violence during the 1890s seems to have spurred efforts at reform by the government and by industry between about 1896 and 1904, but the effect was short-lived.[6] The great turning point in the struggle of the American working class was the enactment in the 1930s of legislation that guaranteed workers the right to organize and to bargain collectively, and this turning point was followed by a "sharp decline in the level of industrial violence."[7] But I'm not aware of any evidence that the legislation was *motivated* by a desire to prevent labor violence.

The data support the conclusion that labor violence was damaging to labor unions and counterproductive in relation to the workers' immediate goals.[8] On the other hand, it seems clear that labor violence could not have been ended except by addressing the grievances of the working class.[9] Thus, the threat of violence could have impelled the government to enact legislation guaranteeing the workers' right to organize and to bargain collectively. But, again, I don't know of any evidence that this was actually what happened.

Be that as it may, we can dispense with the labor movement for present purposes. The revolt of American black people (the "civil rights movement") of the 1950s and 1960s can serve to illustrate the points I tried to make on page [345 note 121] and pages [322–323] of this book. And it's easy to give other examples of cases in which popular revolt, short of revolution, has forced governments to pay attention to people's grievances. Thus, the Wat Tyler Rebellion in England (1381) failed as a social revolution, but it impelled the government to refrain from enforcing the poll tax that was the immediate cause of the revolt.[10] The Sepoy Mutiny in India (1857–58) was ruthlessly crushed, but it caused the British to drop their effort to impose westernizing social changes upon Hindu civilization.[11]

NOTES

[1] *Encycl. Britannica*, 2003, Vol. 20, article "Germany," page 96.

[2] The fact that Prussia's access was to the Baltic Sea rather than directly to the Atlantic was not a terribly important factor in the 18th century, when round-the-world voyages were nothing very extraordinary; still less was it important in the 19th century, when sailing ships of advanced design, and later steamships, made voyages to all parts of the world a routine matter. Even the tiny duchy of Courland, situated at the eastern end of the Baltic, made a start at overseas colonization during the 17th century (*Encycl. Britannica*, 2003, Vol. 3, article "Courland," page 683), so there was certainly no physical obstacle to Prussia's doing the same in the 18th and 19th centuries.

[3] John Keegan, *The Second World War*, Penguin, 1990, page 219.

[4] B. H. Liddell Hart, *The Real War, 1914–1918*, Little, Brown and Company, 1964, page 44.

[5] *Encycl. Britannica*, 2003, Vol. 19, article "France," page 521.

[6] Foster Rhea Dulles, *Labor in America: A History*, third edition, AHM Publishing Corporation, Northbrook, Illinois, 1966, pages 166–179, 183–88, 193–99, 204–05.

[7] Hugh Davis Graham and Ted Robert Gurr (editors), *Violence in America: Historical and Comparative Perspectives*, Signet Books, New York, 1969, pages 343–45, 364–65.

[8] *Ibid.*, pages 361–62.

[9] *Ibid.*, pages 364–66.

[10] *Encycl. Britannica*, 2003, Vol. 9, article "Peasants' Revolt," pages 229–230.

[11] *Ibid.*, Vol. 6, article "Indian Mutiny," pages 288–89.

BIBLIOGRAPHY

Aboujaoude, E. et al. 2006. Potential Markers for Problematic Internet Use. *CNS Spectrums*, 11(10).Har

Akai, L. 1997. In Search of the Unabomber. *Anarchy: A Journal of Desire Armed.*

Aleksiuk, M. 1996. *Power Therapy*. H & H Publishers.

Allen, N. et al. 2008. Animal Foods, Protein, Calcium and Prostate Cancer Risk. *British Journal of Cancer*, 98.

Axtell, J. 1985. *The Invasion Within*. Oxford University Press.

Barclay, H. 2002. Letter to the editor. *Anarchy: A Journal of Desire Armed*, Spring/Summer, pp. 70-71.

Barcott, B. 2002. From Tree-Hugger to Terrorist. *New York Times Sunday Magazine*. (April 7; pp. 56-59).

Black, B. 1998. Primitive Affluence. In *The Abolition of Work/Primitive Affluence: Essays Against Work.* Green Anarchist Books (London).

Bolívar, S. 1990. *Simon Bolívar: Escritos Politicos*. (G. Soriano, ed.) Alianza Editorial (Madrid).

Bonvillain, N. 1998. *Women and Men: Cultural Constructs of Gender*, 2nd edition. Prentice Hall.

Carlyle, T. 1837/2002. *The French Revolution*. The Modern Library.

Carr, N. 2008. Is Google Making us Stupid? *The Atlantic* (July/August).

Cashdan, E. 1989. Hunters and Gatherers: Economic Behavior in Bands. In S. Plattner (ed.), *Economic Anthropology*. Stanford University Press.

Chao, A. et al. 2005. Meat Consumption and Risk of Colorectal Cancer. *Journal of the American Medical Association*, 293(2).

Christakis, D. et al. 2004. Early Television Exposure and Subsequent Attentional Problems in Children. *Pediatrics* 113(4): 708-713.

Coatimundi. 1998. Unabomber Cops a Plea. *Fifth Estate* 33(1): 2.

Coon, C. 1971. *The Hunting Peoples*. Little, Brown and Company.

Corey, S. 2000. On the Unabomber. *Telos*, 118: 157-181.

Davidson, H. 1990. *Gods and Myths of Northern Europe*. Penguin Books.

Davis, G. 2006. *Means Without End*. University Press of America.

Debo, A. 1976. *Geronimo: The Man, His Time, His Place*. University of Oklahoma Press.

De Camp, L. 1974. *The Ancient Engineers*. Ballantine.

Diamond, J. 2005. *Collapse: How Societies Choose to Fail or Succeed*. Viking.

Divan, H. et al. 2008. Prenatal and Postnatal Exposure to Cell Phone Use and Behavioral Problems in Children. *Epidemiology*, 19(4): 523-529.

Dynarski, M. et al. 2007. Effectiveness of Reading and Mathematics Software Products: Findings from the First Student Cohort. *US Department of Education*, Institute of Education Sciences.

Dyson, F. 2005. The Bitter End. *New York Review* (April 28).

Egeberg, R. et al. 2008. Meat Consumption, N-acetyl Transferase 1 and 2 Polymorphism, and Risk of Breast Cancer. *European Journal of Cancer Prevention*, 17.

Elkin, A. 1964. *The Australian Aborigines*, 4th edition. Anchor Books, Doubleday.

Ellul, J. 1954/1964. *The Technological Society*. Knopf.

Ellul, J. 1969/1971. *Autopsy of Revolution*. Knopf.

Ellul, J. 1991. *Anarchy and Christianity*. (G. Bromiley, trans.) W. B. Eerdmans.

Ehrlich, P. 1968. *The Population Bomb*. Ballantine.

Evans-Pritchard, E. 1972. *The Nuer*. Oxford University Press.

Fernald, M. and A. Kinsey. 1996. *Edible Wild Plants of Eastern North America*, Revised Edition. Dover.

Ferris, W. 1940. *Life the Rocky Mountains*. (P. Phillips, ed.) Old West Publishing.

Finnegan, W. 1998. Defending the Unabomber. *New Yorker* (March 16).

Fulano, T. 1996. The Unabomber and the Future of Industrial Society. *Fifth Estate* 31(2): 5-9.

García López, J. 1959. *Historia de la literatura espanola*. (5th ed.) Las Americas Publishing.

Gershon, E. and R. Rieder. 1992. Major Disorders of Mind and Brain. *Scientific American* (September).

Gibbons, E. 1972. *Stalking the Wild Asparagus*, Field Guide Edition. David McKay Company.

Gibbs, N. 1996. Tracking Down the Unabomber. *Time* (April 15).

Goodman, P. 1960. *Growing Up Absurd*. Random House.

Graham, H. and T. Gurr. 1979. *Violence in America*. Sage Publications.

Grant, B. et al. 2004. Prevalence, Correlates, and Disability of Personality Disorders in the U.S. *Journal of Clinical Psychiatry*, 65: 948-958.

Greaves, M. 2000. *Cancer: The Evolutionary Legacy*. Oxford University Press.

Hardell, L. et al. 2007. Long-Term Use of Cellular Phones and Brain Tumors: Increased Risk Associated With Use for Greater Than 10 Years. *Occupational Environmental Medicine*, 64.

Haviland, W. 1999. *Cultural Anthropology*, 9th edition. Harcourt Brace College Publishers.

Hoffer, E. 1951/2002. *The True Believer*. Harper Perennial Modern Classics.

Hollander, P. 1988. *The Survival of Adversary Culture*. Transaction Books.

Holmberg, A. 1969. *Nomads of the Long Bow: The Siriono of Eastern Bolivia*. The Natural History Press.

Hunter, J. 1957. *Manners and Customs of Several Indian Tribes Located West of the Mississippi*. Ross and Haines.

Huxley, A. 1932/2004. *Brave New World*. HarperCollins.

Illich, I. 1973. *Tools for Conviviality*. Harper and Row.

Illich, I. 1974. *Energy and Equity*. Harper and Row.

Jones, S. 2006. *Against Technology: From the Luddites to the Neo-Luddites*. Routledge.

Joy, B. 2000. Why the Future Doesn't Need Us. *Wired* (April): 238-262.

Kaplan, D. (ed.) 2004. *Readings in the Philosophy of Technology*. Rowman and Littlefield.

Kaplan, J. and L. Weinberg. 1998. *The Emergence of a Euro-American Radical Right*. Rutgers University Press.

Keegan, J. 1990. *The Second World War*. Penguin.

Keniston, K. 1974. *The Uncommitted*. Dell.

Kessler, R. et al. 2004. Prevalence, Severity, and Unmet Need for Treatment of Mental Disorders in the World Health Organization "World Mental Health" surveys. *Journal of the American Medical Association*, 291.

Keyfitz, N. 1993. Review of "Only One World" (by G. Piel). *Scientific American* (February).

Kirkham, J. et al. 1969. *Assassination and Political Violence*. US Government Printing Office.

Kubey, R. and M. Csikszentmihalyi. 2002. Television Addiction is No Mere Metaphor. *Scientific American* (February).

Kurzweil, R. 2003. Promise and Peril. In *Living with the Genie* (A. Lightman et al., eds.) Island Press. (See also Kurzweil 2006).

Kurzweil, R. 2006. *The Singularity is Near*. Penguin.

Leach, D. 1978. Colonial Indian wars. In *Handbook of North American Indians*, W. Sturtevant (ed.), Vol. 4: *History of Indian-White Relations*. Smithsonian Institution.

Leakey, R. 1981. *The Making of Mankind*. E. P. Dutton.

Lee, M. 1995. *Earth First!: Environmental Apocalypse*. Syracuse University Press.

Leuchtenburg, W. 1963. *Franklin D. Roosevelt and the New Deal, 1932-1940*. Harper and Row.

Li, V. 2006. *The Neo-Primitivist Turn*. University of Toronto Press.

Luke, T. 1996. Re-reading the Unabomber Manifesto. *Telos*, 107: 81-94.

Marcuse, H. 1964. *One Dimensional Man*. Beacon Press.

Marquis, T. 1967. *Wooden Leg: A Warrior Who Fought Custer*. Bison Books, University of Nebraska Press.

Massola, A. 1971. *The Aborigines of South-Eastern Australia: As They Were*. Heinemann.

McGinn, R. 1991. *Science, Technology, and Society*. Prentice-Hall.

McKibben, W. 1989. *End of Nature*. Random House.

Mello, M. 1999. *The USA versus Theodore John Kaczynski*. Context Publications.

Mercader, J. (ed.) 2003. *Under the Canopy: The Archaeology of Tropical Rain Forests*. Rutgers University Press.

Morris, D. 1969. *The Human Zoo*. Cape.

Mumford, L. 1966. *Knowledge Among Men*. Smithsonian Institution.

Mumford, L. 1967-1970. *The Myth of the Machine* (2 volumes). Volume 1: *Technics and Human Development*. Volume 2: *The Pentagon of Power*. Harcourt, Brace, and World.

Naess, A. 1989. *Ecology, Community, and Lifestyle*. Cambridge University Press.

Nakazawa, T. et al. 2002. Association Between Duration of Daily VDT Use and Subjective Symptoms. *American Journal of Industrial Medicine*, 42: 421-426.

Nietzsche, F. 1888/1990. *Twilight of the Idols / The Antichrist*. Penguin Classics.

Nitzberg, J. and J. Zerzan. Back to the Future Primitive. *Mean*, April: 68-78.

Norat, T. et al. 2005. Meat, Fish, and Colorectal Cancer Risk. *Journal of the National Cancer Institute*, 97(12).

Nothlings, U. et al. 2005. Meat and Fat Intake as Risk Factors for Pancreatic Cancer. *Journal of the National Cancer Institute*, 97(19).

Orwell, G. 1937/1958. *The Road to Wigan Pier*. Harcourt, Brace.

Packard, V. 1957. *The Hidden Persuaders*. D. McKay.

Perera, F. et al. 2009. Prenatal Airborne Polycyclic Aromatic Hydrocarbon Exposure and Child IQ at Age 5 Years. *Pediatrics*, 124(2).

Pfeiffer, J. 1969. *The Emergence of Man*. Harper & Row.

Pfeiffer, J. 1977. *The Emergence of Society*. McGraw-Hill.

Poncins, G. de. 1980. *Kabloona*. Time-Life Books.

Posner, R. 2004. *Catastrophe: Risk and Response*. Oxford University Press.

Powers, S. et al. 2003. Quality of Life in Childhood Migraines. *Pediatrics* 112(1-5): E1-E5.

Ramos, S. 1934/1982. *El perfil del hombre y la cultura en Mexico*. Espasa-Calpe Mexicana.

Rani, F. et al. 2008. Epidemiologic Features of Antipsychotic Prescribing to Children and Adolescents. *Pediatrics*, 121(5).

Rees, M. 2003. *Our Final Century*. Heinemann.

Reichard, G. 1990. *Navaho Religion: A Study of Symbolism*. Princeton University Press.

Rice, C. 2007. Prevalence of Autism Spectrum Disorders. *Morbidity and Mortality Weekly Report*, US CDC, 56.

Rousseau, J-J. 1762/1979. *Emile, or On Education*. (A. Bloom, trans.) Basic Books.

Russell, O. 1965. *Journal of a Trapper*. Bison Books.

Ruthen, R. 1993. Strange Matters: Can Advanced Accelerators Initiate Runaway Reactions? *Scientific American* (August).

Sahlins, M. 1972. *Stone Age Economics*. Aldine-Atherton.

Sale, K. 1995. Unabomber's Secret Treatise: Is There Method in his Madness? *Nation* 261(9): 305-311.

Sale, K. 1996. *Rebels Against the Future*. Basic Books.

Sale, K. 2006. *After Eden: The Evolution of Human Domination*. Duke University Press.

Sallust. ca. 50 BCE/2007. *Catiline's War, the Jugurthine War, Histories*. (A. Woodman, trans.) Penguin.

Sarmiento, D. 1990. *Facundo: Civilizacion y barbarie*. Catedra (Madrid).

Scarce, R. 2006. *Eco-Warriors: Understanding the Radical Environmental Movement* (revised edition). Left Coast Press.

Scharff, R. and V. Dusek. 2003. *Philosophy of Technology: The Technological Condition*. Blackwell Publishing.

Schebesta, P. 1938-1941. *Die Bambuti-Pygmäen vom Ituri*, Bands I and II. Institut Royal Colonial Belge (Brussels).

Seligman, M. 1975. *Helplessness: On Depression, Development, and Death*. W. H. Freeman.

Skolimowski, H. 1983. *Technology and Human Destiny*. University of Madras (India).

Smelser, N. 1971. *Theory of Collective Behavior*. Macmillan.

Split, W. and W. Neuman. 1999. Epidemiology of Migraine Among Students from Randomly Selected Secondary Schools in Lodz. *Headache* 39: 494-501.

Stefansson, V. 1951. *My Life with the Eskimo*. Macmillan.

Stivers, R. 1999. *Technology as Magic*. Continuum.

Tan, C. 1971. *Chinese Political Thought in the 20th Century*. Doubleday.

Thomas, E. 1989. *The Harmless People*, 2nd edition. Random House.

Thurston, R. 1996. *Life and Terror in Stalin's Russia, 1934-1941*. Yale University Press.

Trotsky, L. 1932/1980. *History of the Russian Revolution*. (M. Eastman, trans.) Pathfinder Press.

Turnbull, C. 1961. *The Forest People*. Simon and Schuster.

Turnbull, C. 1965. *Wayward Servants: The Two Worlds of the African Pygmies*. The Natural History Press.

Turnbull, C. 1983. *The Mbuti Pygmies: Change and Adaptation*. Harcourt Brace.

Vacca, R. 1973. *The Coming Dark Age*. (J. Whale, trans.) Doubleday.

Vestal, S. 1989. *Sitting Bull, Champion of the Sioux: A Biography*. University of Oklahoma Press.

Von Laue, T. 1971. *Why Lenin? Why Stalin?* J. B. Lippencott, Co.

Waldman, M. et al. 2006. Does Television Cause Autism? Unpublished. Available online at: http://www.johnson.cornell.edu/faculty/profiles/waldman/autpaper.html

Wissler, C. 1989. *Indians of the United States*, Revised Edition. Anchor Books, Random House.

Wright, R. 1995. Evolution of Despair. *Time* (August 28).

Xu, W-H. et al. 2007. Nutritional Factors in Relation to Endometrial Cancer. *International Journal of Cancer*, 120.

Zakaria, R. 1989. *The Struggle Within Islam*. Penguin.

Zerzan, J. 1994. *Future Primitive and Other Essays*. Autonomedia.

Zerzan, J. (Undated). Whose Future? *Species Traitor*, 1.

Zimmerman, F. et al. 2007. Associations Between Media Viewing and Language Development in Children Under Age 2 Years. *Journal of Pediatrics* (July).

Zimmermann, G. 1902. *Das Neunzehnte Jahrhundert*. Zweite Hälfte. Druck und Verlag von Geo. Brumder (Milwaukee).

ABOUT THE AUTHORS

Theodore J. Kaczynski attended Harvard University, received a PhD in mathematics from the University of Michigan, taught at the University of California, Berkeley, and then moved to Montana where he attempted to live a self-sufficient life. Called a "domestic terrorist" and the "Unabomber" by the FBI, Kaczynski was convicted for illegally mailing bombs, resulting in the deaths of three people and the injury of 23 others. He now serves a life sentence in the supermax prison in Florence, Colorado.

David Skrbina (born 1960) is currently a Lecturer in Philosophy at the University of Michigan at Dearborn, where he created and teaches that university's first-ever course in Philosophy of Technology. He has also served as a guest lecturer at Eastern Michigan University, and the University of Ghent (Belgium). In addition to philosophy of technology, his research interests include philosophy of mind, metaphysics, and environmental ethics. Skrbina has published two books in philosophy, and has written numerous articles. Among his many current projects, he is preparing a collection of readings in technology criticism, for use as a college text.

www.FeralHouse.com